The Weather and Climate of Southern Africa

M

Th

The Weather and Climate of Southern Africa

PD TYSON AND
RA PRESTON-WHYTE

SECOND EDITION

OXFORD

UNIVERSITY PRESS

OXFORD
UNIVERSITY PRESS

Great Clarendon Street, Oxford OX2 6DP

Oxford University Press is a department of the University of Oxford.
It furthers the University's objective of excellence in research, scholarship,
and education by publishing worldwide in

Oxford New York

Athens Auckland Bangkok Bogotá Buenos Aires Calcutta
Cape Town Chennai Dar es Salaam Delhi Florence Hong Kong Istanbul
Karachi Kuala Lumpur Madrid Melbourne Mexico City Mumbai
Nairobi Paris São Paulo Singapore Taipei Tokyo Toronto Warsaw

with associated companies in Berlin Ibadan

Oxford is a registered trademark of Oxford University Press
in the UK and certain other countries

Published in South Africa
by Oxford University Press Southern Africa, Cape Town

The Weather and Climate of Southern Africa

ISBN 0 19 571806 2

First published 1988 as *The Atmosphere and Weather of Southern Africa*
by R. A. Preston-Whyte and P. D. Tyson

Editor: Arthur Attwell
Designer: Mark Standley
Cover photographs: Jennifer Tyson and Peter Tyson
Indexer: Jeanne Cope

Published by Oxford University Press Southern Africa
PO Box 12119, N1 City, 7463, Cape Town, South Africa

Set in 9.5 pt on 12 pt Adobe Caslon by Scan Shop, Cape Town
Reproduction by Scan Shop
Cover reproduction by The Image Bureau, Cape Town
Printed and bound by Creda Communications, Eliot Avenue, Epping Industry II, Cape Town

Contents

Preface to the second edition

The provision of a textbook with a southern-hemisphere perspective for students of meteorology, climatology and cognate disciplines in southern Africa has proved highly successful. First published in 1988 as *The Atmosphere and Weather of Southern Africa*, the book has been prescribed in almost all institutions of higher learning in the region teaching atmospheric sciences in one form or another and has been recommended for general use in many more.

Since first appearing, climatology has assumed an ever-increasing importance in the pantheon of sciences as a consequence of public interest and national concerns over the issue of greenhouse warming of the atmosphere and global change. The international Intergovernmental Panel on Climate Change (IPCC), which first reported in 1990, has ensured that climatology and climatic change are issues high on the agendas of all governments concerned with planning for sustainable development.

The Atmosphere and Weather of Southern Africa has been revised to take account of these developments and the considerable research into the climatology of southern Africa that has taken place since the mid 1980s when work on the first edition began. This revision has necessitated a change in the title of the book to *The Weather and Climate of Southern Africa*. New chapters have been added on climatology and modern science, atmosphere–ocean interactions and the transport of aerosols and trace gases over the region. The chapters on climatic change and variability and the prediction of future conditions (formerly modelling) have been substantially modified to incorporate the latest findings on these subjects. Likewise, the chapter on boundary-layer processes has been adapted and updated. Elsewhere, modifications have been made to improve the clarity and correctness of the text. Wherever appropriate, the findings arising out of the IPCC process have been incorporated. As in the first edition, each chapter ends with selected recommendations for further reading, some introductory and some advanced.

In preparing this edition the authors have benefited from the help and advice of many people, including Johann Lutjeharms, Deon Terblanche, Simon Mason and Alec Joubert. Their constructive comments are much appreciated. Wendy Job, with the assistance of Wendy Phillips, has provided excellent illustrative material in the many new and updated figures. Family support has continued too, not only in the revising of the book, but also in the undertaking of much of the research included in the regional aspects of the work. This support has been, and continues to be, greatly appreciated. To our nearest and dearest we say a heartfelt thanks.

Finally, the continued work in climatology in South Africa has been made possible by the financial support of the National Research Foundation, the Water Research Commission, ESKOM and the Universities of the Witwatersrand and Natal. Their generosity is much appreciated and will continue to benefit students and others throughout South Africa, the southern African region and beyond.

P. D. TYSON
University of the Witwatersrand,
Johannesburg

R. A. PRESTON-WHYTE
University of Natal, Durban

Preface to the first edition

Almost every textbook of meteorology or climatology is written for the northern-hemisphere reader. Few books are directed towards an understanding of the atmospheric processes affecting the weather and climate of the southern hemisphere; fewer still deal specifically with the atmospheric environment of southern Africa. *The Atmosphere and Weather of Southern Africa* has been written to meet the latter need and to provide a textbook for students of earth, atmospheric, environmental and related sciences. It is a book in which the examples of the phenomena being considered are not from the northern hemisphere or areas remote from the experience of students, but are drawn from the region in which they live and are familiar to them. By providing material in which the illustration and explanation of meteorological and climatological phenomena can be part of a student's experience, it is hoped to encourage a greater interest in the general field of atmospheric science.

Some of the most pressing problems of international collaborative research and some of the most stimulating advances in science at present are concerned with the geosphere and biosphere and the question of global change. Central to an understanding of such change, is an appreciation of the atmosphere and its meteorology, of the earth and its climatology. Before students of the atmospheric sciences become involved at the intellectual interfaces between disciplines and in solving some of the problems at the forefront of knowledge, they need to be well grounded in the basic fundamentals of meteorology and climatology. The aim of this book is to provide an introduction to these fundamentals.

Physical meteorology, the principles of atmospheric hydrostatics and thermodynamics, clouds and atmospheric heat transfer are introduced in Chapters 2 to 6. Chapters 7 and 9 cover the dynamics of horizontal motion and large-scale weather-producing systems. Material from both physical and dynamic meteorology is combined in Chapter 8 in the consideration of vertical motion and cumulus convection over southern Africa. The general circulation of the southern hemisphere is considered in Chapter 10 and the circulation patterns and weather-producing disturbances over southern Africa are examined in Chapter 11. In the final three chapters climatic change and variation over southern Africa, boundary-layer meteorology and modelling are handled. A glossary of technical terms has been appended, together with tables of important symbols, constants, Systeme Internationale (si) units and some conversions to other units. At the end of each chapter further reading is suggested with texts being listed in the chronological order of their dates of publication.

Much of Chapters 10, 11 and 12 is derived in large measure from *Climatic Change and Variability in Southern Africa* by P. D. Tyson (Oxford University Press, 1986), except that detailed referencing has been omitted in the interests of easier reading. Full references are available in the parent work.

It is impossible to gain a comprehensive knowledge of atmospheric science without the use of mathematics. For some students this presents a problem. Except in Chapter 14 on modelling (and which is intended for advanced students), mathematics have been kept to a minimum and the book has been written to be understandable even to those without appropriate backgrounds. It is also designed to include material that may be offered in courses ranging from first year to Honours.

In writing this book we have benefited from the generous assistance of many people. Philip Stickler, Wendy Job, Jenny McDowell and Hemraj Hurrypursad have produced outstanding illustrative material with great professionalism. Jenny Tyson, Debbie Melville and Glenda Venn have helped with the cartography. Trish Parrott prepared the manuscript on the University of the Witwatersrand mainframe computer and was a tower of strength in every way; to her we owe special thanks. Janette Lindesay gave invaluable and much-appreciated help with the checking of the final manuscript. The writing of books is always made easier by the support given by one's nearest and dearest; to them we acknowledge our indebtedness.

Finally, the production of this book has been made possible by the generosity of the Council for Scientific and Industrial Research, the Water Research Commission and the Universities of the Witwatersrand and Natal. We believe generations of students will benefit from the stimulation by these institutions of research and learning in atmospheric science. To them we record our great appreciation.

R. A. PRESTON-WHYTE
University of Natal, Durban

P. D. TYSON
University of the Witwatersrand,
Johannesburg

1

CLIMATOLOGY AND MODERN SCIENCE

Climatology has moved to centre stage in the pantheon of modern sciences. The authoritative pronouncement by the international Inter-governmental Panel on Climate Change (IPCC) in 1996 that 'the balance of evidence suggests that there is a discernible human influence on global climate' (IPCC 1996, 5) has re-empha-sized the importance of modern climatology in the affairs of humankind. Issues such as global warming and environmental change, sustainable development, biodiversity and effective regional policy formulation for sustainable development all require an understanding of climatology. Long gone are the days when climatology was a fringe science concerned mainly with the description of the climates of places and different regions of the globe. Now it is a discipline central to modern scientific endeavour. It is grounded in atmos-pheric physics and chemistry, in analysis, hypo-thesis testing and modelling.

Fields of climatological study range from the study of palaeoclimates, through the thermo-dynamics of the atmosphere, the understanding of weather processes and storm systems, the investigation of boundary layer phenomena, to the transport of aerosols and trace gases in the atmosphere and the simulation of future climates and their impacts. It is a discipline of particular relevance and applicability in modern life. One of the core issues underlying much of what clima-tology does concerns the consequences of both deliberate and inadvertent anthropogenic modi-fication of the atmosphere in present times and in future.

In many regions of the earth, change and variability in weather and climate impact significantly on societies. Excluding deaths caused by famine in droughts, of all deaths caused by natural disasters, over 60 per cent are weather or climate related. Droughts, the effects of which occur more slowly, are probably the most damaging disasters of all. This is certainly the case in Africa. During the 1980s, of the recorded natural disasters affecting the conti-nent, those that were weather related dominated (Fig. 1.1). Droughts left most victims dead or homeless. Over the years, drought-induced crop failures are the single greatest reasons for loss in agricultural production in South Africa. Floods

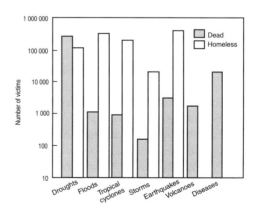

Fig. 1.1 The numbers of people left dead and homeless by recorded disasters in Africa during the 1980s (after World Meteorological Organization, 1990).

and hail damage are also major factors affecting crop production. Small advances in predicting droughts, floods, tropical cyclones and storms will spare the lives of many and bring about great savings to national economies in southern Africa. The improved prediction of El Niño events by understanding better the coupling of ocean–atmosphere variability in advanced general circulation models is a case in point.

It is necessary to emphasize that sustainable development, striven for by all countries, is impossible without a detailed understanding of climatology and how climates may vary in future. Climate and food production, climate and water resources, climate and energy production and, increasingly, climate and health all illustrate how changes in climate are of the greatest relevance to nation states. A recent development of great significance has been the establishment of the IPCC to assess the scientific aspects of global change and the impacts of such change. Major reports were issued in 1990, 1992 and 1996. Their effect in scientific, private-sector and government policy and planning circles has been considerable.

In South Africa, as in many other countries, the regional assessment of global-change impacts has become a priority. Regional environmental-change scenarios have been developed to aid the planning of sustainable development. Major issues of concern include the regional effect of global change on water resources in a region that experiences a high degree of natural climatic variability in any case. Likewise, the impact of environmental change on regional food production may be profound in future. Similarly, forestry, fishing and animal husbandry in southern Africa will not remain unaffected by the global change that many predict is to occur. Diseases sensitive to climate change will also affect animal and human health. The spread of malarial areas is just one of the concerns for the future. The implications for biodiversity also need to be considered when planning for a sustainable future in the southern African region. All planning will have to be based on the prediction of future conditions.

In all scientific disciplines, prediction and assessment have to be based on a sound understanding of the basic scientific underpinning of the subject. This is no less so in climatology. Thus an understanding of the physics and chemistry of the atmosphere, its interaction with the oceans and the terrestrial biosphere (particularly in the boundary layer), the processes controlling past and present circulation patterns, the natural variability of the system and its anthropogenic modification are all subjects that need to be understood before the simulation and prediction of future conditions may be attempted. This is as necessary in a textbook on regional meteorology and climatology as in any other context. Thus *The Weather and Climate of Southern Africa* has been structured accordingly to lead the student, the scientist seeking information on the region or the policy maker or planner to prediction via a basic understanding of the discipline.

ADDITIONAL READING

HOUGHTON, J. T., et al. 1996. *Climate Change 1995, The Science of Climate Change.* The Contribution of Working Group I to the Second Assessment Report of the Intergovernmental Panel on Climate Change. Cambridge: Cambridge University Press.

WATSON, R. T., et al. 1996. *Climate Change 1995, Impacts, Adaptations and Mitigation of Climatic Change: Scientific-Technical Analyses.* The Contribution of Working Group II to the Second Assessment Report of the Intergovernmental Panel on Climate Change. Cambridge: Cambridge University Press.

BRUCE, J. P., et al. 1996. *Climate Change 1995, Economic and Social Dimensions of Climatic Change.* The Contribution of Working Group III to the Second Assessment Report of the Intergovernmental Panel on Climate Change. Cambridge: Cambridge University Press.

2

THE COMPOSITION AND STRUCTURE OF THE ATMOSPHERE

The earth's atmosphere was not always composed of gases in the proportions and quantities found today. It is generally agreed that when the earth was formed about five thousand million years ago, high temperatures precluded the formation of an atmosphere. These high temperatures caused molecular speeds to reach and exceed the *escape velocity* required by molecules to break free from the earth's gravitational attraction. This caused the original gases, probably composed of a mixture of *methane* and *ammonia*, to escape to space. As the earth began to cool, however, gases dissolved in the molten rock and released at the surface by the process of *outgassing* were retained as an atmosphere by the earth's gravitational attraction.

Judging from present-day gases found in volcanoes, it appears likely that the early atmosphere consisted of large amounts of *water vapour, carbon dioxide, sulphur* and *nitrogen*, together with much smaller quantities of *argon, chlorine* and *hydrogen*. As the earth continued to cool, much of the water vapour condensed and accumulated to form the *hydrosphere* (oceans and lakes) and in so doing diminished the amount of water vapour in the atmosphere. Carbon dioxide reacted with calcium and magnesium to produce limestone and, as vast deposits of this rock were precipitated into the sea, so the carbon dioxide content of the atmosphere became greatly reduced. It is estimated that by the end of the Precambrian (600 million years ago) the carbon dioxide content was close to its present atmospheric level with some 99.83

per cent of the carbon stored in carbonate rocks, shales and fossil fuels. Sulphur and its compounds were *scavenged* out of the atmosphere by precipitation particles and accumulated in sulphate deposits in the process of sedimentation. By the end of the Precambrian nitrogen was the predominant gas in the atmosphere – and remains so today.

Oxygen was largely absent from the primordial atmosphere and, when it was formed by the dissociation of water by ultraviolet radiation, was rapidly removed by the weathering of surface minerals. However, by about three thousand million years ago it began to appear with the development of early oxygen-releasing life forms, particularly blue-green algae. For a very long time after this event virtually all available oxygen was taken up in the oxidation of ferrous iron. Estimations of atmospheric oxygen content at two thousand million years are placed at 0.1 per cent of present levels. Gradually photosynthesis and possibly photo-dissociation of water to hydrogen and oxygen were able to produce oxygen at a rate greater than absorption by oxidation. However, this process was slow and by the end of the Precambrian, the oxygen content of the atmosphere is estimated to have been only about 1 per cent that of today. Thereafter it increased to modern-day levels.

The present composition of the atmosphere is given in Table 2.1. Nitrogen and oxygen are the most abundant components of the mixture of fixed and variable quantities of gases, liquids and solids which constitute the earth's atmosphere. Below about 80 km, in a layer known as the *homosphere* (also called the *turbosphere*), the atmospheric constituents are well mixed by turbulence and convection (Fig. 2.1). This causes an almost constant distribution of each gas through the layer (with the exception of water vapour and ozone). Above the homosphere vertical mixing is controlled by molecular diffusion and the stratification of gases occurs in accordance with their molecular weights. This layer is known as the *heterosphere*, with the level of transition from turbulent mixing to molecular diffusion taking place at the top of the *mesosphere*.

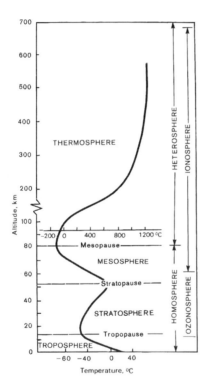

Fig. 2.1 Lapse rates and the vertical structure of the atmosphere (adapted from Miller and Thompson, 1970).

THE VERTICAL TEMPERATURE DISTRIBUTION

Temperature variations with height provide another means of structuring the atmosphere.

Thus the vertical temperature profile may be divided into four layers called the *troposphere, stratosphere, mesosphere* and *thermosphere.* The tops of these layers are respectively called the *tropopause, stratopause, mesopause* and *thermopause* (Fig. 2.1).

The troposphere varies in thickness by latitude and season. On average, the elevation of the tropopause decreases from 15–16 km over the tropics to 5–6 km over the poles. More than 80 per cent of the air in the atmosphere is found in this layer and is heated from below by infrared radiation from the sun-warmed earth's surface and the upward transfer of sensible and latent heat by turbulence. The air nearest the surface is warmed most and temperature decreases with height at an average *lapse rate* of 6.5 °C km^{-1}. Almost all the water vapour in the atmosphere is contained in the troposphere, with the gradient of water vapour density decreasing with height above the surface. The strong vertical currents that occur in this layer cause condensation into cloud with subsequent precipitation.

The tropopause marks a change in the temperature profile from a *lapse* condition (temperature decrease with height) in the well-mixed troposphere to an *isothermal* condition (temperature constant with height). Above the troposphere, in the stratosphere, a temperature *inversion* (temperature increases with height) persists to the stratopause. Temperature inversions cause the atmosphere to become *stable* and suppress vertical motion. Consequently little vertical mixing takes place beyond the tropopause. This is evidenced by the sharp decrease in water vapour concentrations at this level and by the increase in ozone in the stratosphere. The temperature increase that occurs in the stratosphere is due to the absorption of ultraviolet radiation by the ozone.

In the mesosphere, between 50 and 80 km above the earth's surface, temperature once again decreases with height. This layer is transitional between the energy-absorbing ozonosphere and the region of dissociation and ionization in the ionosphere, which are discussed below. With an average lapse rate of 3 °C km⁻¹ vertical mixing does take place with the rare occurrence of *noctilucent* clouds. The composition of these clouds is unknown; they may be formed from cosmic dust.

The temperature increase in the *thermosphere* is a function of the energy of individual particles rather than of any external heating effects. Consequently, the apparently high temperatures in this region are misleading. The density of individual gas molecules at this elevation is so low that the number of inter-molecule, heat-producing collisions is very small in comparison to the molecular collisions in air near the ground. The air is consequently cooler even though the energy of individual particles is high. Above 500 km the density of gaseous molecules is so low that collisions are infrequent, and molecules may leak into space by escaping the earth's gravitational pull. Known as the *exosphere*, this zone marks the outer limit of the earth's atmosphere.

In the heterosphere above 80 km very short waves of ultraviolet and X-ray radiation from the sun cause photochemical reactions in which particles are ionized (electrons ejected from the atom) or molecules are dissociated by splitting into two parts. While molecular nitrogen is readily ionized by short waves, molecular oxygen undergoes dissociation to produce a large number of positively-charged nitrogen and oxygen atoms and free electrons in the *ionosphere*.

The earth's magnetic field considerably influences the motion of electrons above about 150 km, with charged particles and ions being constrained by the earth's magnetic field in what is called the *magnetosphere*. In this region particles behave as individual ballistic particles. The magnetosphere provides a barrier to the stream of ionized particles which emanate from the sun in the *solar wind*, with the *magnetopause* separating solar wind particles from charged particles within the magnetosphere. The idealized dipole shape of the magnetic field is distorted by the solar wind. Field lines are compressed on the daylight side and elongated downwind on the night side (Fig. 2.2).

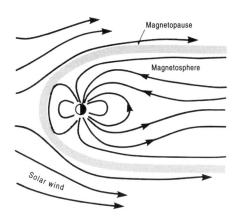

Fig. 2.2 Deformation of the earth's magnetic field by the solar wind.

Solar wind particles which penetrate the magnetosphere are guided toward the earth's magnetic poles where ionization is accompanied by *auroral* displays (*aurora borealis* in the northern hemisphere; *aurora australis* in the southern hemisphere). These displays accompany solar outbursts with concomitant increases in solar wind particle density.

THE CHEMICAL COMPOSITION OF THE ATMOSPHERE

Air is not a chemical compound, but a mixture of gases. Three of these, nitrogen, oxygen and argon, account for more than 99 per cent of the atmosphere by volume (Table 2.1). Despite their scarcity, the remaining trace gases play an important role in the thermodynamics of the atmosphere – by virtue of their properties as *greenhouse gases* – by trapping outgoing long-wave, infrared radiation from the earth and atmosphere to produce the *greenhouse effect*. The changing composition of natural and anthropogenic trace gases in the atmosphere has important consequences for *global warming*.

The nitrogen, oxygen and carbon dioxide in the atmosphere represent the gaseous phase of biogeochemical cycles in which the elements or compounds are converted into a gaseous form before being re-used by the biosphere. Although nitrogen comprises 78 per cent of the atmosphere by volume, most living organisms cannot use it in its gaseous form. Nitrogen is, therefore, cycled through the biosphere after being converted by fixing processes into nitrate salts suitable for plant growth. The nitrogen-containing protein molecules in plants are then transferred through the food web and finally converted back into nitrogen or oxides of nitrogen by the decomposition of organic material. In the oxygen and carbon cycle, plants utilize solar energy through the process of photosynthesis to take up carbon dioxide and release oxygen. Carbon dioxide and water are converted by

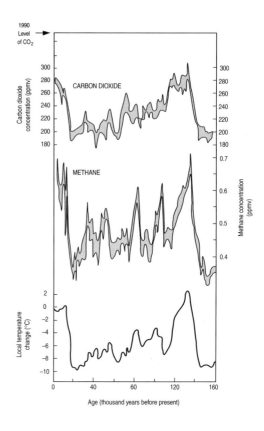

Fig. 2.3 The variation of carbon dioxide, methane and temperature over the past 160 000 years at Vostok, Antarctica. The thickness of the lines for carbon dioxide and methane indicate the degree of uncertainty associated with the measurements made from an ice core (after Reynaud et al., 1993).

Table 2.1 The composition of the homosphere (after Miller and Thompson, 1970).

Constituent	Symbol	% by vol (dry air)
Nitrogen	N_2	78.084
Oxygen	O_2	20.946
Argon	A	0.934
Carbon dioxide[a]	CO_2	0.033
Neon	Ne	18.18×10^{-4}
Helium	He	5.24×10^{-4}
Krypton	Kr	1.14×10^{-4}
Xenon	Xe	0.09×10^{-4}
Hydrogen	H_2	0.05×10^{-4}
Methane[a]	CH_4	2.0×10^{-4}
Nitrous oxide[a]	N_2O	0.5×10^{-4}
Radon	Rn	6.0×10^{-18}
Water vapour[a]	H_2O	< 4.0
Carbon monoxide[a]	CO	$\sim 2.0 \times 10^{-5}$
Sulphur dioxide[a]	SO_2	$< 1.0 \times 10^{-4}$
Nitrogen dioxide[a]	NO_2	$< 2.0 \times 10^{-6}$
Ozone[a]	O_3	$< 7.0 \times 10^{-6}$
Particles (dust, salt)[a]		$< 1.0 \times 10^{-5}$
Water (liquid, solid)[a]		< 1.0

a Variable constituent.

this process into biomass, whence the carbon and oxygen compounds pass along the food web from herbivore to carnivore. At each step some of the stored organic molecules are broken down to produce energy for the organism. Through the process of respiration, carbon dioxide is returned to the atmosphere. The remaining carbon and oxygen are returned to the air when plants and animals die and their tissues decompose.

The concentration of carbon dioxide (CO_2) has varied considerably in the past and has increased steadily over the last century (Fig.

2.3). This is attributed in part to the burning of fossil fuels, which releases stored carbon, and in part to the clearing of forests, where carbon was previously stored in the forest biomass. Carbon dioxide is also produced by volcanic activity, soil processes, ocean evaporation and the respiration of biota.

Methane (CH_4) is another important greenhouse gas the levels of which have varied considerably in the past (Fig. 2.3). It is produced primarily through anaerobic (i.e., oxygen deficient) processes by natural wetlands and agricultural systems (e.g., rice paddies), as well as by enteric fermentation in animals, by termites, through coal and oil extraction, biomass burning and from fermentation in landfills. It is formed by the reaction of carbon dioxide and hydrogen gas (H_2):

$$CO_2 + 4H_2 \rightarrow CH_4 + 2H_2O \qquad (2.1)$$

More than two-thirds is anthropogenically produced. Methane is extracted from the atmosphere by a complex photochemical process:

$$CH_4 + O_2 + 2x \rightarrow CO_2 + 2x\,H_2 \qquad (2.2)$$

where x denotes any methane-destroying chemical species such as H (hydrogen), OH (hydroxyl radical), NO (nitric oxide), Cl (chlorine) or Br (bromine).

While the amount by volume of nitrogen and oxygen in the atmosphere remains largely unchanged over time, the anthropogenic and biogenic emissions of oxides of nitrogen have important consequences both locally and globally. Oxides of nitrogen are now known to play a role in producing and maintaining the stratospheric ozone hole. Nitrous oxide (N_2O) is produced by biological activity in the oceans and soils, by combustion in industry, automobiles and aircraft, and by chemical fertilizers. It is unreactive in the troposphere, but is destroyed in the stratosphere by reactions with ozone. In so doing it contributes to the formation of the ozone hole.

Ozone is formed mainly in the stratosphere.

In the troposphere it is emitted as an anthropogenic pollutant and naturally in volcanic eruptions. It is also formed as a consequence of biomass burning.

Chlorofluorocarbons (CFCs) are entirely anthropogenic in origin. They originate as refrigerants and propellants in aerosol cans. They have been present in the atmosphere since the 1930s, and their usage is now limited by international convention. They destroy ozone in the stratosphere and have lifetimes of 65–130 years. Hydrogenated hydrocarbons (HFCs and HCFCs) are also entirely anthropogenic and their levels have increased rapidly in the atmosphere in recent years following their use as substitutes for CFCs.

Among the trace gases, water vapour is the most important from the human viewpoint. It is the most important greenhouse gas. Although the average concentrations of global water vapour do not undergo large changes, considerable variations occur over time (seasonally) and space (from place to place). Water is the only substance that can exist as a gas, a liquid and a solid at the temperatures normally found in the lower atmosphere. During the *phase changes* between states of water, heat is taken up in the process of evaporation from a liquid to a gaseous state and released when water vapour condenses into water droplets. This *latent heat* may, therefore, be transferred considerable distances to play a vital role in maintaining such phenomena as thunderstorms and hurricanes.

In addition to the greenhouse gases, traces of *reactive gas species* are produced by the biogeochemical cycles of nitrogen, sulphur and chlorine. These play important roles in acid rain formation and in stratospheric ozone destruction. Unlike nitrous oxide (N_2O), nitric oxide (NO) and nitrogen dioxide (NO_2) are reactive. Collectively they are termed NO_x. This is produced by fossil-fuel combustion, lightning and biomass burning as well as being produced biogenically. Nitrogen is removed from the troposphere as dilute nitric acid in precipitation and as ammonium in aerosol fallout.

Reactive sulphur species are in the oxidized

form of sulphur dioxide (SO_2) and sulphur trioxide (SO_3), together with the reduced forms of hydrogen sulphide (H_2S) and dimethyl sulphide (DMS). Atmospheric sulphur is mainly anthropogenic and originates primarily from coal and oil combustion and copper smelting. DMS is produced locally by biological activity near the surface of the ocean. Despite the short lifetime of SO_2, sulphur may be transported over long distances by being converted into sulphate adhering to small aerosols. Sulphur is precipitated from the atmosphere as dilute sulphuric acid in rain, snow, fog and dew as well as in aerosol deposition.

Although ozone occurs in very small quantities in the atmosphere, its importance lies in its absorption of ultraviolet radiation in the stratosphere. Concentrations of ozone peak at about 25 km, and the *ozonosphere* (Fig. 2.1), therefore, shields the earth from radiation which is harmful to life. Ozone is formed by photochemical processes in which an oxygen molecule, O_2, absorbs a *photon* of energy, hv, and dissociates into highly reactive atomic oxygen, O. Thus

$$O_2 + hv \rightarrow 2O \tag{2.3}$$

where h is Planck's constant and v is the frequency of the radiation. Atomic oxygen then reacts with molecular oxygen and a third molecule, m, in a three-body collision

$$O_2 + O + m \rightarrow O_3 + m \tag{2.4}$$

to form ozone, O_3. Equilibrium is maintained by the dissociation of ozone under the influence of ultraviolet light with wavelengths larger than 0.2 μm according to the reaction

$$O_3 + hv \rightarrow O_2 + O \tag{2.5}$$

The depletion of stratospheric ozone takes place in reaction cycles summarized by

$$X + O_3 \rightarrow XO + O_2 \tag{2.6}$$
$$O_3 + hv \rightarrow O + O_2 \tag{2.7}$$
$$O + XO \rightarrow X + O_2 \tag{2.8}$$

where the catalyst X and its oxidized derivative XO are highly reactive free radicals (molecular fragments). Three such radicals are involved in the destruction of ozone. The first involves the hydroxyl radical resulting from the photochemical dissociation of water vapour molecules in the ozonosphere. The second is by nitric oxide resulting from the reaction of atomic oxygen with nitrous oxide transported from the lower troposphere or directly from aircraft exhausts. The third is the chlorine radical originating from the transport of CFCs and small amounts of methyl chloride from algal photosynthesis into the stratosphere. In addition, the conversion of ozone to oxygen is accelerated during

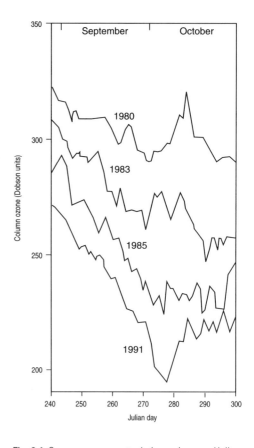

Fig. 2.4 Ozone measurements during spring over Halley Bay, Antarctica, for 1980, 1983, 1985 and 1991 showing the normal spring decline in ozone and the annual deepening of the ozone hole (after Jiang et al., 1996).

the polar night winter. The consequence is to produce a spring ozone hole until solar radiation and ozone production increase again during the polar day summer. However, the hole fails to fill to its previous level, and annual ozone depletion continues (Fig. 2.4).

A great variety of *aerosols* from natural and anthropogenic sources are carried into the atmosphere by wind (Table 2.2). These include dust from exposed soil, salt from the oceans, soot from natural fires, a variety of aerosols from industrial plants and emissions from volcanoes. Aerosols may also form by being converted into particles from inorganic trace gases, such as sulphates from SO_x, ammonium salts from ammonia (NH_3), nitrates from NO_x and phosphates from PO_x (oxides of phosphorus, P). Usually the concentrations of aerosols decrease with height, although occasionally they may be forced to high levels by volcanic eruptions or in the strong convection associated with large cumulonimbus clouds. The residence time of larger aerosols in

Table 2.2 Aerosol production estimates for aerosols less than 5 µm radius (10^9 kg/year) and typical concentrations near the surface (µg m^{-3}) (after Barry and Chorley, 1998).

		Concentration	
	Production	Remote	Urban
Natural			
Primary production:			
Sea salt	1 300	5–10	
Mineral particles	1 500	0.5–5[a]	
Volcanic	33		
Forest fires and biological debris	50		
Secondary production (gas → particle):			
Sulphates from H_2S	100	1–2	
Nitrates from NO_x	22		
Converted plant hydrocarbons	75		
Total natural	3 080		
Anthropogenic			
Primary production:			
Industrial dust	100		
Combustion (soot)	8	100–500[b]	
Biomass burning (soot)	5		
Secondary production (gas → particle):			
Sulphate from SO_2	140	0.5–1.5	10–20
Nitrates from NO_x	30	0.2	0.5
Biomass combustion (organics)	80		
Total anthropogenic	363		

a 10–60 µgm^{-3} during dust episodes from the Sahara over the Atlantic.
b Total suspended particles.

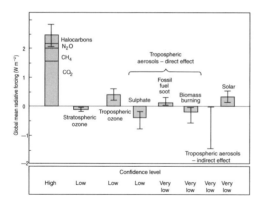

Fig. 2.5 Estimates of the annual, globally averaged anthropogenic forcings (W m^{-2}) due to changes in greenhouse gases and aerosols from pre-industrial times to the present (1992) and to natural changes in solar forcing from 1850 to the present. The error bars indicate estimates of uncertainty (after IPCC, 1996).

the atmosphere may be only a few days; small aerosols may remain suspended for many weeks to even years. The atmosphere is cleansed of aerosols by dry deposition as well as wet, in which the particulate material is rained out by precipitation. These effects are particularly noticeable over cities where dustfall is often high during fine, calm weather and where the scavenging effect of rainfall causes short periods of relatively pollution-free air.

The changing composition of the atmosphere at the present time is leading to a modification of the greenhouse effect. The increase in trace gases is leading to an enhanced heating effect. The increasing aerosol load of the atmosphere is leading to a reduction in the amount of radiation reaching the earth's surface and so acts against the increased greenhouse effect. The exact magnitudes of these different radiative forcings are not known owing to all the uncertainties in the estimates (Fig. 2.5). The greatest uncertainty is associated with indirect aerosol effects. These include the effect fine aerosols may have in altering the precipitation-producing mechanisms of the atmosphere and may lead to a diminution of rainfall in some regions. Whichever way aerosols are considered, their effect on the atmosphere is profound.

ADDITIONAL READING

BARRY, R. G. and CHORLEY, R. J., 1998. *Atmosphere, Weather and Climate*, 7th Edition. London: Routledge.

GRAEDEL, T. E. and CRUTZEN, P. J. 1997. *Atmosphere, Climate and Change*, Scientific American Library. New York: W. H. Freeman and Company.

SCHNEIDER, S. H. ed. 1996. *Encyclopaedia of Weather and Climate*, Vols 1 and 2. Oxford: Oxford University Press.

MCILVEEN, R. 1992. *Fundamentals of Weather and Climate*. London: Chapman and Hall.

MILLER, A. and THOMPSON, J. C. 1979. *Elements of Meteorology*, 3rd Edition. Columbus: Charles E. Merrill Publishing Company.

WALLACE, J. M. and HOBBS, P. V. 1977. *Atmospheric Science, an Introductory Survey*. New York: Academic Press.

3

PRESSURE, TEMPERATURE AND DENSITY RELATIONSHIPS

In seeking to understand the processes controlling the weather and climate it is essential to consider some of the fundamental laws of atmospheric physics that link air pressure, temperature and density. It is these quantities, together with water vapour and air motion, that define the state of the atmosphere; it is their variation in time and space that provides the basis for predicting weather and for understanding climate.

PRESSURE

Atmospheric pressure is defined as the mass of a column of air of unit cross section or the force acting on a unit area. Since force is defined as the product of mass and acceleration, it follows that pressure, p, is given by

$$p = \frac{ma}{\text{area}} \qquad (3.1)$$

or

$$p = ma \qquad (3.2)$$

per unit area where m is mass and a is acceleration. The unit of force applied to a mass of 1 kg to produce an acceleration of 1 m s^{-2} is a newton (N). Atmospheric pressure is thus measured as

$$\frac{N}{m^2} = Nm^{-2} = kgms^{-2}m^{-2} = kgm^{-1}s^{-2} \qquad (3.3)$$

Conventionally, newtons per square metre are defined as pascals, Pa, such that

$$100Nm^{-2} = 1hPa \qquad (3.4)$$

where the hectopascal (formerly millibar, mb) is the standard measure of atmospheric pressure. Both hPa and mb are used in the meteorological and climatological literature. The mb unit is increasingly falling into disuse.

An isopleth of equal pressure at a constant height is an *isobar*. Pressure at sea level is usually given by isobars. Above the surface it is more convenient to specify the *geopotential* or height at which a specific pressure is observed, such as the height (in geopotential metres) of the 850 hPa surface over South Africa (see Chapter 8 for a fuller discussion of geopotential).

TEMPERATURE

Whereas heat measures total kinetic energy (energy of movement) of molecular motion and is thus directly proportional to the mass of air being considered, temperature measures the average energy of molecular motion and is independent of mass. Temperature is measured on the scale 0 (melting point of ice) to 100 (boiling point of water) as °C on the Celsius scale or above the absolute zero of temperature, -273°C, on the Kelvin scale such that

$$T \text{ K} = 273 + t \text{ }^\circ\text{C} \tag{3.5}$$

By convention the degree symbol is not used for degrees Kelvin.

THE EQUATION OF STATE

The relationships between temperature, pressure and volume are expressed in laws developed by Boyle, Charles, and Gay-Lussac. Boyle found that if the temperature of a gas was kept constant while its volume varied, the pressure varied in a manner so as to keep the product of pressure, p, and volume, v, the same. Thus, the temperature and mass remain unchanged,

$$pv = c_1 \tag{3.6}$$

where c_1 is a constant. Equation (3.6) shows that if volume is doubled pressure is halved. This is in accordance with Boyle's law which states that the pressure of a given mass of gas at constant temperature varies inversely with the volume.

Boyle's law describes the effect of varying pressure on a volume at constant temperature. In contrast, Charles and Gay-Lussac investigated the effect of varying temperature on a volume at constant pressure. It was determined that with each degree Celsius rise or fall in temperature, the volume increased or decreased by 1/273 of its volume. This so-called volume-expansion coefficient is the same for all gases. In terms of the absolute temperature scale, the volume-expansion coefficient is such that 273 ml of gas at 273 K becomes 274 ml at 274 K and so on. Therefore, according to Charles's law, the volume occupied by a given mass of gas at different temperatures is directly proportional to the absolute temperature, if the pressure is kept constant. Thus

$$v = c_2 T \tag{3.7}$$

where c_2 is a constant. A definition of a perfect gas is that it obeys these two laws. The mixture of gases that makes up the atmosphere obeys

these laws only approximately, but for most practical purposes the air may be regarded as a perfect gas.

Pressure and temperature usually vary together so that a formula connecting temperature and volume changes at constant mass may be deduced by combining Boyle's and Charles's law as follows:

$$\frac{pv}{T} = c_3 \tag{3.8}$$

or

$$pv = c_3 T \tag{3.9}$$

where c_3 is a constant.

By Avogadro's law, one gram atomic weight of any substance contains 6.02×10^{23} molecules. This quantity is called a *mole*. One mole of gas molecules occupies 22.414 litres at a pressure of one atmosphere (1 013.25 hPa) and a temperature of 273 K. This applies to all gases. Avogadro was therefore able to show that gases containing the same number of molecules occupy the same volumes at the same pressure and temperature. For one *kilomole* of any gas the value of the constant in equation (3.9) is the same and is called the *universal gas constant*. The gas constant for air is the universal gas constant divided by the molecular weight of air. Thus equation (3.9) may be rewritten as

$$pv = RT \tag{3.10}$$

where R is the gas constant for dry air: 287 J kg^{-1}K^{-1}. The *density* of air is defined as

$$\rho = \frac{m}{v} \tag{3.11}$$

If unit mass is considered, the specific volume, α, is given by

$$\alpha = \frac{1}{\rho} \tag{3.12}$$

and equation (3.10) becomes

$$p = R\rho T \tag{3.13}$$

or

$$p\alpha = RT \tag{3.14}$$

In either form this is the *equation of state* and is of fundamental importance in atmospheric science, since it governs the relationships between air density and temperature. From

$$\rho = \frac{p}{RT} \tag{3.15}$$

it follows that the greater the temperature of a parcel of air, the lower will be its density (and consequently the greater its buoyancy and tendency to rise) and vice versa.

A problem arises when an attempt is made to accommodate the effect of water vapour in the atmosphere. The gas constant for dry air is less than the gas constant for water vapour, R_w. Since the density of water vapour, ρ_w, is given by

$$\rho_w = \frac{p}{R_w T} \tag{3.16}$$

it follows that the presence of water vapour at a given temperature and pressure lowers the density of the air. Rather than use the gas constant for moist air in computational work, a better solution is to determine the temperature that dry air must have in order to have the same density and pressure as a volume of moist air. This fictitious temperature is called the *virtual temperature*, T_v. Its effect is to increase the temperature by an amount corresponding to the lowered density caused by water vapour. Virtual temperature may be calculated from

$$T_v = \frac{T}{1 - \dfrac{e(1 - \epsilon)}{p}} \tag{3.17}$$

where e is water vapour pressure and $\epsilon = mw_w/mw_d = 18.106/28.99 = 0.622$ and where

mw_w and mw_d are the molecular weights of water vapour and dry air respectively. By incorporating the effect of moisture into the definition of virtual temperature, the equation of state may be expressed in an alternative form as

$$p = R\rho T_v \tag{3.18}$$

If, like water, the atmosphere were incompressible, pressure would decrease upward at a uniform rate. The compressibility of atmospheric gases, however, is demonstrated by their ability to expand with decreasing pressure and become compressed with increasing pressure. From equation (3.18) it is apparent that in a gas kept at constant temperature, the specific volume is inversely proportional to the pressure; that is, if the pressure is doubled, specific volume is reduced by half. Under these conditions, pressure is directly proportional to density, with falling pressure causing density to decrease. This is apparent in Table 3.1 which shows density to decrease from 1.2 kg m^{-3} at sea level to 0.6 kg m^{-3} at 6 000 m (472.2 hPa). Put another way, approximately half the mass of the atmosphere is shown to lie below 6 000 m. This is a sobering thought when the pollution of this vital natural resource is brought to mind.

The variation of atmospheric pressure is much greater vertically than horizontally. It is useful, therefore, to define the *Standard Atmosphere* which describes the vertical structure of the atmosphere as a function of height and against which departures can be observed. By international agreement the Standard Atmosphere at sea level is defined as the pressure to support a mercury column 76 cm high at latitude 45° with a surrounding temperature of 15 °C. This pressure is 1 013.2 hPa. Selected values from the Standard Atmosphere are given in Table 3.1.

Table 3.1 Selected values from the Standard Atmosphere.

Altitude, m	Temperature, °C	Pressure, hPa	Density, kg m^{-3}
0	15.0	1 013.2	1.2250
6 000	−24.0	472.2	0.6601
11 000	−56.4	227.0	0.3648
20 000	−56.5	55.3	0.0889
50 000	−2.5	0.8	0.0010
90 000	−92.5	0.002	0.000003

THE HYDROSTATIC EQUATION

In the absence of an atmosphere, a free-falling object would move downward under the influence of the earth's gravitational acceleration, g, at a rate of 9.8 m s^{-2}. A volume of air also has mass and should be similarly affected. Yet air shows no inclination to accelerate downward and, in general, the atmosphere appears to be in a state of equilibrium. This can only be explained by the existence of an equal and opposite force to that of gravity. This force is an upward-directed pressure gradient force acting from higher pressure near the surface to lower pressure above. The balance of forces that exists in an atmosphere free of vertical acceleration is known as *hydrostatic equilibrium*. Consider a thin slice of air of unit cross-sectional area, a, and thickness, dz, which exerts an incremental downward pressure, dp, within a column of air (Fig. 3.1, *upper left*). Since

$$\rho = \frac{m}{v} \tag{3.19}$$

and

$$m = \rho v \tag{3.20}$$

the mass of the slice will be

$$m = \rho dza = \rho dz \tag{3.21}$$

By definition

$$p = mg \tag{3.22}$$

The pressure contributed by the air extending to height z is p and that to height $z + dz$ is $p + dp$, where dp is the incremental pressure contributed by the slice. Since pressure decreases with height in the atmosphere, dp must be negative and thus the incremental pressure contribution will be

$$- dp = \rho dzg \tag{3.23}$$

Equation (3.23) is termed the *statical* or *hydrostatic equation* and is usually expressed in the form

$$\frac{dp}{dz} = -\rho g \tag{3.24}$$

THE PRESSURE–HEIGHT RELATIONSHIP

Substituting

$$\rho = \frac{p}{RT} \tag{3.25}$$

from equation (3.15) into equation (3.24), it is seen that

$$- \frac{dp}{dz} = \frac{pg}{RT} \tag{3.26}$$

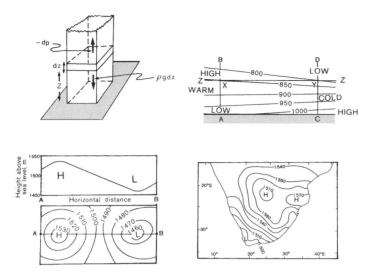

Fig. 3.1 *Upper left:* the hydrostatic balance of forces when no vertical accelerations are present. Arrows represent the downward weight of the air and the upward force due to the vertical pressure gradient (*dp* is negative because pressure decreases with height); *upper right:* the effect of warming and cooling on the production of horizontal pressure gradients on constant pressure surfaces; *lower left:* a cross-section from A to B through a constant pressure surface; and *lower right:* an example of contours (gpm) of the 850 hPa surface.

Consequently in warm air (higher T) the decrease of pressure with height ($-dp/dz$) will be less rapid than in cold air (lower T). The spacing between pressure surfaces will thus be wider in warm air than in cold air, as is seen in Figure 3.1 (*upper right*), where at height z the pressure at X (850 hPa) exceeds that at Y (800 hPa). The difference in height, z_2-z_1, between any two pressure levels in the atmosphere is called the *thickness* of the layer. The finite thickness between pressure surfaces may be determined by integrating equation (3.23) from z_0 to z and p_0 to p. This gives

$$\Delta z = \frac{R\overline{T}}{g} \ln \frac{p}{p_0} \qquad (3.27)$$

where Δz is the thickness in metres between the lower pressure surface p_0 and the upper pressure surface p and $\overline{T}$ is the mean temperature (in K)

between the pressure surfaces. This equation simplifies to

$$\Delta z = 29.27\overline{T}\,(\log_e p_0 - \log_e p) \qquad (3.28)$$

or

$$\Delta z = 67.4\overline{T}\,(\log_{10} p_0 - \log_{10} p) \qquad (3.29)$$

depending on whether natural logarithms or logarithms to the base ten are used respectively. A more precise solution is obtained using the mean virtual temperature, $\overline{T}_v$, defined previously in equation (3.17).

From equation (3.27) it is apparent that the thickness of a layer bounded by two pressure surfaces is directly proportional to the mean temperature of the layer. Although this equation assumes an isothermal atmosphere, it can be used without serious error provided the thickness is not large. For example, if the mean temperature of the atmospheric layer bounded by the 950 hPa and the 900 hPa pressure surfaces in Figure 3.1 (*upper right*) is 20 °C along the vertical section AB and 0 °C along CD, the difference in thickness calculated from equation (3.27) is 32 m.

The variation of pressure with height means that in mountainous terrain surface pressure recorded at individual stations will fluctuate with differences in elevation. In order to identify pressure fields associated with the passage of weather systems, it is necessary to reduce pressure to a common reference level. This datum may be either mean sea level or a standard pressure level.

Whereas the reduction of pressure to sea level is a relatively simple matter over low-lying areas, difficulties occur when the surface pressure of higher-altitude inland stations is reduced to sea level. Problems arise in the determination of the correct temperature of the layer between the recording station and mean sea level (which is below-ground). A fictitious temperature has to be used in the reduction process and this seldom produces satisfactory results. Southern Africa, with its high-standing plateau surrounded by a narrow coastal margin, presents a particular problem. This has been overcome in the past by using two levels on South African weather maps. Over the sea, pressure has been represented by hectopascals at the constant height of mean sea level; over the land, surface pressure is represented by means of contours of the 850 hPa pressure surface (which approximates the height of the plateau). This convention is followed throughout in this book.

On a constant pressure surface variations are represented by undulations in the surface such that topographic highs represent pressure highs and topographic lows represent pressure lows (Fig. 3.1, *lower left*). An example of the pressure map drawn from the topography of the 850 hPa surface is shown in Figure 3.1 (*lower right*). The variations in height of the surface are represented by contours in geopotential metres above mean sea level with the topographic gradient of the surface increasing toward the north-east from the south-west. Equation (3.27) reveals that the mean temperature between sea level and the 850 hPa surface will increase in proportion to the contours of the 850 hPa surface.

ADDITIONAL READING

HOUGHTON, J. T. 1997. *The Physics of Atmospheres*. Cambridge: Cambridge University Press.

PEIXOTO, J. P. AND OORT, A. H. 1992. *Physics of Climate*. New York: American Institute of Physics.

MCILVEEN, R. 1992. *Fundamentals of Weather and Climate*. London: Chapman and Hall.

MCINTOSH, D. H. and THOM, A. S. 1978. *Essentials of Meteorology*. London: Wykeham Publications.

WALLACE, J. M. and HOBBS, P. V. 1977. *Atmospheric Science, an Introductory Survey*. New York: Academic Press.

BYERS, H. R. 1974. *General Meteorology*, 4th Edition. New York: McGraw-Hill.

4

THE ADIABATIC PROCESS

Heat is added to or taken away from the atmosphere in many different ways, each resulting in warming or cooling of the air. Absorption of radiation causes warming; emission results in cooling. Conduction plays a minor role and convection a major role in heat transfer. Large amounts of heat may be transported by horizontal advection of air. In the vertical, as buoyant air rises it expands and cools; sinking air contracts and warms. Considerable amounts of latent heat may be released in the atmosphere when the condensation of water vapour occurs in cloud formation. In this chapter adiabatic thermodynamic heating and cooling effects associated with vertically moving air will be examined.

SENSIBLE AND LATENT HEAT

Changes in atmospheric heat that are accompanied by no changes of state involve *sensible heat* that can be sensed by temperature-measuring devices. The heat required to effect a change from a lower to a higher state (such as solid to liquid or liquid to gas) or released in the reverse process is *latent heat*. In an isobaric process, when pressure is held constant, the amount of heat, dh, required to raise the temperature of a parcel of air of unit mass by dT is

$$dh = C_p\, dT \tag{4.1}$$

where C_p is the specific heat of air at constant temperature. If the volume of the parcel is held constant then

$$dh = C_v dT \tag{4.2}$$

where C_v is the specific heat of air at constant volume. The two specific heats are related such that

$$C_p - C_v = R \tag{4.3}$$

The changes in molecular configuration that accompany the fusion of atmospheric water into ice or condensation of vapour into liquid are accompanied by a release of latent heat without any change of temperature. Likewise, changes of phase involving melting and evaporation require the input of latent heat (extracted from the environment), but also without involving a temperature change. For example, if heat is supplied to ice at normal atmospheric pressure and 0 °C, the temperature will remain constant until all the ice has melted. The latent heat of fusion is 3.34×10^5 J kg^{-1}. The latent heat of vaporization, which is the amount of heat required to convert unit mass of material from the liquid to the vapour phase without a change of temperature, is 2.50×10^6 J kg^{-1}. The latent heat of condensation has the same numerical value.

THE FIRST LAW OF THERMODYNAMICS

The first law of thermodynamics governs the changes that take place when air is heated in the atmosphere. A closed parcel of air (i.e., one of constant volume) has an *internal energy* that is made up of both the kinetic and potential energy (i.e., energy of motion and of position) of its constituent molecules that may be defined by equation (4.l). If the parcel is heated, its internal energy will increase as the molecular motion of the air becomes more vigorous. If at the same time the air is allowed to expand, then in so doing it must perform *work* against the ambient environmental air. This work, dW, will depend on the volume expanded, dv, as well as the constricting pressure exerted by the surrounding air, p, such that

$$dW = pdv \tag{4.4}$$

The amount of heat added in the heating process, or *enthalpy*, dh, must be balanced by both the increase in the internal energy of the parcel and the work it has to do on the environment. This is a statement of *the first law of thermodynamics* in its general form. Thus

heat added = change in internal energy + work done

which is

$$dh = C_v dT + pdv \tag{4.5}$$

In a more convenient form, meteorologically, the first law of thermodynamics may be expressed as

$$dh = C_p dT - vdp \tag{4.6}$$

in which heat changes are related to pressure and temperature variations rather than to temperature and volume.

ADIABATIC PROCESSES AND LAPSE RATES

Consider dry air (for this purpose air in which no condensation occurs). An adiabatic process is one in which no heat enters or leaves the system, so in equation (4.6) dh is zero, and

$$C_p dT = vdp \tag{4.7}$$

This equation expresses the relationship between adiabatic changes in temperature and pressure. Usually it is more convenient to express temperature changes with height rather than pressure, namely as *lapse rates*. This can be done in equation (4.7) by converting changes of pressure to changes in height using the statical equation and by replacing volume by the ratio of mass to density. In so doing, it becomes apparent that in an adiabatic process the lapse rate of temperature of the parcel undergoing vertical motion will be

$$-\frac{dT}{dz} = \frac{g}{C_p} = \Gamma \tag{4.8}$$

where Γ is the *dry adiabatic lapse rate*. Notice that equation (4.8) is read as the decrease of temperature with height (hence the negative sign before dT/dz). Γ is a constant approximating to 9.8 K km⁻¹, or more usually 1 °C 100 m⁻¹. A parcel of air rising dry-adiabatically will cool at this rate; correspondingly, a parcel sinking (subsiding) warms at this rate.

Even though moisture may be present in the atmosphere, providing this does not condense, the air may be considered to be dry in an adiabatic process. If condensation occurs, the latent heat so released will retard the cooling of the rising air and the adiabatic lapse rate will begin to alter depending on the variation of the saturated humidity mixing ratio with height. The new lapse rate is the *saturated* or *wet adiabatic lapse rate*, Γ', and is defined as

$$-\frac{dT}{dz} = \Gamma + \frac{L}{C_p}\frac{dx_s}{dz} = \Gamma' \qquad (4.9)$$

where L is the latent heat of condensation. Since the change in the saturated mixing ratio, dx_s, is negative, owing to the drying of the air as vapour is converted to liquid in the condensation process, it follows that the second term on the right in equation (4.9) is negative and that $\Gamma' < \Gamma$. Notice also that Γ' is not a constant, but varies as x_s varies with height. Near the surface the difference between the dry and wet adiabatic lapse rates is large, but in the upper troposphere it may be quite small owing to the dryness of the atmosphere.

POTENTIAL TEMPERATURE

If the temperature, T, at any pressure level, p, is reduced adiabatically to a standard reference level of 1 000 hPa, then the resulting temperature is known as the potential temperature, θ. The equation for doing this is

$$\theta = T\left(\frac{1000}{p}\right)^{R/C_p} \qquad (4.10)$$

in which the constant R/C_p for dry air is 0.286. Potential temperature provides a useful label for an air parcel because it remains unchanged no matter how many adiabatic processes occur; that is, it conserves its properties. Consequently it may be used to facilitate the identification of different parcels of air and their movement.

STABILITY AND INSTABILITY

The atmosphere is in a constant state of temperature change caused by such factors as surface heating, large-scale subsidence of air and advection. These changes produce spatial as well as temporal departures from the normal temperature lapse rate and influence the ability of the atmosphere to promote or inhibit the vertical motion of air.

The word *stability* is used to indicate a condition of equilibrium. A state of stable equilibrium has occurred when a mass of air, uplifted by some outside force, tends to return to its original position. In a state of unstable equilibrium, a mass of air given an upward or downward impetus continues to rise or sink of its own accord. If the air is unsaturated, it will cool at the dry adiabatic lapse rate with lifting, and warm at the same rate with sinking. Saturated air cools and warms at the saturated adiabatic lapse rate. In each case the equilibrium state of the atmosphere is determined by referring to the temperature of the ambient environmental air and that of adiabatically cooled air at the same height. In the *parcel method* of determining stability and instability, it is important to distinguish between the behaviour of discrete parcels of air undergoing lifting and adiabatic cooling and the surrounding air, which is assumed to be undergoing no compensatory vertical motion. Parcels are also assumed to undergo minimal mixing with the surrounding air and to be able to retain their identity.

The thermal structure of the atmosphere is described by the lapse rate of temperature. If temperature decreases with height, a *lapse* condition is said to prevail; if it increases with height, a temperature *inversion* occurs. With no variation of temperature with height *isothermal* conditions prevail (Fig. 4.1, *upper*). Both inversions of temperature and isothermal stratifications are stable. Lapse conditions may be either stable or unstable depending on the magnitude of the environmental lapse rate and on whether the air is saturated or not.

Unsaturated air

A condition of stable equilibrium exists when the lapse rate of environmental air, α, is similar to that depicted in Figure 4.1 (*lower left*). If air at level z_0 is lifted by some external forcing mechanism, orographically or by convergence in the wind field (see Chapter 8), for instance, then it will cool at the dry adiabatic lapse rate, and at

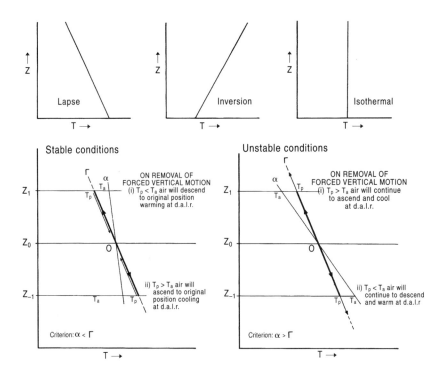

Fig. 4.1 *Upper:* types of lapse rates; *lower:* temperature–height diagrams and the definition and occurrence of stability and instability in the atmosphere.

height z_1 it will be cooler and, therefore, denser than the environmental air. In this case $T_p < T_a$ where T_p is the temperature of the parcel and T_a the temperature of the ambient air. With the removal of the upward force, the parcel of air will sink back to its original position.

If the parcel is forced downward from z_0, it will warm at the dry adiabatic lapse rate. At a height, z_{-1}, below that from which it has been displaced, it will be warmer and thus less dense than the surrounding air; that is, $T_p > T_a$. The positive buoyancy of the parcel will cause it to return to its original position once the downward force is removed. The buoyancy force is such as to cause the parcel to overshoot its original displacement level and to become immediately cooler than its environment, whereupon it will sink, overshoot the reference level, become warmer to rise again, and so on in a rapidly diminishing oscillation about the origi-

nal level of displacement. Such behaviour is common in stable air and is responsible for lenticular wave clouds forming over mountains (Fig. 4.2, *upper left*). Such waves, which form in the vertical, are rapidly damped out with distance away from the source of uplift. Lenticular wave clouds are of great beauty, and may maintain their positions for hours. They are common over the Western Cape mountains and the Drakensberg (see Figure 10.15).

The criterion by which *stable conditions* are recognized is the magnitude of the lapse rate, α. If the environmental lapse rate is less than the dry adiabatic lapse rate ($\alpha < \Gamma$) then the air is stable. Using a similar argument, if the environmental lapse rate exceeds the dry adiabatic lapse rate ($\alpha > \Gamma$), then the air will be *unstable* (Fig. 4.1, *lower right*). In an unstable environment, given an upward impetus, an air parcel ascending from z_0 will be warmer than the surround-

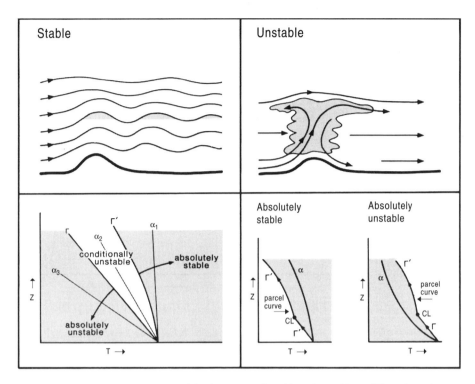

Fig. 4.2 *Upper:* characteristic vertical motion and cloud occurrence in stable and unstable conditions; *lower:* lapse rates and conditional and absolute states of instability.

ing air at z_1, and will continue to ascend. Conversely, a downward impetus causes the parcel to become progressively cooler and denser than its surroundings so that the parcel continues to sink. Under such conditions the development of strong vertical motions, both upward and downward, may be facilitated as, for example, in a storm forming over a mountain where orographic uplift may provide the initial vertical forcing necessary to trigger the instability (Fig. 4.2, *upper right*). When the environmental lapse rate and the dry adiabatic lapse rate are equal ($\alpha = \Gamma$), the condition is known as *neutral stability* and is one in which vertical motion is neither favoured nor resisted.

Saturated air

Once a parcel of rising air is saturated it cools at the saturated adiabatic lapse rate, Γ'. Using a

similar argument to that relating to unsaturated air, it follows that if $\alpha < \Gamma'$ the atmosphere will be stable, when $\alpha > \Gamma'$ it will be unstable, and when $\alpha = \Gamma'$ it will be neutral.

Since the dry adiabatic lapse rate is greater than the wet adiabatic lapse rate, it follows that $\alpha > \Gamma$ defines the most extreme unstable conditions, called *absolute instability*. Similarly, $\alpha < \Gamma'$ denotes *absolutely stable* conditions. Such situations may be illustrated graphically as in Figure 4.2 (*lower left*). Absolutely stable conditions occur before inversion conditions form. An inversion is absolutely stable, but absolutely stable air is not necessarily associated with an inversion. Absolutely stable layers are highly restrictive to the upward transfer of aerosols and trace gases; they are considered further in Chapters 12, 14 and 15.

In the region between the two adiabatic lapse rates, the presence or absence of instability will be conditional upon the moisture content of the

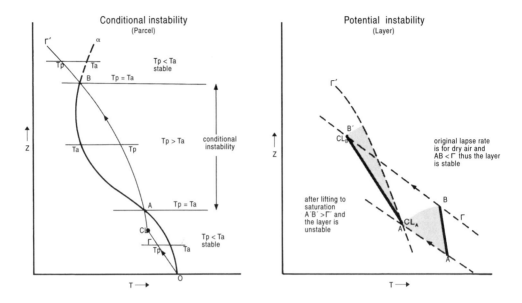

Fig. 4.3 Temperature–height diagrams and the occurrence of conditional and potential instability (T_p denotes parcel temperature and T_a ambient air temperature).

air. If air is lifted to condensation level, it will cool dry-adiabatically until saturation and condensation occurs; thereafter it will cool wet-adiabatically. If the combined parcel ascent curve is such that at all levels the parcel is cooler than the ambient environmental air, then the air will be absolutely stable (Fig. 4.2, *lower right*). If the ascent curve is such that the parcel is everywhere warmer than the ambient air, then it will be absolutely unstable. Usually the situation is more complicated than this since, depending on its moisture content, the atmosphere may show *conditional instability*.

CONDITIONAL INSTABILITY

If a parcel of unsaturated air is forced to ascend, it will cool at the dry adiabatic lapse rate until it becomes saturated with respect to a plane surface of pure water. The height at which this occurs is known as the *lifting condensation level* (CL in Fig. 4.3, *left*). The line from the surface

at 0 through A to B and beyond is referred to as the *ascent curve* of the rising parcel of air. Below condensation level the parcel is everywhere cooler than its environment: $T_p < T_a$, and the air is stable. This means that if the lifting force is removed, the negative buoyancy of the air will immediately cause it to sink back to its original position. Above condensation level further ascent of air takes place at the wet adiabatic lapse rate until a point is reached when $T_p = T_a$ at A. Thereafter, as the air continues to ascend at the wet adiabatic lapse rate, the parcel will be warmer than the ambient air, $T_p > T_a$. The air is now unstable and will rise freely of its own accord without any forced lifting. This will continue until point B is reached after which the air again becomes stable. The layer AB is said to be conditionally unstable and point A is referred to as the level of *free convection*. The height at which this occurs depends upon the moisture content of the rising air as well as the magnitude of the environmental lapse rate.

POTENTIAL (CONVECTIVE) INSTABILITY

In addition to the localized ascent of small parcels of air, the atmosphere may also undergo large-scale vertical movements which may extend over thousands of square kilometres. Embedded within such a rising mass of air there frequently exists a layer that has a vertical distribution of moisture in which the lapse rate of wet-bulb temperature exceeds the wet adiabatic lapse rate. If lifted sufficiently, such a layer becomes unstable and is said to be *potentially unstable*. An alternative term for such instability is *convective instability*. The term is a misnomer, however, since it refers to the characteristic of a layer of air and has little reference to cumulus (parcel) convection in the generally accepted sense. Only the term *potential instability* will be used in this book. The instability is potential because the air is stable until lifted by the appropriate amount. By contrast, conditional instability refers to the condition of the air before parcels are lifted through it. It is possible for both types of instability to be present together or for a layer to be potentially unstable, while showing no conditional instability. With potential instability, the initial lapse rate in layer AB in Figure 4.3 (*right*) is stable. The air is dry (i.e., has not reached saturation) and must be compared with the dry adiabatic lapse. When this is done it is seen that $\alpha < \Gamma$ and the air is consequently stable. Provided the moisture content at the bottom of the layer, A, is much higher than at the top, B, when the layer is lifted, the base of the layer reaches condensation level before the top. Once the air in the layer has

reached saturation at condensation level and beyond, further ascent will take place at the wet adiabatic lapse rate. The new lapse rate of the layer at and above condensation level is A′B′, which now needs to be compared to the wet adiabatic lapse rate in order for its state of stability to be assessed. It is seen that with the new lapse rate A′B′ > Γ. Consequently, the air is now unstable and further vertical motion will be enhanced as a result of the realization of the potential instability.

A final type of instability, called *latent instability*, may be present in the atmosphere. It is most appropriately defined on a thermodynamic diagram called a tephigram. This will be done in Chapter 6 where the discussion of stability and instability will continue.

ADDITIONAL READING

HOUGHTON, J. T. 1997. *The Physics of Atmospheres*. Cambridge: Cambridge University Press.

MCILVEEN, R. 1992. *Fundamentals of Weather and Climate*. London: Chapman and Hall.

PEIXOTO, J. P. and OORT, A. H. 1992. *Physics of Climate*. New York: American Institute of Physics.

MCINTOSH, D. H. and THOM, A. S. 1978. *Essentials of Meteorology*. London: Wykeham Publications.

WALLACE, J. M. and HOBBS, P. V. 1977. *Atmospheric Science, an Introductory Survey*. New York: Academic Press.

BYERS, H. R. 1974. *General Meteorology*, 4th Edition. New York: McGraw-Hill.

5

MOISTURE AND PRECIPITATION

Moisture in the atmosphere is fundamental to all life forms on earth. The moisture content of the atmosphere governs the degree of stability in the atmosphere and the occurrence of precipitation. Before precipitation can take place, condensation must occur, clouds must form and large enough water drops or ice crystals must develop to precipitate out of clouds. The study of clouds provides useful information on both the type of weather that can be expected at the earth's surface and the nature of weather-producing systems themselves. Clouds have long featured in weather lore due to the perceptive observation that the shape, structure, texture and patterns of clouds express the kinds of air movements responsible for their development. Recent analyses of cloud characteristics using satellite pictures have also greatly enhanced the understanding of weather systems and concomitant rainfall patterns. (A full discussion of cloud types and classification appears in *The Atmosphere and Weather of Southern Africa*, the first edition of this book, and it should be consulted for details. See 'Additional reading' at the end of this chapter). In this chapter, water vapour in the atmosphere, the processes of cloud-droplet and ice-crystal formation, and cloud-droplet growth and precipitation mechanisms will be considered.

WATER VAPOUR IN THE ATMOSPHERE

Water may be present in the air as a solid, a liquid or a gas and may change phase into a gaseous state by the process of *evaporation* (liquid to gas) or *sublimation* (solid to gas directly) at temperatures commonly observed in the atmosphere. Conversely it may change from gas to liquid by *condensation* and from gas to solid or liquid to solid by *crystallization*. Water vapour plays a fundamental role in regulating atmospheric processes, not only in the formation of precipitation, but also in heat exchange by affecting the transmission of radiation and through heat released or taken up in phase changes. Water is supplied to the atmosphere from the surface through evaporation and returns to the surface as precipitation. In general, for an entire continent precipitation exceeds evaporation, whereas over the ocean the reverse is true. The amount of water present in the atmosphere as a gas is enormous and far exceeds that present as a liquid or solid. Over South Africa the water vapour content of the air is significantly greater than the volume of water transported by all rivers combined. Only a small fraction of this water vapour is actually precipitated.

Unlike temperature and rainfall, there is no single measure of water vapour content of the atmosphere and different measures are used for different purposes.

Vapour pressure

Atmospheric pressure is the total pressure exerted by all the gases in the atmosphere. It is, therefore, the sum of the partial pressures of the constituent gases, of which the evaporated water vapour is one. The partial pressure exerted by the water vapour is called the *vapour pressure, e*. Like atmospheric pressure it is expressed in hectopascals. The amount of water vapour that can remain in a gaseous state is a function of temperature alone. Evaporation takes place initially from a plane water surface to the surrounding air. A state of equilibrium is reached when, at a particular temperature, no further exchange takes place between the air and water. At this stage the air is said to be *saturated*. The pressure exerted by this saturated air is called the *saturation vapour pressure, e*$_s$. The magnitude of the saturation vapour pressure increases with temperature (Fig. 5.1). At 100 °C the saturation vapour pressure reaches a pressure of one atmosphere. Boiling point can therefore also be defined as the temperature at which the saturation vapour pressure is the same as the Standard Atmosphere sea level pressure. This explains why at lower pressures associated with higher altitudes water boils at lower temperatures.

The difference between e_s and e is known as the *saturation deficit* and is a measure of the dryness of the air. The variation of saturation deficit over South Africa in January shows highest values over the north-western interior (Fig. 5.2, *upper left*). Since both e_s and e vary with temperature, spatial comparison of saturation deficit at different times of day becomes meaningless. Hence comparisons are usually made for fixed times, for example 14:00. In most cases $e_s > e$. However, it is possible for water vapour to remain a gas at vapour pressures exceeding the saturated water vapour pressure. Such a state is known as *supersaturation*. If two equal masses of air, A and B in Figure 5.1 (*left*), each unsaturated but close to saturation, are mixed, then the resulting mixture will have a vapour pressure, C, lying halfway along the straight line joining A and B. The resultant mixture will be supersaturated by the amount CD and the excess moisture usually will precipitate as cloud or fog in order to restore the air to the equilibrium saturation vapour pressure, D. The position of C along AB is proportional to the masses of air involved in the mixing. For fog or cloud of this kind to form, the air masses initially must be close to saturation. If this is not so, then the straight line joining their respective temperature-dependent vapour pressures may not intersect the equilibrium saturation vapour pressure curve.

Changes in the phase of water are important in the atmosphere and vary with vapour pressure and temperature (Fig. 5.1, *right*). Along the evaporation curve, water and vapour are in equilibrium; to the left of the curve, only water exists and to the right only gas. The evaporation curve ends at the so-called triple point (e = 6.11 hPa, T = 273 K) where the distinction between liquid and vapour ceases. The sublimation curve represents the equilibrium conditions between vapour and air, whereas the melting curve represents those between water and ice.

Water-drops in the atmosphere may readily exist when they are cooled below freezing. The equilibrium between such *supercooled* (or *undercooled*) water and its saturated vapour is

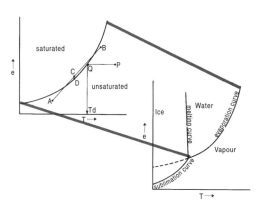

Fig. 5.1 Phase diagram for water showing equilibrium *e*, *T* curves separating vapour, liquid and solid states. *Left:* the saturation vapour-pressure curve for a plane water surface.

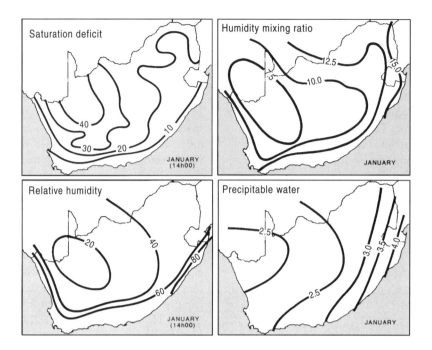

Fig. 5.2 Mean January measures of moisture content of the air over southern Africa. *Upper left:* saturation deficit (hPa) at 14:00; *upper right:* humidity mixing ratio (g kg⁻¹); *lower left:* relative humidity (per cent); and *lower right:* precipitable water (cm) (after Jackson, 1961; Schulze, 1984; McGee, 1986).

given by the dashed extended evaporation curve in Figure 5.1.

As can be seen in the phase diagram of Figure 5.1, air may be saturated with respect to an ice surface, in which case the saturation vapour pressure differs from that over water. Table 5.1 shows that at 0 °C (i.e., 273 K) the saturation vapour pressures are the same. At temperatures below 0 °C, however, the saturation vapour pressure with respect to ice is less than that with respect to water. This has important implications for the growth of ice crystals in clouds. Vapour pressure may be determined from

$$e = e_s - Ap(T - T_w) \qquad (5.1)$$

where T is the dry-bulb temperature, T_w is the wet-bulb temperature, e_s is the saturation vapour pressure over water at the temperature of the wet bulb, p is the atmospheric pressure and A is

a constant which relates to ventilation conditions and instrument characteristics.

Dry-bulb temperature is determined from a ventilated or aspirated thermometer. *Wet-bulb temperature* is the temperature of a ventilated thermometer whose bulb is surrounded by water evaporating freely from a saturated wick dipped into pure (i.e., distilled) water. As evaporation of water vapour occurs, so the latent heat of vaporization causes cooling. The wet-bulb temperature is defined, therefore, as the lowest temperature to which air can be cooled by evaporating water into it.

Mixing ratio

The amount of water vapour in a given volume of air that is measured as the ratio of the mass of water vapour, m_w, to the mass of dry air, m_d, is called the *mixing ratio, x*. It is expressed as

Table 5.1 Selected saturation vapour pressures (hPa) over water and ice.

T K	253	263	273	283	293	303
Water	1.26	2.87	6.11	12.28	23.38	42.43
Ice	1.04	2.60	6.11			

$$x = \frac{m_w}{m_d} \tag{5.2}$$

in grams of water vapour per kilogram of dry air. It is usually more convenient to determine the mixing ratio from measurements of vapour pressure and total pressure, in which case x can be determined from

$$x = \frac{mw_w}{mw_d} \frac{e}{p-e} \tag{5.3}$$

Since the ratio of the molecular weights of dry air, mw_d, and water vapour, mw_w, is 0.622 and e is significantly less than p, for most practical purposes

$$x = \frac{0.622e}{p} \tag{5.4}$$

Typical values of surface humidity mixing ratio vary between about 7.0 and 15.0 g kg^{-1} over South Africa in January (Fig. 5.2, *upper right*) to between about 7.5 and less than 5.0 g kg^{-1} in July. The great value of the humidity mixing ratio and specific humidity (to follow) is that they are measures of moisture content that are independent of temperature and vary only with the addition and removal of water vapour.

The *saturated mixing ratio*, x_s, is defined as the mass of water vapour in a given volume of air, saturated with respect to a plane surface of pure water, to the mass of dry air. Like equations (5.3) and (5.4) it may be expressed as,

$$x_s = \frac{0.622e_s}{p - e_s} \tag{5.5}$$

or

$$x_s = \frac{0.622e_s}{p} \tag{5.6}$$

Specific humidity

Specific humidity, q, is defined as the mass of water vapour, m_w, contained in a unit mass of moist air (dry air plus water vapour), m_a, expressed as grams of water vapour per kilogram of moist air:

$$q = \frac{m_w}{m_a} \tag{5.7}$$

Alternatively

$$q = \frac{0.622e}{p - 0.378e} \tag{5.8}$$

Relative humidity

Relative humidity is the ratio of the observed humidity mixing ratio to that which would saturate the air at the same temperature. Expressed as a percentage it is

$$RH = \frac{x100}{x_s} \tag{5.9}$$

Alternatively

$$RH = \frac{e100}{e_s} \tag{5.10}$$

or

$$RH = \frac{q100}{q_s} \qquad (5.11)$$

Relative humidity is a highly variable quantity and is influenced by both temperature and pressure. Consequently any comparisons of relative humidity must be made at similar times of day to eliminate the effect of the diurnal inverse variation of temperature. An example of relative humidity variation over South Africa at 14:00 is given in Figure 5.2 (*lower left*).

Dew-point temperature

Dew-point temperature is the temperature to which air at constant pressure and water vapour content must be cooled in order to become saturated and for dew to precipitate. In Figure 5.1 if air at P is cooled, saturation will occur at Q with temperature T_d the dew-point temperature.

Precipitable water

Precipitable water is a measure of the total water vapour content of the atmosphere over a given place and is obtained from

$$w = \frac{1}{g} \int_{p_2}^{p_1} x \, dp = \frac{\bar{x}(p_1 - p_2)}{g} \qquad (5.12)$$

where $\bar{x}$ is the mean humidity mixing ratio within the layer extending from pressure levels p_1 to p_2. The quantity w is the precipitable water and is measured in units of 10^{-1} kg m^{-2} which is equivalent to a depth of water in units of 10^{-1} mm. More usually precipitable water is given in centimetres and refers to the whole atmospheric column. Over South Africa precipitable water in January varies from about 4.0 cm in the east to less than 2.5 cm over the western areas (Fig. 5.2, *lower right*).

THE CONDENSATION PROCESS

If particle-free air is rapidly expanded (and therefore cooled) so that relative humidity is increased by several hundred per cent, embryonic water droplets form by *spontaneous nucleation* as water vapour molecules come together by random fluctuations. However, the atmosphere is not free of airborne *aerosol* particles and relative humidities seldom exceed 101 per cent. In order for cloud droplets to form, therefore, a process is required that combines the presence of aerosols in the atmosphere with low supersaturations. A discussion of this process must begin with the nature and characteristics of aerosols.

Aerosol characteristics

When an unsaturated parcel of air is cooled and/or acquires additional moisture to the point where it becomes saturated, water vapour condenses onto aerosol nuclei which then grow in size to become cloud drops. This process was demonstrated by Aitken at the end of the nineteenth century. If a known volume of saturated air is expanded rapidly in a cloud chamber so that it becomes highly supersaturated, the vapour will condense onto virtually all the nuclei in the chamber to form a cloud of water droplets. When counted, the number of droplets approximates the number of *Aitken nuclei* available for nucleation. Near the earth's surface Aitken nucleus counts have been shown to vary widely, with average values of 10^3 cm^{-3} over oceans, 10^4 cm^{-3} over rural land and 10^5 cm^{-3} over urban areas.

Condensation nuclei sizes range from a diameter of 10^{-4} μm to tens of micrometres and are broadly grouped into three size categories. Where diameters are less than 0.2 m (1 μm = 10^{-6} m), the nuclei are termed *Aitken nuclei*; those between 0.2 and 2 μm are defined as *large aerosols* and those larger than 2 μm, *giant aerosols* (Fig. 5.3). These aerosols originate from a variety of sources. Combustion from both anthropogenic and natural sources contributes

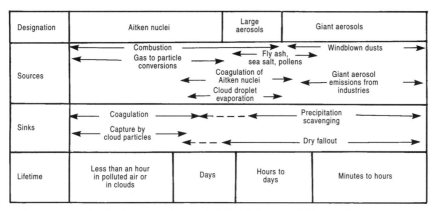

Designation	Aitken nuclei	Large aerosols	Giant aerosols
Sources	← Combustion → ← Gas to particle → conversions ← Coagulation of Aitken nuclei → ← Cloud droplet → evaporation	← Fly ash, → sea salt, pollens	Windblown dusts → ← Giant aerosol → emissions from industries →
Sinks	← Coagulation →← – – – ← ← Capture by → cloud particles ← – – – ←		Precipitation scavenging → Dry fallout →
Lifetime	Less than an hour in polluted air or in clouds	Days	Hours to days	Minutes to hours

Fig. 5.3 Principal sources and sinks of atmospheric aerosols (after Wallace and Hobbs, 1977).

primarily to Aitken nuclei and large aerosols, while windblown sources contribute dust and pollen in the form of giant aerosols (Fig. 5.4). Chemical reactions between water vapour, oxygen and nitrogen and various trace gases (sulphur dioxide, chlorine, ammonia, ozone, oxides of nitrogen) cause gas-to-liquid or gas-to-particle conversions, such as the reaction of sulphur dioxide and sulphur trioxide with water to produce sulphurous and sulphuric acids and the reaction of ammonia and water to produce ammonium sulphate. These reactions produce aerosols which fall mainly into the Aitken nucleus range. Sea-salt provides another important source of nuclei with the ejection of particles into the air with the bursting of bubbles in breaking waves at the surface.

Growth of cloud droplets

The difference between the vapour pressure, e, and the saturation vapour pressure with respect to a plane surface of pure water, e_s, (and hence the relative humidity) is an important consideration in the growth history from condensation nucleus to cloud droplet. On some nuclei water vapour condenses at vapour pressures lower than the saturation vapour pressure (i.e., $e < e_s$). The incipient cloud droplets formed on these *hygroscopic* nuclei make up a *solution* rather than consisting of pure water. However, these droplets are *curved* rather than being plane surfaces and cannot grow into cloud droplets until the air becomes *supersaturated* (i.e., until $e > e_s$). The implications of the effect of both droplet-solution and curvature effects need to be discussed separately, prior to a description of the growth of a cloud drop.

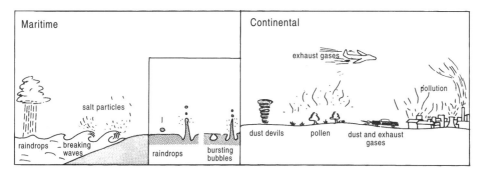

Fig. 5.4 Origins of individual nuclei.

The saturation vapour pressure for a solution is less than that for pure water. The solution, caused by the condensation of water vapour onto a hygroscopic nucleus, therefore acquires an equilibrium vapour pressure at relative humidities less than those required for pure water. The difference between the saturation vapour pressure over the solution and that over pure water depends on the strength of the solution. In terms of this *solute effect*, a hygroscopic nucleus will attract water vapour at relative humidities less than 100 per cent with respect to a plane surface of pure water until it reaches a state of stable equilibrium with the ambient air. Any further growth would cause the vapour pressure over the droplet to rise above that of the surrounding air, resulting in its evaporating back to equilibrium size. Similarly, evaporation would cause the vapour pressure over the droplet to fall below that of the ambient air and it would grow by condensation back to its equilibrium size. As the relative humidity increases, the droplet would grow to new equilibrium sizes along the curve DN in Figure 5.5 (*left*). It is

apparent from this curve that as the droplet size increases so the solute effect diminishes until the droplet grows as if the water were pure.

Due to the *curvature effect* work must be done to overcome the surface tension of the curved droplet. This causes the droplet to resist the advance of water vapour molecules until the vapour pressure is greater than the saturation vapour pressure for a plane surface (i.e. $e > e_s$). This means that for a small droplet to be in a state of equilibrium with the surroundings, the air would need to be supersaturated. If a droplet of pure water were introduced into air with the necessary supersaturation, the effect of drop curvature would cause the droplet to grow along the curve PW in Figure 5.5 (*left*). As the droplet increases in size, smaller and smaller supersaturations are required until the droplet grows as if the surface were plane.

The actual growth of a droplet is a function of both solute and curvature effects. A droplet at an initial stage of stable equilibrium is shown at point A in Figure 5.5. Progressive cooling would increase the relative humidity and the

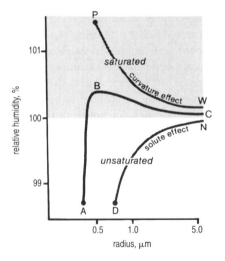
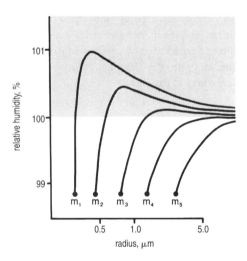

Fig. 5.5 *Left*: the theoretical control of the growth of a pure water drop by the curvature effect (PW) and that of a drop forming about a hygroscopic nucleus by the solute effect (DN), together with an example of actual drop growth (ABC); *right*: the effect of increasing mass (and size) of the condensation nucleus on the growth of drops ($m_1 < m_2 < m_3$, where m is mass).

droplet would grow to new stable equilibrium, first at subsaturation and then at supersaturation levels to point B. Droplets in the stage of growth between A and B are known as *haze* droplets and their formation can severely restrict visibility. The droplet is said to be *activated* at point B. This means that it is no longer in a state of stable equilibrium and would continue to grow along the curve to C provided that $e > e_s$. As the drop size increases, smaller and smaller supersaturations are required until it grows as if the surface were plane and the water pure.

Each unique drop radius has its own growth curve. The smaller the condensation nucleus, the larger the supersaturation required for droplets to grow to the point of activation (Fig. 5.5, *right*). This means that condensation on large nuclei, and their subsequent growth to cloud droplets, occurs at the expense of small nuclei, since water vapour is abstracted from the air by condensation to levels below the critical supersaturation for smaller droplets. The haze droplets are the first to evaporate, leaving the activated droplets to grow by condensation.

The rate of growth of a droplet by condensation is rapid while the droplet is small, but slows as the drop size increases, taking about 3 hours to reach a radius of 1 mm (Fig. 5.6). The comparative sizes of droplets relative to an initial condensation nucleus is shown in Figure 5.7. Taken together, Figures 5.6 and 5.7 indicate that condensation alone cannot account for the development of raindrops, particularly when these occur only a few hours after the formation of cloud. Other processes are required to produce the rapid growth of cloud drops.

WARM-CLOUD PRECIPITATION

Droplets fall through the air at a terminal velocity, V, which increases with the size of the droplet according to *Stokes' Law*:

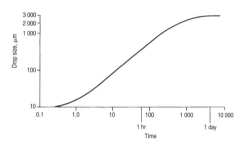

Fig. 5.6 Droplet growth by condensation (after Barry and Chorley, 1998).

$$V = \frac{2g(\rho_d - \rho_a)r^2}{9\eta} \tag{5.13}$$

where g is the acceleration due to gravity, r is the drop radius, ρ_d and ρ_a are the densities of the droplet and atmosphere respectively and η is the viscosity of the air. The rate of fall relative to drop size is given in Table 5.2.

Coalescence occurs as drops overtake their neighbours and collide. In the process droplet size increases rapidly. Small droplets may also be pulled into the wake of the larger drop by *wake capture* as shown in Figure 5.8. The collision efficiency of droplets increases with droplet size when the droplet acquires a radius in excess of

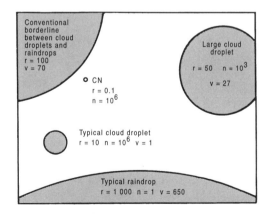

Fig. 5.7 Comparative drop sizes where r is radius in μm, n is number of drops per litre and v is terminal velocity in cm s^{-1} (after Mason, 1962).

Table 5.2 Terminal velocities of raindrops and cloud droplets in still air.

Diameter, μ	Fall rate, ms^{-1}	Drop type
5 000	8.9	Large raindrop
1 000	4.0	Small raindrop
500	2.8	Fine rain
200	1.5	Drizzle
100	0.3	Large cloud drop
50	0.076	Medium cloud drop
10	0.003	Small cloud drop
2	0.00012	Incipient drop
1	0.00004	Large nucleus

20 μm. Once started, the process accelerates and enables the droplet to grow rapidly by coalescence at the stage when growth by condensation becomes slow (Fig. 5.9).

The process of collision and coalescence takes place in *warm clouds* in which the ambient temperature is greater than 0 °C and the condensation particles consist of liquid water. The terminal velocities given in Table 5.2 are for still air conditions which seldom occur in the free atmosphere. Updraughts will keep cloud particles in suspension for as long as they exceed the terminal velocity. Only when the terminal velocity exceeds the updraught velocity will droplets begin to fall out of the cloud and precipitate as rain.

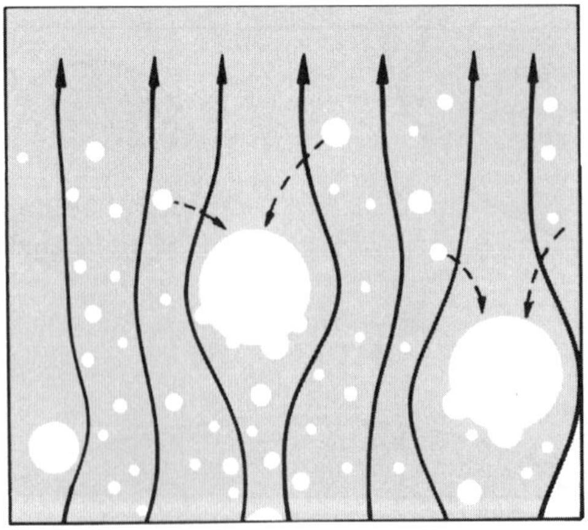

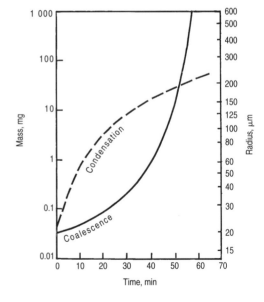

Fig. 5.8 Droplet growth by wake capture and coalescence.

Fig. 5.9 Comparative rates of droplet growth by condensation and coalescence (after East and Marshall, 1954).

COLD-CLOUD PRECIPITATION

Clouds with temperatures lower than 0 °C are called *cold clouds*. At temperatures between 0 °C and −40 °C water droplets can exist as such in a cloud and are termed *supercooled droplets*. Usually supercooled droplets co-exist with ice crystals in this temperature range, giving rise to a *mixed cloud*. At temperatures below −40 °C only ice crystals occur in cold clouds. These ice crystals play an important role in the precipitation process from such clouds.

Ice-crystal formation

Ice particles in clouds can form in various ways:
- A pure water droplet with a radius between 20 and 60 μm freezes by *homogeneous* (or *spontaneous*) *nucleation* at temperatures below −40 °C.
- A droplet containing a *freezing nucleus* freezes by *heterogeneous nucleation* at temperatures determined by the drop size, but generally warmer than −40 °C. The larger the droplet, the lower the freezing temperature. Freezing nuclei are derived mainly from dust particles carried aloft by winds. The most efficient freezing nuclei are those that are insoluble in water and contain a crystallographic arrangement similar to the hexagonal structure of ice.
- Contact between a freezing nucleus and a droplet causes freezing by *contact nucleation*.
- The direct conversion of vapour to ice on *deposition nuclei* occurs when air is supersaturated with respect to ice. This process of sublimation is particularly effective when temperatures are low. Table 5.1 showed, for instance, that at −10 °C air saturated with respect to water is supersaturated with respect to ice by 10 per cent and at −20 °C by 20 per cent.

Ice-crystal growth

In cold, layer-type clouds ice crystals grow into snowflakes; in deep convective clouds they develop into hailstones. Precipitation particles may be of different sizes and shapes, particularly when in the form of ice crystals (Fig. 5.10). Ice-crystal growth processes take place by sublimation and collision in much the same way as the growth of liquid drops by condensation and coalescence in warm clouds. However, the growth of ice crystals is complicated by the non-spherical shape of ice crystals and each individual mechanism needs some elaboration.

Sublimation
The high supersaturations that occur with respect to ice enable ice nuclei, whether formed by freezing nuclei, contact nuclei or deposition nuclei, to grow by sublimation more rapidly than water droplets. Thus ice crystals tend to grow at the expense of liquid water drops in a mixed cloud and may do so until the cloud is completely converted to ice particles. Ice crystal growth is complicated, however, by the various crystal *habits* (i.e., shapes), with each habit determined by the temperature at which it grows.

Collisions with ice crystals
Ice crystals grow by collision with other ice crystals due to stirring within the cloud and the differences in the terminal velocities of the crystals. This growth by *aggregation* occurs when collisions take place between crystals with intricate habits which allow them to become entangled and entwined. Temperature also plays a role in the aggregation process with the adhesion between ice crystal surfaces becoming particularly effective at temperatures near about −5 °C.

Collisions with supercooled droplets
Mixed clouds provide a suitable environment for the growth of ice crystals by collision with supercooled droplets. When this occurs it leads to growth by *riming* as the supercooled droplet freezes onto the crystal. Many such impacts obscure the original shape of the ice crystal at which point the particle becomes *graupel* (soft, small hail particles). The extreme form of this type of growth is the *hailstone*. The alternate clear and opaque layers apparent in large

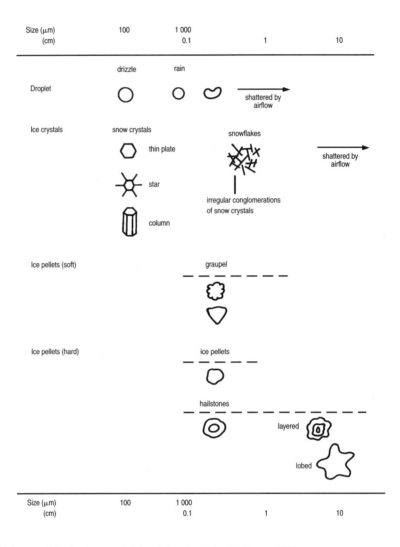

Fig. 5.10 Various precipitation forms and their relative sizes (after McIlveen, 1986).

hailstones depict the growth history of the stone as it falls (or rises) in vigorous convective clouds. The hailstone may rise and fall many times within the cloud while trapped within up-draughts, with each cycle producing one clear and one opaque layer (Fig. 9.17, *right*). The opaque layers form when the stone collides with ice crystals. The aggregation that results is most likely to occur at low temperatures and great heights. At higher temperatures (and possibly in the descending phase of the stone) collision

with supercooled droplets is more common. A clear layer occurs when the stone collects super-cooled droplets so rapidly that its surface temperature rises to 0 °C allowing the stone to become covered with a layer of liquid water which freezes as clear (bubble-free) ice.

Precipitation

The growth of cloud crystals by sublimation and aggregation and of water droplets by collision and coalescence ultimately reaches a stage where particles in the form of ice particles (snow, graupel, hail) or liquid droplets (rain) can no longer be sustained by updraughts. Precipitation then occurs from the cloud. According to the *Bergeron-Findeison theory*, all precipitation from cold clouds relies initially upon the growth of ice crystals by aggregation to form snowflakes. These precipitate to the surface when the air temperature between the cloud base and the ground is below 0 °C. However, if the temperature within this layer is above 0 °C, the snow-flake melts into a water drop and precipitates as rain (Fig. 5.11).

The Bergeron-Findeison theory does not account for precipitation from warm clouds, the tops of which occur in air with temperatures above 0 °C. In this case droplet growth by coalescence is thought to be the primary mechanism producing precipitation. With cumulus clouds, a limited number of droplets form on giant nuclei with diameters in excess of 20 μm. These are carried aloft in the updraft and grow by coalescence. Near the top of the cloud the

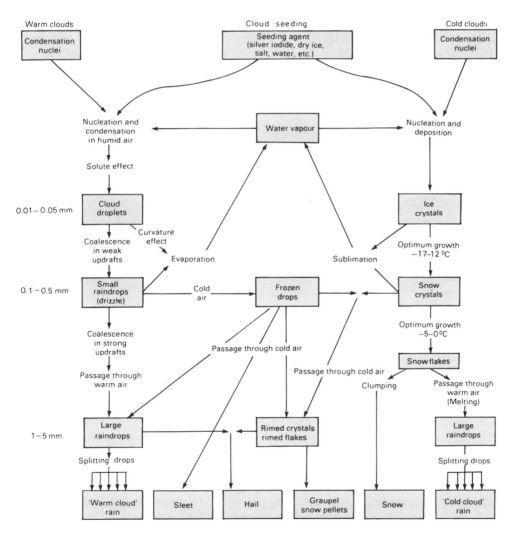

Fig. 5.11 Various precipitation paths in warm and cold clouds (after Eagleman, 1980).

droplets reverse direction either because their terminal velocity exceeds the updraught velocity or because they are blown out of the updraught by a vertical wind shear. Further growth by coalescence occurs as the droplet falls through the cloud with some drops reaching a diameter exceeding 5 mm. Droplets of this size become unstable and fragment into smaller particles. These are again carried aloft to repeat the coalescence procedure. Ultimately, the chain reaction of raindrop multiplication is responsible for shower precipitation.

CLOUD MODIFICATION AND RAINFALL AUGMENTATION

Successful cloud seeding depends on a thorough understanding of the microphysics of the clouds in specific regions. In particular it is important to distinguish between maritime-type clouds and those of continental origin. Maritime clouds are characterized by low drop concentrations with a relatively broad drop-size distribution whereas continental clouds have high drop concentrations and a narrow drop-size distribution. Each type responds differently to different types of seeding agents. A seeding agent that works effectively in clouds with a uniform drop-size distribution, for instance, will not necessarily be effective in a cloud with a non-uniform distribution. Over the last five decades the seeding of clouds to stimulate rainfall or suppress hail has been attempted in many countries. South Africa is no exception, and for a number of years cloud-seeding experiments were undertaken at Bethlehem and Nelspruit to assess the potential for rainfall stimulation.

The numerous seeding techniques adopted worldwide vary in accordance with their objective and the nature of the clouds concerned. In warm clouds the growth of raindrops by the collision-coalescence mechanism requires the seeding of the cloud by large hygroscopic particles, such as potassium chloride, whereas in cold clouds it is necessary to seed the cloud with artificial ice nuclei, such as silver iodide. If the

dissipation of the cloud (or fog) or the suppression of hail is the objective, overseeding of the cloud is generally practised.

Warm-cloud modification

In warm-cloud modification, warm clouds may be seeded with water droplets or hygroscopic nuclei, either into the top of the cloud or into the cloud base. Traditionally, hygroscopic seeding has used fairly large (> 10 μm) salt particles that almost immediately become active in the coalescence process of rainfall formation, with the particles themselves acting as collectors. The most effective modern technique, developed in South Africa, involves flare seeding with small (~5 μm) hygroscopic nuclei of potassium chloride from aircraft flying just below the cloud base. These particles are then carried upward in updraughts into the cloud where they compete with natural cloud condensation nuclei for the available moisture, condense into cloud droplets and then grow, initially by condensation and then by the collision-coalescence mechanism to the point of precipitation. The idea behind this approach is to change the drop-size spectrum of the clouds to resemble more closely more efficient rain-producing maritime clouds. Significant success has been achieved in South Africa in the seeding of continental clouds using this technique. The immediate consequences of such seeding is to broaden the spectrum of droplet size and the mass density of droplets in seeded clouds and shift them significantly in the direction of larger, heavier drops (Fig. 5.12, *left*). At the same time, seeded clouds show significantly greater rates of increase in cloud top heights, storm mass above 6 km, heights of the 45 dBZ radar reflectivity contour and peak radar reflectivity than do unseeded storms.

In a five-year randomized cloud-seeding experiment in South Africa, in which 120 storms were treated, it was found that seeded storms, on average, produce 30 per cent more radar-measured rainfall than their natural counterparts. The increase in rainfall is the result of

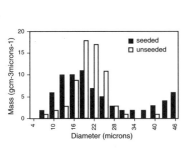
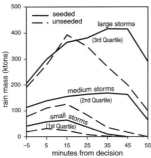
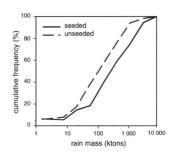

Fig. 5.12 Differences between unseeded (natural) and seeded Highveld thunderclouds; *left*: the variation of mass with water droplet size (diameter); *middle*: the variation of rain mass in small, medium and large storms; *right*: the difference in radar-measured rain mass (after Mather et al., 1997).

longer storm lifetimes and a more efficient conversion of the available moisture to precipitation in the seeded clouds. Small storms produce small increases in rain mass about 20–30 minutes after seeding, whereas medium and large storms show highly significant differences 30–40 minutes after seeding in comparison to control unseeded clouds. Seeded storms last 30–40 minutes longer than those unseeded (significant at $p < 0.01$) (Fig. 5.12, *middle*). They also produce 30 per cent more rainfall (significant at $p < 0.01$) (Fig. 5.12, *right*). It appears likely that hygroscopic seeding causes small storms to transform into larger, more effective ones by accelerating their growth rates.

Whereas it is clear that precipitation can be increased from individual storms, it remains to be demonstrated in a statistically significant sense that rainfall may be augmented over a large area. The signs that it may be possible to do this are encouraging. A pilot river-basin rainfall-augmentation project in an often drought-ravaged part of the Northern Province of South Africa shows that positive rainfall anomalies over the basin coincided with combined tracks of seeded storms. Hygroscopic seeding over the catchment area appears to have contributed meaningfully to enhancing area-averaged rainfall.

Cold-cloud modification

In cold-cloud modification, most experiments on rainfall stimulation have been based on seeding cold clouds with artificial ice nuclei to induce ice-crystal growth by sublimation and aggregation and thence rainfall in the manner suggested by the Bergeron-Findeison theory. Silver iodide is one of the most common seeding agents in use owing to the similarity of its crystallographic structure to that of ice and its ability to act as an ice nucleus at temperatures as high as −4 °C. Solid carbon dioxide (dry ice) is also a common seeding agent.

In general, cold-cloud modification has not proved as successful as that of warm clouds using potassium chloride. A successful cold-cloud rainfall augmentation programme is run in Israel, where a statistically significant increase in rainfall of up to 15 per cent has been achieved.

Fog dissipation

The principle of overseeding is used to dissipate troublesome cold cloud or fog at airports. A successful technique often used is the introduction of dry ice into the cloud. Dry ice does not by itself act as an ice nucleus. Instead, its low temperature (−78 °C) causes ice crystals to form by homogeneous nucleation in the wake of the falling dry ice particles and causes the cloud to glaciate. The ice crystals thus formed are very small and, in the absence of supercooled droplets, soon evaporate, causing the cloud or fog to dissipate.

Hail suppression

Suppression of hail operates on the principle of seeding cold, hail-producing clouds with artificial ice nuclei with the objective of increasing the number of ice crystals competing for supercooled droplets. If successful, this increases the number of small hailstones and reduces the number of large, damaging hailstones. This may also be achieved by overseeding hail-producing clouds. In this case glaciation would reduce the number of supercooled droplets and thereby slow the growth of hailstone aggregation. Care has to be taken, however, not to over-seed for fear of stimulating too much extra cloud growth as a result of excessive latent heat release.

ADDITIONAL READING

BARRY, R. G. and CHORLEY, R. J. 1998. *Atmosphere, Weather and Climate*, 7th Edition. London: Methuen.

STURMAN, A. S. and TAPPER, N. 1996. *The Weather and Climate of Australia and New Zealand.* Melbourne: Oxford University Press.

PRESTON-WHYTE, R. A. and TYSON, P. D. 1988. *The Atmosphere and Weather of Southern Africa.* Cape Town: Oxford University Press.

LUDLAM, F. H. 1980. *Clouds and Storms.* University Park: Pennsylvania State University Press.

WALLACE, J. M. and HOBBS, P. V. 1977. *Atmospheric Science, an Introductory Survey.* New York: Academic Press.

MASON, B. J. 1962. *Clouds, Rain and Rainmaking.* Cambridge: Cambridge University Press.

6

THE TEPHIGRAM

Temperature–height diagrams of the kind shown in Chapter 4 do not provide a method of representing atmospheric data in a form that satisfies the analytic requirements of meteorologists. As a consequence, a number of thermodynamic diagrams have been designed to provide for the graphic display of many atmospheric processes regularly encountered. Such diagrams include the tephigram, emagram and skew *T-log p* diagram. Different national weather services use different diagrams. In South Africa and other southern African countries the tephigram is used. In this chapter its construction and use will be considered in some detail.

The tephigram is constructed by overlaying five sets of lines, namely *isobars, isotherms, saturated mixing ratio* (or *dew-point*) *lines, dry adiabats* (each one representing the dry adiabatic lapse rate) and *wet adiabats* (each representing the wet adiabatic lapse rate) (Fig. 6.1). Only one set of lines is curved (wet adiabats); the others are nearly straight (isobars) or linear (dry adiabats, isotherms and mixing ratio or dew-point) isopleths, with isotherms and dry adiabats crossing at right angles. An important characteristic of the tephigram is that it allows changes in energy or work done during a cyclic process to be represented by the area enclosed by lines representing the process. This characteristic greatly enhances the usefulness of the tephigram, particularly in the evaluation of the thermal instability of the atmosphere.

Pressure, temperature and moisture data obtained from *radiosonde* instruments carried aloft by meteorological balloons are recorded by ground telemetry stations and are plotted on tephigrams to provide a vertical sounding of the atmosphere at standard times. Radar tracking of the balloons allows the variation of wind speed and direction throughout the troposphere and lower stratosphere to be determined at the same time.

TEPHIGRAM APPLICATIONS

Plotting the data

The South African Weather Bureau issues the results of balloon soundings in terms of dry-bulb temperature, T, dew-point temperature, T_d, and pressure, p, at 50 hPa intervals from the surface to 800 hPa, at 100 hPa intervals up to 300 hPa, and then at decreasing intervals of 50 hPa or less up to the balloon ceiling at about 10 hPa. In addition data are provided for all *significant levels*, which are characterized by a change in lapse rate. At each pressure level corresponding values of T and T_d are plotted on the diagram. Successive values of T are joined by a solid line to identify the environmental lapse rate, α, and by a pecked line for the dew-point curve. An example of a typical sounding taken at 06:00 at Durban shows a lapse rate in which nocturnal radiational cooling has produced a surface inversion, above which an elevated inversion separates subsided air above from unmodified air below (Fig. 6.2).

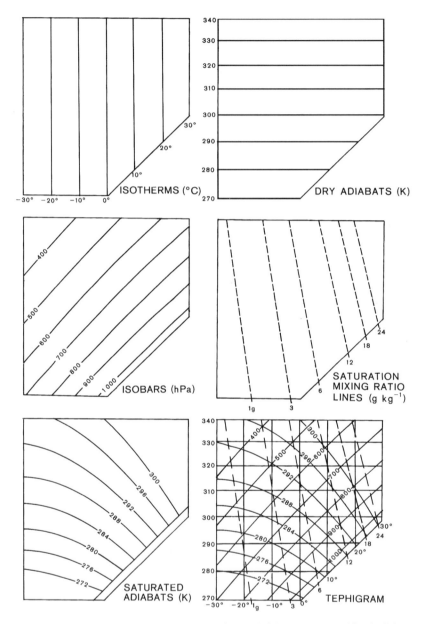

Fig. 6.1 Isotherms, dry adiabats, isobars, saturation mixing ratio (dew-point) lines and saturated (wet) adiabats on tephigrams, together with a composite tephigram.

Normand's proposition

Normand was the first to show that the dry adiabat through the dry-bulb temperature, the wet adiabat through the wet-bulb temperature and the dew-point line through the dew-point temperature intersect at a point which gives the level at which condensation of water vapour will occur when air is lifted dry adiabatically (Fig. 6.3). The direct consequence of this simple theorem is the facilitation of a number of computations on the tephigram.

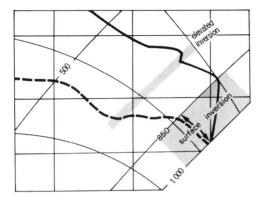

Fig. 6.2 Typical early-morning lapse rates of dry-bulb (solid line) and dew-point temperature (broken line) on a tephigram to show surface and elevated inversions of temperature.

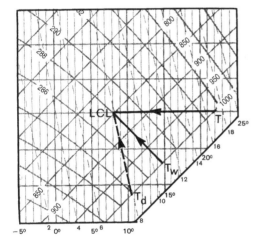

Fig. 6.3 The determination of lifting condensation level on a tephigram.

Using the tephigram

The *lifting condensation level*, LCL, may be determined from dry-bulb temperature, T, and dew-point temperature, T_d, at level p by drawing a dry adiabat through T and a dew-point line through T_d. The level (in pressure units) of intersection of the two lines is the LCL (Fig. 6.4a). Wet-bulb temperature, T_w, is fixed by T and T_d. The first step is to determine the lifting

condensation level and then to drop a wet adiabat back through the condensation point to the original level (Fig. 6.4b). The intersection of this adiabat and the isobar at that level defines T_w.

The *humidity mixing ratio*, x, is read off as the value of the dew-point line through T_d. Likewise, the *saturated humidity mixing ratio, x_s*, is the value of the line through T (Fig. 6.4c). *Relative humidity* is the ratio $100x/x_s$.

The *cloud base* is defined by the height of the lifting condensation level. The process during the lifting of the air to the cloud base is dry adiabatic (since no condensation has occurred). The height of the cloud base, Δz, in metres above the ground may be determined, therefore, using the dry adiabatic lapse rate:

$$\Delta z = \Delta T 100 \tag{6.1}$$

where ΔT is the temperature difference between T and LCL (Fig. 6.4d).

Vapour pressure, e, may be approximated as

$$e = \frac{8xp}{5} \tag{6.2}$$

Graphically, the value of the mixing ratio when T_d is altered isothermally to 620 hPa gives the vapour pressure (Fig. 6.4e). *Dry-bulb potential temperature, θ*, is the dry-bulb temperature reduced dry adiabatically to 1000 hPa (Fig. 6.4f). Similarly, *wet-bulb potential temperature, θ_w*, is the wet-bulb temperature reduced wet adiabatically to 1000 hPa.

Equivalent temperature, T_e, is the temperature attained if, at constant pressure, all the water vapour in the air is condensed and the latent heat so released is added to the air:

$$T_e = T + \frac{L_v x_{sw}}{C_p} \tag{6.3}$$

where L_v is the latent heat of condensation, x_{sw} is the saturated mixing ratio at dry-bulb temperature, T, and C_p is the specific heat of air at constant pressure. On the tephigram equivalent

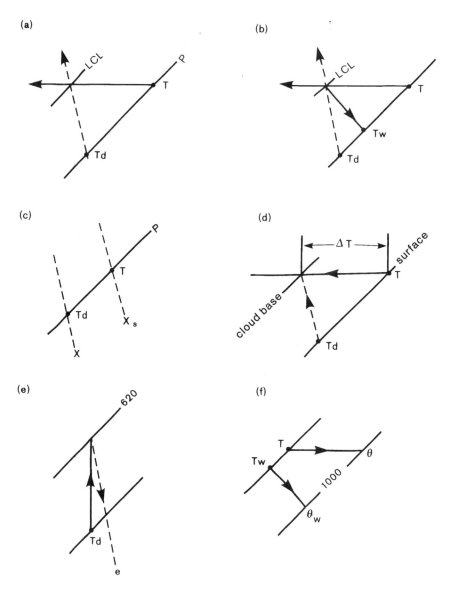

Fig. 6.4 The determination of (a) condensation level, (b) wet-bulb temperature, (c) humidity and saturation humidity mixing ratios, (d) height of cloud base, (e) vapour pressure, and (f) potential and wet-bulb potential temperature.

temperature is determined by extending the ascent curve in Figure 6.5a to A, the point where all moisture in the air has condensed and which is defined by the asymptotic dry adiabat at that point. Where this dry adiabat cuts the original pressure level, p, the equivalent temperature is read off.

Equivalent potential temperature, θ_e, is a useful parameter for studying cumulus convection. It is defined in the same way as dry-bulb potential temperature (equation 4.10) and is the equivalent temperature reduced dry adiabatically to the standard level of 1000 hPa (Fig. 6.5a). If $d\theta_e/dz < 0$, the atmosphere is potentially unstable.

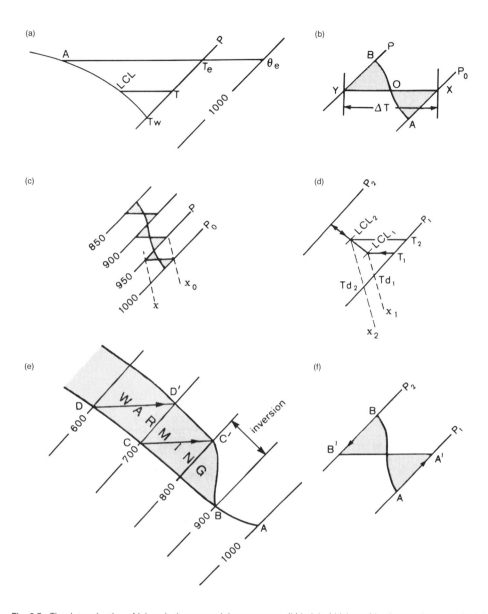

Fig. 6.5 The determination of (a) equivalent potential temperature, (b) height (thickness) by the equal-area method, (c) precipitable water, (d) the föhn effect, (e) warming by subsidence, and (f) the effect of turbulent mixing on the lapse rate.

The *thickness of a layer*, Δz, may be determined approximately by the method of equal areas. In order to calculate the thickness between pressure levels p_0 and p, given the environmental lapse rate, AB, a line XY is drawn through AB (Fig. 6.5b) in such a way that area XAO approximately equals area YOB. The temperature difference between the points X and Y then defines ΔT from which the thickness in metres is

$$\Delta z = \Delta T 100 \qquad (6.4)$$

For an accurate determination, the pressure–

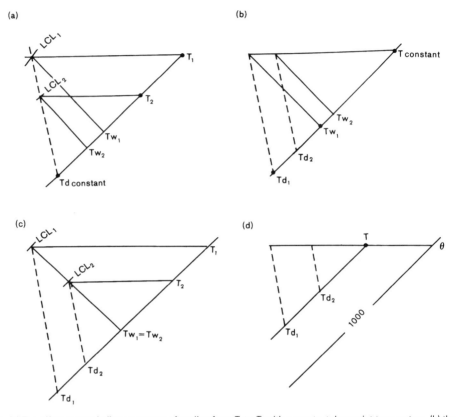

Fig. 6.6 (a) The effect on wet-bulb temperature of cooling from T_1 to T_2 with a constant dew-point temperature, (b) the effect of increasing dew-point temperature on wet-bulb temperature providing dry-bulb temperature remains constant, (c) the effect of evaporation on dry-bulb and dew-point temperatures, and (d) the effect of changing dew-point temperature on dry-bulb temperature and potential temperature providing no condensation takes place.

height equation must be used in which the mean temperature, $\overline{T}$ K, between the levels p_0 and p is

$$\left(\frac{T_A + T_B}{2}\right) + 273 \tag{6.5}$$

The height difference between the layers in metres is then given by

$$\Delta z = 29.27\,\overline{T}\,(\log_e p_0 - \log_e p) \tag{6.6}$$

Precipitable water, w, is defined as

$$w = \frac{\overline{x}(p_0 - p)}{g} \tag{6.7}$$

where $\overline{x}$ is the mean humidity mixing ratio between any two levels, p_0 and p, having mixing ratios x_0 and x (Fig. 6.5c). The calculation is performed for the layer p_0 to p and thence for successive layers above. Finally, the total precipitable water is determined by integrating (adding) the values for all layers.

The *föhn effect* of air ascending a mountain, precipitating and then descending to reach the other side of the mountain warmer than at the same height on the windward side may also be demonstrated on a tephigram. Air with properties T_1, T_d, and x_1 at p_1 on the windward side of the mountain is forced to rise orographically to p_2 in its ascent over the barrier (Fig. 6.5d).

Condensation occurs at LCL_1 and Δx humidity is lost through precipitation. In ascending above LCL_1, the air, being saturated, cools at the wet adiabatic lapse rate until it reaches p_2 and begins to descend. On descent it continues to cool until it reaches a condensation level now defined in terms of the diminished humidity mixing ratio, LCL_2. After evaporating at this level the air continues cooling but now at the dry adiabatic lapse rate. It reaches p_1 on the leeward side with temperatures T_2 and T_{d_2}; that is, it has become warmer and drier, and the condensation level on this side of the mountain is higher than it was on the windward side.

Subsiding air undergoes adiabatic heating and may produce an absolutely stable layer and/or a subsidence inversion as illustrated in Figure 6.5e, where subsidence by 100 hPa to the 800 hPa level changes the original lapse rate ABCD to the modified lapse rate ABC'D' with an inversion BC' between the subsided and unmodified air.

Turbulent mixing of air acts to produce a uniform heat distribution in the atmosphere, and lapse rates tend to the dry adiabatic as a result. Hence the mixing of air in the layer p_1p_2 in Figure 6.5f will alter the lapse rate from AB to A'B'.

In general, dry-bulb, wet-bulb and dew-point temperatures undergo changes under conditions imposed by radiation, evaporation, condensation, ascent, subsidence and turbulent mixing. For example, in the case of radiational cooling, the dry-bulb temperature in Figure 6.6a might decrease from T_1 to T_2 and the lifting condensation level descends from LCL_1 to LCL_2. Similarly, wet-bulb temperature will change from T_{w_1} to T_{w_2}. If, instead of radiational changes, an increase occurs in water-vapour content without any alteration in dry-bulb temperature, T_{d_1} increases to T_{d_2} with concomitant wet-bulb temperature changes from T_{w_1} to T_{w_2} (Fig. 6.6b).

There are, however, certain exceptions to the generalization that temperature changes will occur with changing atmospheric conditions. Some properties remain constant when acted upon. These are said to be *conservative*. For instance, the definition of wet-bulb temperature shows it to be conservative with respect to changes in evaporation and condensation. In Figure 6.6c dry-bulb temperature, T_1, and dew-point temperature, T_{d_1}, represent conditions before evaporation occurs. The effect of evaporation is to increase the water-vapour content of the air to T_{d_2} and lower the dry-bulb temperature to T_2. The wet-bulb temperature remains unchanged and is so conserved during this process. Likewise, dry-bulb potential temperature is conservative with respect to changes in moisture (providing no condensation occurs). Thus in Figure 6.6d, as moisture increases and dew-point temperature rises from T_{d_1} to T_{d_2} (or falls from T_{d_2} to T_{d_1}), θ remains unchanged. Conservative properties are particularly useful in tracing the origin of air and in the classification of different masses of air. Table 6.1 shows how different properties are conserved for different processes. Wet-bulb potential temperature, θ_w, is shown to be suitable as an air mass indicator since it is conservative with respect to evaporation and condensation and to ascent and descent. Likewise, the humidity mixing ratio, x, is conservative during subsidence and may be used as an indicator of such subsiding air.

STABILITY AND INSTABILITY ON TEPHIGRAMS

Absolute stability and instability

The more rapidly temperature decreases with height in the atmosphere, the less stable will be the air. Thus in Figure 6.7a lapse rate $\alpha_1 > \alpha_2$ and $\alpha_3 > \alpha_4$. Any lapse rate less than the wet adiabatic lapse rate (i.e., $\alpha < \Gamma'$) will be absolutely stable. Conversely, any lapse rate greater than the dry adiabatic lapse rate (i.e., $\alpha > \Gamma$) will be absolutely unstable. Neutral

Table 6.1 Processes for which certain properties of the atmosphere are conserved.

Property	Conservative for the process of				
	Radiational heating and cooling	Evaporation and condensation	Ascent and descent	Turbulent Mixing	
				Heat	Water vapour
T	no	no	no	no	yes[a]
T_d	yes[a]	no	no	yes[a]	no
T_w	no	yes	no	no	no
θ	no	no	yes[a]	no	yes[a]
θ_w	no	yes	yes	no	no
θ_e	no	yes	yes	no	no
χ	yes[a]	no	yes[a]	yes[a]	no
RH	no	no	no	no	no

a Provided condensation does not occur.

conditions are defined by $\alpha = \Gamma$ and $\alpha = \Gamma'$ for dry (or moist) air and saturated air respectively. That temperatures are decreasing more rapidly along α_1 than α_4 is seen by comparing the decreases ΔT_{OA} and ΔT_{OD} along the lapse rates between pressure levels p_1 and p_2 (Fig. 6.7a).

The effect of ascent and descent on lapse rates

If dry air with lapse rate AB (Fig. 6.7b) is lifted from level p_1 to p_3 and p_2 to p_4 (i.e., by the same amount throughout the layer), then, since the isobars on a tephigram are increasingly spaced apart with decreasing pressure, the new lapse rate CE will exceed the original one AB (which is equal to CD). Consequently, with lifting, the air has become less stable. Conversely, air having an initial lapse rate CE, and which is forced to subside from E to B and C to A, will have a new lapse rate AB, which is less than CE. The sub-sided air has thus become more stable. The latter process is common over southern Africa and has the immediate effect of clearing bad weather and maintaining fine-weather conditions. Subsidence likewise causes a deterioration in the ability of the atmosphere to disperse air pollution and may produce adverse air pollution conditions because of the marked stability of the atmosphere.

Conditional instability

It is not possible to say whether air having a lapse rate between the dry and wet adiabatic lapse rates is stable or not, unless the moisture content of the air is known. The instability of the air is conditional upon the amount of moisture present and must be determined using the parcel method, the observed lapse rate of temperature and surface moisture (dew-point temperature) (Fig. 6.8). If a parcel of air is displaced upward (as a result of surface heating or low-level convergence in the wind field, for instance), it will cool at the dry adiabatic lapse rate until condensation occurs at G in Figure 6.8. If the upward displacement is maintained, the parcel of now-saturated air would rise at the

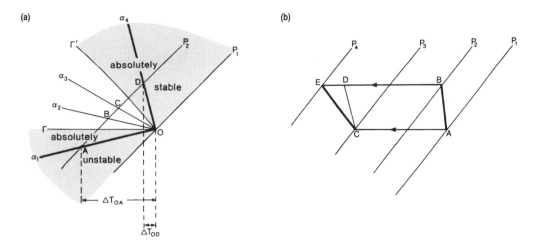

Fig. 6.7 (a) Absolute stability and instability on a tephigram, and (b) the effect of ascent from AB to CE (or descent from CE to AB) on the lapse rate.

wet adiabatic rate along curve GC. At level BH the parcel is cooler than the ambient air (i.e., the parcel temperature, T_p, will be less than the air temperature, T_a or $T_H < T_B$), and the air is stable. At point C the parcel and ambient air temperatures are equal. Beyond this point no further external lifting is required and the level of free convection has been reached. Above the level of free convection $T_p > T_a$ (i.e., $T_F > T_D$), the parcel will be warmer than the ambient air and will rise of its own accord. At level E the parcel and ambient air temperatures are equal (i.e., $T_p = T_a$) and upward motion will cease.

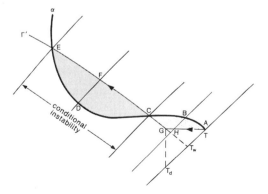

Fig. 6.8 The determination of conditional instability.

Between levels C and E conditional instability is said to occur.

Because of the thermodynamic properties of the tephigram, the area CDEFC is directly proportional to the convective energy released on the realization of the conditional instability. Similarly the area AGCBA is proportional to the amount of energy required to lift the air from the surface to the point where it becomes unstable. Visual inspection of the respective areas allows a judgement to be made as to the likelihood of the instability being realized (Fig. 6.9a and b). Thus the larger the area of possible convective energy release vis-à-vis the amount required to realize the release, the more likely it is that deep convection will occur. In a strongly conditionally unstable atmosphere little external lifting is necessary for strong uplift to be initiated and sustained. In a less unstable atmosphere stronger initial thermal and dynamic forcing will be required. Even in a stable atmosphere vertical motion is possible. However, in such a case the forcing is unlikely to have a thermal component and uplift is invariably associated with strong dynamic forcing (such as widespread low-level convergence or upper-level divergence) in the wind-field (see Chapter 10).

The shape (and hence slope) of the environ-

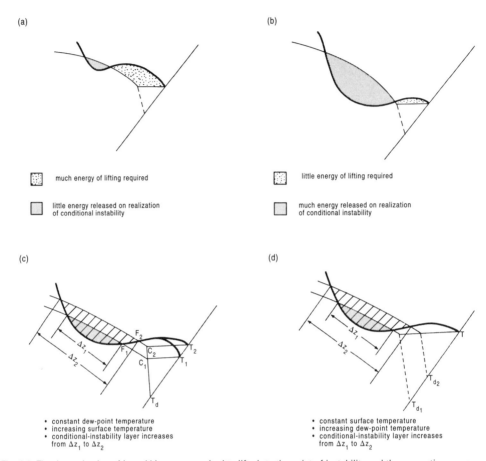

(a)

much energy of lifting required

little energy released on realization
of conditional instability

(b)

little energy of lifting required

much energy released on realization
of conditional instability

(c)

- constant dew-point temperature
- increasing surface temperature
- conditional-instability layer increases
  from Δz_1 to Δz_2

(d)

- constant surface temperature
- increasing dew-point temperature
- conditional-instability layer increases
  from Δz_1 to Δz_2

Fig. 6.9 The determination of (a and b) energy required to lift air to the point of instability and the convective energy released by the realization of conditional instability, (c) the effect of increasing surface temperature, and (d) the effect of increasing dew-point temperature (moisture content) on conditional instability.

mental lapse rate is important in determining the degree of conditional instability. Likewise, the manner in which surface heating and moisture advection may change ground-level, dry-bulb and dew-point temperatures affects the degree of instability. If dew-point temperature remains constant, then an increase in surface temperature will result in an increase in instability (Fig. 6.9c). As the temperature rises from T_1 to T_2, so too does the condensation level from C_1 to C_2. At the same time the level of free convection drops from F_1 to F_2. When F_2 and C_2 coincide, the corresponding surface temperature is the minimum required to generate cumulus clouds by thermal forcing (buoyancy) alone.

If temperature is held constant and surface T_d is allowed to increase, then as T_d increases so the amount of conditional instability increases (Fig. 6.9d). It is for this reason that the more moist the air mass, the more easily convection is initiated and sustained. Thus storms develop much more readily in moist equatorial regions than in dry subtropical areas. Likewise, as a dry airstream is replaced by a moist one over South Africa, so the probability of showers or thunderstorms increases noticeably.

The type of convective cloud likely to result from conditional instability depends on the degree of instability prevailing (Fig. 6.10, *upper*). Clearly the greater the degree of free

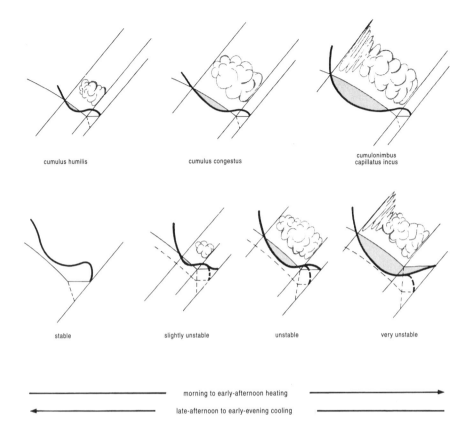

cumulus humilis   cumulus congestus   cumulonimbus
                                                    capillatus incus

stable   slightly unstable   unstable   very unstable

morning to early-afternoon heating

late-afternoon to early-evening cooling

Fig. 6.10 *Upper:* the variation of cloud type with the amount of conditional instability; *lower:* the growth of convective clouds during the day as surface temperature increases and the opposite with afternoon cooling.

thermal convective energy that may be released on the realization of instability, the deeper and stronger the cumuliform cloud. Providing that no change in air mass occurs (i.e., that the moisture content remains the same), morning and early-afternoon heating and late-afternoon cooling of the surface will produce a clear cycle of cumuliform cloud development and may produce a mid- to late-afternoon peak in the occurrence of showers and storms (Fig. 6.10, *lower*).

Latent instability

In the above discussion conditional instability has been defined for a parcel of air at the surface, for which dry-bulb and wet-bulb temperatures are given. If the parcel cooling curve cuts the environmental lapse rate, then, it was shown, the parcel is conditionally unstable. This is the same as saying that the *wet adiabat* drawn through the *wet-bulb temperature* of the air must cut the environmental curve at a higher level (see Fig. 6.8). If this extension does not hold at the surface, but holds for air above ground level (i.e., any other wet adiabat through any other part of the wet-bulb curve cuts the environmental curve at a higher level), then the air is said to possess latent instability. Note that for this to be determined, both dry- and wet-bulb temperature curves must be plotted on the tephigram, and not just dry-bulb and dew-point temperatures.

Given dry-bulb and wet-bulb temperature lapse rates as shown in Figure 6.11, it is clear that the air at the surface is stable, since the parcel ascent curve TXY does not cut the environmental curve at all. Air at level AB, however, is unstable if forced to rise, since parcels starting from this level cool along BCDE, and between D and E free convection will occur. Such instability is said to be latent and latent instability obtains at all levels between points G and H, since any parcels being initiated between these levels will be unstable. Parcels being forced to rise from levels between the surface and H will be stable. The tangent wet adiabatic lapse rate FGH drawn through the lowest point F on the dry-bulb temperature curve defines the layer GH in which the latent instability resides.

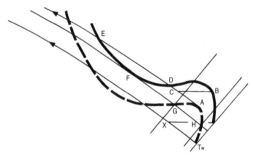

Fig. 6.11 Latent instability in the layer HG.

Potential instability

Consider a segment, BC, of a continuous dry-bulb temperature lapse rate, ABCD, and a similar segment, QR, of a wet-bulb temperature lapse rate, PQRS (Fig. 6.12). Since the dry-bulb temperatures B and C do not equal the wet-bulb temperatures Q and R, the air is not saturated. Lapse rate AB is less than the dry adiabatic lapse rate (AB < Γ) and the layer is stable. If the entire layer is lifted to saturation, then the condensation point for the air at the bottom of the layer having dry- and wet-bulb temperature B and Q is B'. Similarly, the condensation point for the top of the layer is C' and the condensation points for all other sub-layers of air between B and C lie along the straight line joining B' and C'. This line is the new lapse rate of the now saturated (condensed) air. Since the

layer is saturated, its lapse rate must now be compared to the wet adiabatic lapse rate (i.e., to the wet adiabat continued through QB'). Lapse rate B'C' exceeds the wet adiabatic lapse rate (B'C' > Γ') and the whole layer between levels QB and RC is said to be potentially unstable. For the instability to be realized, the layer must be lifted through a height CC' which is $\Delta T_{CC'}$100 m deep.

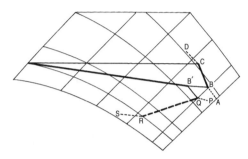

Fig. 6.12 Potential instability in the layer BC (or QR) as defined by the lapse rate of wet-bulb temperature, QR, exceeding the wet adiabatic lapse rate.

The criterion for recognizing the occurrence of potentially unstable air is that *the lapse rate of wet–bulb temperature must exceed the wet adiabatic lapse rate.* After determining the lapse rate of wet-bulb temperature from the dry-bulb and dew-point temperature curves, that lapse rate is inspected and those layers in which the criterion is satisfied are designated as potentially unstable. Alternatively, potential instability may be defined as the layer in which potential temperature decreases with height (i.e., $d\theta/dz < 0$). Where $d\theta/dz > 0$ the atmosphere is stable. The instability refers to the lifting of the whole layer and, unlike conditional instability (and its subtype latent instability), not just to the lifting of parcels within it.

In most instances forced ascent is required to trigger thermal instability of any type. The forced ascent may be on a convective scale, may be mesoscale due to local air circulations or orographic lifting, or may be at the synoptic scale due to fronts or large-scale convergence and divergence in the wind field. Often more than one lifting mechanism may prevail. If a lifting mechanism is absent, then any potential instability present may be unrealizable.

Changes in instability

Any process raising wet- and dry-bulb temperatures at the surface favours the development of conditional instability. Any process raising the wet-bulb temperature in a lower layer, while lowering the dry-bulb temperature in a higher layer, enhances the development of latent instability. Any process increasing the wet-bulb temperature in a lower layer while decreasing it in a higher layer, favours the development of potential instability.

At or near the surface wet-bulb temperature increases when heat is added to the air as part of the normal diurnal heating cycle and is mixed into the lower layers by turbulence. Local surface evaporation may also increase wet-bulb temperatures, but seldom enough to alter large-scale stability. If the atmosphere has been moistened by rainfall at a previous time (e.g., a few hours or the day before), this effect, combined with heating, may produce strong convection. Advection of low-level, moist, warm air enhances conditions for the development of instability. Along coasts, moist-air advection over land warmer than the sea will enhance instability, whereas advection of air from the land over a colder sea will decrease wet-bulb temperatures and inhibit the development of instability.

At upper levels over South Africa dry- and wet-bulb temperatures are lowered and instability enhanced by the advection of cold air with a southerly component. Radiational cooling from the top of a cloud layer will have the same effect. Any combination of low- and upper-level tropospheric circulation that favours airflow with a northerly component at the surface and a southerly component aloft favours the development of latent and potential instability. The opposite circulation inhibits these types of instability.

Given its topography, coastal mountain ranges and escarpment, a process which acts to enhance latent, but not potential, instability in South Africa is large-scale orographic lifting of whole layers of air. Consider air with a lapse rate ABC and a wet-bulb temperature curve DEF being forced to rise throughout by 100 hPa in its ascent over a coastal mountain range or the Escarpment (Fig. 6.13). The parcel method shows this air to be stable. Lifting by 100 hPa (dry-bulb temperatures are lifted dry adiabatically; wet-bulb wet adiabatically) results in the new lapse rates GHJ and KLM. Conditional (latent) instability is now present and needs little forced lifting for the realization of free convection.

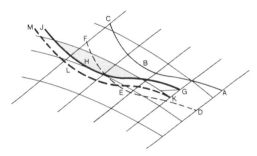

Fig. 6.13 The effect of large-scale lifting of an air mass.

Stability indices

It is often useful to be able to express the degree of instability by a single index. Several indices have been developed for this purpose, the most commonly used of which is the *Showalter Index*. In South Africa the original definition has been modified for local use and the Index is determined from the 09:00 South African Standard Time (SAST) radiosonde ascent. Using 850 hPa dew-point and 800 hPa dry-bulb temperatures, a parcel is lifted dry adiabatically to condensation level and thence wet adiabatically to 500 hPa. The difference at this level between the parcel and ambient air temperature, ΔT_{a-p}, is the Showalter Index. The magnitude of the negative temperature difference is the quantitative measure of the degree of instability. For the

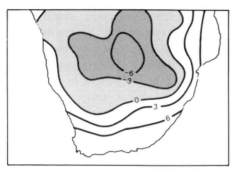 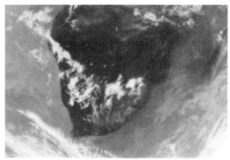

Fig. 6.14 Showalter index ($T_a - T_p$ at 500 hPa) for a spring day (after Poolman, 1986) and the mid-afternoon Meteosat infrared image for the same day to illustrate the cumulus convection that developed in the unstable air indicated by the shaded area on the map (Meteosat image supplied by the European Space Agency).

13:00 SAST radiosonde sounding the 800 hPa dew-point is normally used.

A value of −3 °C is usually taken as defining conditions in which conditional instability will support the development of cumulus clouds. An example of an early summer distribution of the Index and the afternoon occurrence of convection is given in Figure 6.14. The correspondence is close.

ADDITIONAL READING

STURMAN, A. S. and TAPPER, N. 1996. *The Weather and Climate of Australia and New Zealand.* Melbourne: Oxford University Press.

MCINTOSH, D. H. and THOM, A. S. 1978. *Essentials of Meteorology.* New York: John Wiley and Sons.

LONGLEY, R. W. 1970. *Elements of Meteorology.* New York: John Wiley and Sons.

AIR MINISTRY METEOROLOGICAL OFFICE. 1960. *Handbook of Aviation Meteorology.* London: H. M. S. O.

HEWSON, E. W. and LONGLEY, R. W. 1944. *Meteorology Theoretical and Applied.* New York: John Wiley and Sons.

7

ATMOSPHERIC HEAT TRANSFER

Energy may be conducted from one place to another by conduction, convection, advection or radiation. *Conduction* takes place by the transfer of the kinetic energy of atoms and molecules. The application of heat to a body causes the atoms and molecules in the heated area to vibrate faster about their mean position than those in adjoining cooler areas. Collisions between the fast-moving and slow-moving atoms permit the transfer of energy from the heated to the cooler areas. At the earth–atmosphere interface this exchange of energy takes place in a layer less than one centimetre deep. *Convection* in the atmosphere involves the mixing of air by organized motions and may occur through the depth of the atmosphere. Heat may likewise be transported horizontally by *advection*. The transfer of energy by means of an ensemble of waves which travel through a vacuum at the speed of light takes place by electromagnetic *radiation*. Any body that is not at the absolute zero of temperature radiates energy to the space around it.

Each of these modes of heat transfer will be discussed in this chapter, beginning with solar radiation. The planetary circulation systems of the atmosphere and oceans are driven by solar radiation. Nearly all exchanges of matter and energy at the earth–atmosphere interface and almost all life processes are supported by solar radiation. The sun's energy is of fundamental importance in meteorology and climatology.

THE STRUCTURE OF THE SUN AND ITS RADIATION

More than five thousand million years ago the gravitational collapse of stellar matter resulted in the formation of the sun. As the widely dispersed matter, mainly in the form of hydrogen atoms, came together, potential energy (the energy of position) was converted into kinetic energy (the energy of motion). Since the average kinetic energy of atoms and molecules in a gas is a measure of the temperature of the gas, gravitational collapse and the generation of kinetic energy led to very high core temperatures (possibly as high as 20 million degrees Celsius). These were sufficient to trigger the nuclear fusion reaction that today powers the generation and transfer of energy up through the various layers of the sun and from there into space.

Hydrogen and helium are the most abundant chemical elements in the sun. The fuel of the nuclear-fusion process is provided by hydrogen atoms which, under conditions of high temperature and pressure, combine to form the next simplest element, helium. This fusion process converts mass into energy, which is transferred as radiant energy from the sun's core, through the radiation and convection zones, to its surface, which is termed the *photosphere* (Fig. 7.1). The temperature of the sun's surface is about 6 000 °C and indicates a rapid decrease in temperature from the core to the surface. The sun's atmosphere, called the corona, extends great distances into space. This is particularly evident when *solar flares*, caused by sudden violent eruptions in the corona, send floods of solar cosmic rays into space.

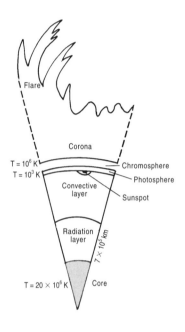

Fig. 7.1 The structure of the sun (after Lockwood, 1979).

The sun's surface is characterized by both hot (bright) and relatively cooler (darker) areas, which give it a granular appearance. The bright granules are areas of violent convective activity, which transfers energy upward into the solar atmosphere, and are separated by darker areas of subsidence. The temperature above the surface rises sharply due to this energy transfer to reach 10^6 °C some 15 000 km above the surface in the transition to the corona known as the *chromosphere*.

Large darker areas on the photosphere are called *sunspots*. These less disturbed regions are cooler by 1 000 °C or more and are characterized by strong magnetic fields. The number of sunspots varies with time. A period of low sunspot activity is followed by a period of high sunspot activity some 3–6 years later. The entire period from sunspot minimum to sunspot minimum lasts for an average of 11 years and has become known as the 11-year sunspot cycle. Longer sunspot cycles have also been identified. These include a 22-year cycle and an 80-year period caused by high and low peaks in the 11-year cycle.

SOLAR RADIATION

The amount of incoming radiation falling perpendicularly on unit area in unit time at the outer limit of the atmosphere, when the distance between the sun and the earth is at its mean value, is called the *solar constant*. This value is estimated at an average of 1.37 kW m^{-2}. It is not a true constant as it can fluctuate by 1.5 per cent about its mean value, with much of the fluctuation being in the ultraviolet portion of the electromagnetic spectrum. Incoming solar radiation (short-wave radiation or *insolation*) represents almost all the energy available to the earth (the energy emanating from the radioactive decay of earth minerals is insignificant by comparison). In Table 7.1 the energy from the sun in one day is compared with other energy sources on the earth or in its atmosphere.

Table 7.1 Total energy of various phenomena relative to total solar energy received per day (after Sellers, 1965).

Solar energy received per day	1
Strong earthquake	10^{-2}
Average hurricane	10^{-4}
Krakatoa explosion of August 1883	10^{-5}
Detonation of thermonuclear bomb	10^{-5}
Average squall line	10^{-6}
Average summer thunderstorm	10^{-8}
Average earthquake	10^{-8}
Average tornado	10^{-11}
Average lightning stroke	10^{-13}

Variations in solar output

One of the enduring unanswered problems in meteorology concerns the link between fluctuations in the solar constant and weather and climate. Variations in the solar constant are likely to occur on a number of time scales that range from short-term variations due to sunspot activity to long-term variations due, first, to changes in activity within the sun, and, secondly, to the changing nature of the earth's orbit around the sun. Sunspots block some radiation to space, but evidence to indicate that this results in short-term variations in the solar constant, in excess of the small fluctuations in ultraviolet levels, is inconclusive at present. However, over the long term it is estimated that the evolving sun has undergone an increase of some 20 per cent in energy output.

The Milankovitch theory

Long-term variations in solar output may also occur as a result of changes in the distance of the earth from the sun. At present the earth follows an elliptical orbit around the sun, being nearest to the sun on 2–3 January (perihelion) and furthest away on 5–6 July (aphelion). These changes in proximity to the sun cause a January increase in the solar beam over the average solar constant of about 3.5 per cent; a similar decrease occurs in July.

Over thousands of years, changes occur in the earth's orbit around the sun and cause subsequent variations in the strength of the solar beam reaching the earth. Known as the *Milankovitch theory*, three different cycles can be identified in Figure 7.2 as resulting from the following:

The eccentricity of the earth's orbit, whereby over a quasi-cyclic period of about 100 000 years the earth's orbit undergoes a change from almost circular to highly elliptical. At present the distance between earth and sun varies from 147×10^6 km at perihelion to 152×10^6 km at aphelion. This is about as small as the variation

gets, but it may increase to as large as 142.5×10^6 km to 156×10^6 km. Clearly the amount of solar radiation reaching the earth will increase and decrease respectively, producing opposite effects in the two hemispheres.

The obliquity of the ecliptic, whereby the tilt of the earth's axis is believed to vary from about 21.8° to about 24.4° over a 42 000-year period (at present it is 23.5°). This slight wobble would influence the severity of the difference between summer and winter. When the obliquity is large so is the range of seasonality.

The precession of the equinoxes, whereby the times of aphelion and perihelion change regularly. At present perihelion is about 2 January (i.e., in the southern-hemisphere summer). In about 11 000 years the sun will be closest to the earth in July, during winter. Winters will then be milder (and summers hotter in the northern hemisphere). The complete cycle produces two periodicities: one at 23 000 years, the other at 18 800 years.

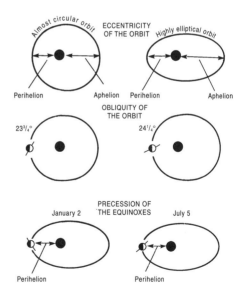

Fig. 7.2 The Milankovitch mechanisms affecting the receipt of solar radiation at the earth's surface.

The hypothesis that variations in energy input caused by the Milankovitch cycles are the cause

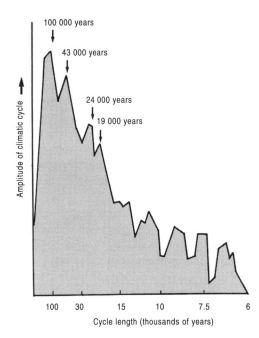

Fig. 7.3 A schematic illustration of the peaks in the spectrum of climatic variation that have occurred over the last half-million years. Variations in eccentricity produce the 100 000-year peak variation, variations in obliquity cause the 43 000-year peak and the precession of the equinoxes results in the 18 000- to 24 000-year peaks (after Imbrie et al., 1984).

of long-term climatic changes is strengthened by the identification of cycles with periods of about 100 000 years, 43 000 years, 24 000 years and 19 000 years in oxygen isotope temperatures from deep-sea sediment cores (Fig. 7.3).

Daily and seasonal variations in solar radiation

The longer the duration of daylight the greater the quantity of radiation received over a given portion of the earth. Due to the axial tilt of 23.5° with respect to the plane of the ecliptic (i.e., to the orbital plane), the revolution of the earth around the sun causes the sun to be vertically overhead over the Tropic of Capricorn at the

time of the southern hemisphere *summer solstice* on 22 or 23 December. At that time the duration of daylight poleward of the tropic is at a maximum. Poleward of the Antarctic Circle the daylight length is 24 hours. When the sun moves north at the time of the *winter solstice* to its overhead position at the Tropic of Cancer on 20 or 21 June, the longest night occurs in the southern hemisphere with 24-hour nights occurring poleward of the Antarctic Circle (Fig. 7.4). At the equinoxes (22 March and 22 September) the sun is vertically overhead at the equator and all places on the earth experience 12 hours daylight.

The altitude of the sun is given in terms of the angular distance between the sun's rays and the local vertical. This *zenith angle*, Z, varies, firstly, with latitude, ϕ, secondly, with the angular distance of the sun north or south of the equator, called the solar declination, δ, and, thirdly, with time of day given by the hour angle, h, (which is defined as zero at solar noon and increases by 15° for every hour after noon). Zenith angle is given by

$$\cos Z = \sin\phi\sin\delta + \cos\phi\,\cos\delta\,\cos h \qquad (7.1)$$

The daily radiation incident on a horizontal surface at the top of the atmosphere, R_t, is

$$R_t = \frac{86400}{\pi} S_0 \left(\frac{\overline{d}}{d}\right) (H\sin\phi\sin\delta + \cos\phi\cos\delta\sin H) \qquad (7.2)$$

where S_0 is the solar constant, $\overline{d}/d$ is a correction factor for variations in the earth–sun distance ($\overline{d}$ is the mean earth–sun distance; d the instantaneous value) and H is a half-day length from sunrise to sunset expressed in radians. The constant in the equation is the number of seconds in a 24-hour period. The variation in solar radiation at the top of the atmosphere with latitude and month is shown in Figure 7.5. Solar radiation at the poles is at a maximum during the polar day at the time of the summer solstice. It is also somewhat higher at the south pole due to perihelion currently occurring close to the middle of summer in the southern hemisphere.

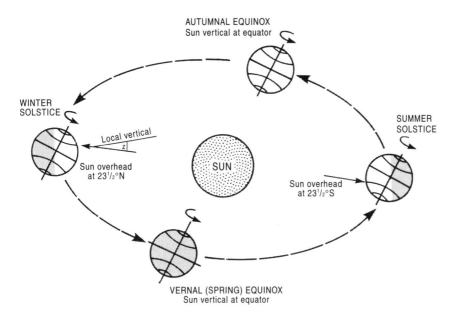

Fig. 7.4   Daily and seasonal variations in solar radiation.

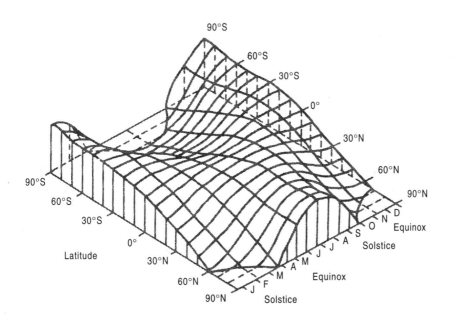

Fig. 7.5   Variations in solar radiation at the top of the atmosphere as a function of latitude and season (Strahler, 1975).

THE ELECTROMAGNETIC SPECTRUM

Radiation is conceived as containing a continuous range of wavelengths, the totality of which is called the *electromagnetic spectrum* (Fig. 7.6). The unit of wavelength commonly used is the micrometre, μm, which is equal in length to 0.001 mm. Wavelengths important in climatology range from very short waves of 0.1 μm to wavelengths of about 100 μm. Within this range the human eye responds only to *visible light* with wavelengths between 0.39 μm and 0.76 μm. Colours in the visible part of the spectrum vary from violet light, which has the shortest, to red, which has the longest wavelengths. Radiation with wavelengths less than 0.39 μm is termed *ultraviolet*; that with wavelengths greater than 0.76 μm is *infrared* radiation. Of the sun's radiation, 9 per cent is in the ultraviolet, 45 per cent in the visible and 46 per cent in the infrared portion of the electromagnetic spectrum.

Radiation is normally unpolarized, which means that it is free to vibrate in all directions (Fig. 7.7, *left*). The relation of wavelength, λ, to frequency, ν, is given by

$$\lambda \nu = c \qquad (7.3)$$

where the speed of light c = 3×10^8 m s^{-1}. Since *c* is constant, any increase in λ will cause a corresponding decrease in ν. Therefore, long waves have a low frequency and short waves a high frequency as shown in Figure 7.7 (*right*).

Not all bodies radiate in all parts of the spectrum. The sun radiates most of its energy between 0.3 μm and 3 μm which explains why solar radiation is often called *short-wave radiation*. The earth–atmosphere system, on the other hand, radiates most of its energy in the band which extends from 3 to 100 μm. This radiation is therefore referred to as *terrestrial* or *long-wave radiation*.

RADIATION LAWS

The rate of energy transfer by electromagnetic radiation is termed the *radiant flux*. When divided by the area through which it passes, it is defined as *irradiance, E*. The rate of energy transfer is specified as energy per unit time and is measured in joules per second or watts obtained as follows:

Force = mass $\times$ acceleration = newton (N)
  = kg m s^{-2}
Energy = force $\times$ distance = N m = joule
  (J) = kg m^2 s^{-2}
Flux = energy per unit time = J s^{-1} = watt
  (W) = kg m^2 s^{-3}
Irradiance = flux per unit area = W m^{-2}
  = kg m s^{-3}

Irradiance per unit wavelength interval is the *monochromatic irradiance, E_λ,* (W m^{-2} μm^{-1}). Radiation laws have been developed for theoret-

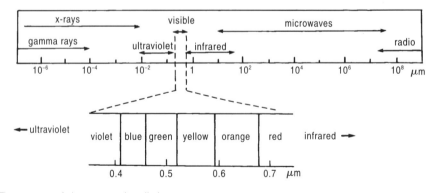

Fig. 7.6 The spectrum of electromagnetic radiation.

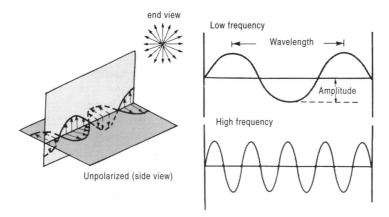

Fig. 7.7 *Left:* the wave nature of radiation in the vertical and horizontal (after Eagleman, 1980); *right:* low- and high-frequency waves.

ically perfect radiators, termed *black bodies*, which absorb all radiation incident upon them and then proceed to emit the maximum possible radiation in all wavelengths and in all directions. These laws govern the emission of radiation from a black body. The first determines the basic shape of the emission curve (Planck's law);

the second gives the relationship between wavelength and the temperature of emission (Wien's law); and the third permits the calculation of the energy of emission as a function of the temperature (the Stefan-Boltzmann law).

The radiation emitted at a particular wavelength, λ, by a black body at temperature T is

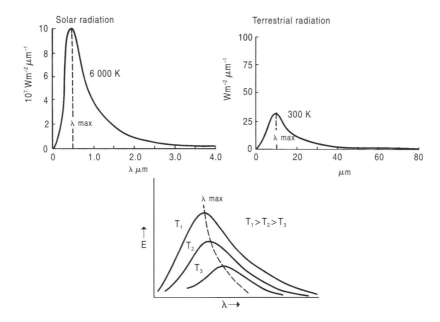

Fig. 7.8 *Upper:* spectra of solar and terrestrial radiation; *lower:* the displacement of the wavelength of maximum radiation toward longer wavelengths with decreasing temperature.

given by *Planck's law*, where

$$E_\lambda = \frac{c_1}{\lambda^5 \left[\exp(c_2/\lambda T) - 1\right]} \qquad (7.4)$$

where E_λ is the amount of energy (W m^{-2} μm^{-1}) emitted from a black body at wavelength λ (μm) and c_1 and c_2 are constants. When E_λ is plotted as a function of wavelength for temperatures characteristic of the sun (6 000 K) and the earth (300 K), the curves appear as in Figure 7.8 (*upper*). The spectral ranges of the sun and earth differ considerably with a small overlap only from about 3.5 to 8 μm. The wavelength of maximum radiation from the sun falls within the visible range of the electromagnetic spectrum and occurs at about 0.5 μm. The wavelength of maximum terrestrial radiation occurs at about 10 μm and within the infrared portion of the spectrum.

In practice, as is shown in Figure 7.9, theoretical and actual solar and terrestrial radiation curves vary significantly, owing to attenuation of radiation by the earth's atmosphere. (The shapes of the curves in Figure 7.8 (*upper*) and Figure 7.9 are different because logarithmic wavelength scales have been used in the latter.)

Wien's law states that the wavelength of maximum emission, λ_{max}, for a black body at temperature T (K) is given by

$$\lambda_{max} = \frac{2897}{T} \qquad (7.5)$$

Given the inverse relationship between wavelength of maximum radiation and temperature of emission, it is clear that the hotter the body, the shorter its radiation. As temperature decreases, so the wavelength of maximum radiation is displaced to longer wavelengths (Fig. 7.8, *lower*), giving rise to an alternative name for the law, namely *Wien's displacement law*.

The *Stefan-Boltzmann law* states that the total amount of energy emitted by a black body, integrated over all wavelengths (i.e., the area under the curves given in Fig. 7.8, *upper*), is

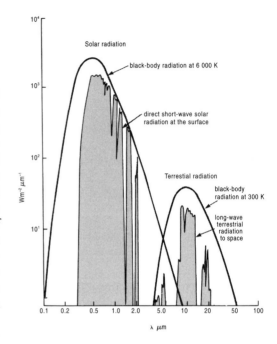

Fig. 7.9 Black-body radiation at 6 000 K and 300 K for the sun and earth respectively, together with the direct solar radiation at the surface and the infrared emission to space from the earth's surface (shaded). Note the logarithmic scales for wavelength and radiation (after Sellers, 1965).

directly proportional to the fourth power of the absolute temperature:

$$E_\lambda = \sigma T^4 \qquad (7.6)$$

where the Stefan-Boltzmann constant σ has the value 5.67×10^{-8} W m^{-2} K^{-4}. The rate of energy emission increases dramatically with an increase in temperature. For example, doubling the temperature increases the rate of emission by 2^4, or 16 times, tripling the temperature by 3^4, or 81 times, and so on.

No natural material radiates or emits as a perfect black body. It is possible, however, to use the amount of energy emitted by a theoretical black body as a basis for comparison with other materials. The fraction of energy emitted by a natural material relative to that of a black body is its *emissivity*, given by

$$\epsilon_\lambda = \frac{E_\lambda \, (material)}{E_\lambda \, (black\ body)} \tag{7.7}$$

The emissivity of a black body is unity at all wavelengths. In contrast, the emissivity of natural materials varies with the wavelength of emission between values of 0 and 1. A corresponding quantity is the *absorptivity*, a_λ, which is the ratio of the incident to the absorbed radiation.

The relationship between absorptivity and emissivity is given by *Kirchoff's law*, which states that for a given wavelength and a given temperature,

$$\epsilon_\lambda = a_\lambda \tag{7.8}$$

This law is important because it indicates that at the same wavelength good absorbers are good emitters and vice versa. This applies to solids, liquids and gases.

THE ATTENUATION OF RADIATION

It is clear from Figure 7.9 that radiation is attenuated (reduced) in its passage through the atmosphere. The processes depleting radiation are absorption, scattering and reflection.

Absorption

When radiation passes through a mixture of gases, it is depleted as a portion of the radiation is transformed into heat by *absorption*. Selective absorption takes place as each gas absorbs in certain wavelengths. The net result is that, over the spectrum as a whole, strong absorption occurs in certain wavelengths with very little in others. The absorption mechanism can best be understood by considering an elementary model of an atom consisting of a nucleus around which electrons move in orbits. According to the quantum theory, only certain orbits are permitted, with each orbit containing a certain energy level. If for some reason an electron moves to a larger permitted orbit, a discrete amount of energy must be supplied to excite the electron to move.

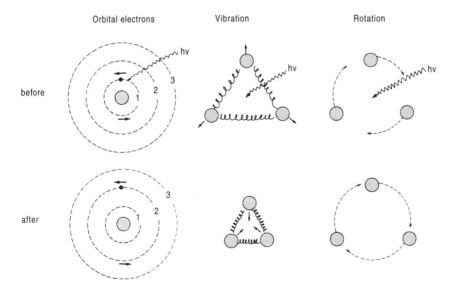

Fig. 7.10 Atomic and molecular energy transitions following the absorption of a quantum of radiation (adapted from Hess, 1959; Gedzelman, 1980).

According to the quantum theory, radiant energy exists in discrete units called *photons*. If the energy of a photon is equal to that required by the electron to move to the next permitted orbit, the electron will absorb this energy in the process of the move (Fig. 7.10). Conversely, electrons which occupy high-energy-level orbits may fall back to lower energy levels and in so doing will emit energy.

The amount of energy, Q, associated with a photon of radiation is given by

$$Q = hv \qquad (7.9)$$

where v is the frequency of radiation and h is Planck's constant. If a beam of radiation with a continuous range of frequencies (and therefore wavelengths) were to fall on a specific gas, absorption would occur in the part of the spectrum that contains photons of energy equal to that required to elevate the electrons contained in the relevant atoms. The absorption characteristics of different gases can be obtained by means of a spectrometer, where the absorption corresponding to Q produces an absorption line in the spectrum of radiation. Similarly, the emission of energy will produce an emission line.

Atmospheric absorption is not restricted to orbital changes in atoms. Radiation may also be absorbed or emitted by molecular transitions, through the process of vibration of the atoms about their mean position in the molecule, or rotation of the molecule around its centre of mass (Fig. 7.10). The type of atomic or molecular transition tends to be related to the wavelength of radiation. Orbital changes, which occur with the highest frequency, occur in the short-wave part of the radiation spectrum occupied by X-rays and ultraviolet radiation. Vibrational changes are associated with the near infrared and the least frequent rotational transitions with infrared and microwave radiation.

Atmospheric gases that are important in the absorption of radiation are oxygen, ozone, water vapour and carbon dioxide. Each gas absorbs energy in narrow wavelength intervals called *spectral absorption lines*, which are grouped to form *absorption bands*. The location of these bands in the radiation spectrum varies with the gas concerned. The amount of absorption that takes place depends upon the quantity and temperature of the gas.

Almost all the ultraviolet radiation in wavelengths less than 0.29 μm is absorbed by oxygen and ozone (Fig. 7.11; see also Fig. 7.9, *left*). The ozone in the ozonosphere is particularly important in this regard and provides a protective shield against the ultraviolet radiation that is so harmful to life. Little absorption takes place in wavelengths between 0.3 and 0.8 μm. The atmosphere is consequently transparent in that part of the spectrum in which the major portion of solar-radiation transmission occurs (Figs. 7.8 and 7.9). However, in the wavelengths in which terrestrial radiation is emitted, water vapour absorbs in several bands between 2.5 μm to 3.5 μm, 5 μm to 7 μm, and beyond 13 μm. Thus it is clear that clouds do not absorb much solar radiation. Individual clouds may absorb 10–20 per cent. On average probably less than 3 per cent is absorbed. Carbon dioxide absorbs terrestrial radiation in bands at 1.6 μm, 2.0 μm, 2.7 μm, 4.3 μm, and 15 μm. Other absorption bands for additional gaseous components of the atmosphere, as well as the total absorption capacity of the atmosphere as a whole, are given in Figure 7.11. Figure 7.9 gives a similar picture of the absorption of radiation in the atmosphere.

Scattering

When a parallel beam of radiation passes through a gaseous medium, the impact of photons with particles, such as air molecules, causes the photons to change their direction of travel. The radiant energy redistributed in all directions by this process of *scattering* is responsible for the brightness of the daytime sky. Without scattering the sky would appear black.

The type of scattering is determined by the size of the scattering particles in relation to the wavelength of incident radiation. For scattering particles having a radius less than 1/10 of the wavelength of the radiation, the amount of *Rayleigh scattering* is inversely proportional to the fourth power of the wavelength of radiation. The sky appears blue because the scattering of blue light (at ~0.4 μm) by air molecules is approximately six times greater than that of red light (at ~0.7 μm), a feature shown for two wavelengths in Figure 7.12 (*upper*). When the sun is low in the sky the solar beam must penetrate a greater thickness of atmosphere. This causes blue light to be almost entirely dispersed from the direct beam, leaving the longer wavelengths in the red and orange portion of the visible spectrum. A reddish sky and sun results. Dust or air pollution enhances this effect.

Mie scattering occurs when the radius of the scattering particles is of a similar size to the wavelength of incident radiation. Cloud drop-

lets and pollution particles are the most common scattering particles of this size and produce scattering principally in the forward direction (Fig. 7.12, *lower*). All wavelengths tend to be scattered in this manner leading to a sky colour more grey to white than blue.

When the size of particles lies between 1/10 and 10 times the wavelength of radiation, then *diffraction* occurs. Diffraction is an edge effect which breaks up and bends a radiation beam as it passes around the edge of an object (particle). The amount of bending varies with wavelength. When particle size is larger than 10 times the wavelength, *refraction* and *reflection* occur. Refraction takes place as radiation is bent in its passage through larger water drops, owing to the differences in the refractive indices of water and air. Reflection from cloud drops and from the ground surface is a major attenuating process.

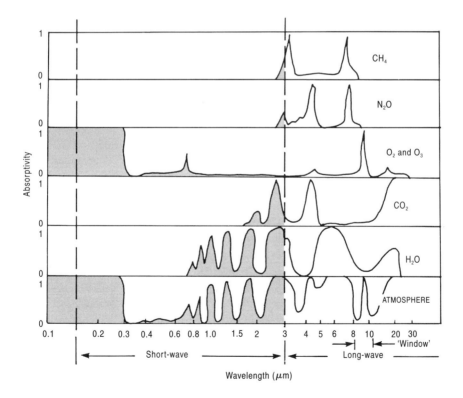

Fig. 7.11 The absorption of short-wave solar radiation by atmospheric constituents (after Fleagle and Businger, 1963).

Reflection

For any wavelength, the radiation incident upon a surface, E_λ, is absorbed and reflected such that

$$E_\lambda \ (absorbed) + E_\lambda \ (reflected) = E_\lambda \ (incident) \quad (7.10)$$

The fraction of incident radiation that is reflected by a surface is called the *albedo*, α, of the surface. If the absorptivity, a, is the fraction of incident radiation which is absorbed by the surface, then

$$a = 1 - \alpha \quad (7.11)$$

which demonstrates that strong absorbers are weak reflectors and vice versa.

Since neither reflection nor scattering involve a transformation to heat through absorption, they may be combined in a measure of the upward and downward radiation streams at the top of the atmosphere to record the *planetary albedo* (Fig. 7.13) and at the surface of the earth to give the *surface albedo*. The albedo of various surfaces in the spectral range for solar radiation is given in Table 7.2.

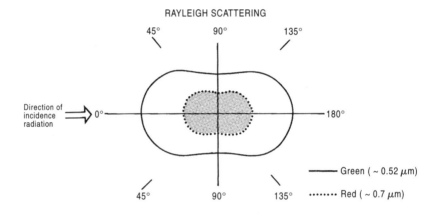

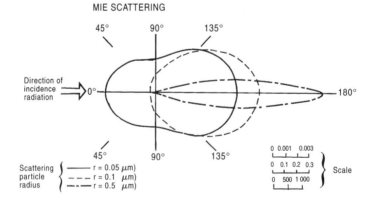

Fig. 7.12   The direction and intensity of Rayleigh scattering for a particle of radius 0.025 μm for two wavelengths of incident radiation and Mie scattering for particles of three different radii with incident radiation, wavelength 0.52 μm (i.e., green light). The polar diagrams give directions of scattering (0° is directly backward and 180° directly forward). The intensity of the scattered beam increases rapidly with increasing particle size (after Henderson-Sellers and Robinson, 1986).

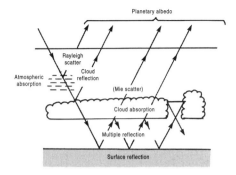

Fig. 7.13 Planetary and surface albedos.

Table 7.2 Some albedos for the short-wave portion of the spectrum (after Sellers, 1965).

Surface type	Per cent
snow, fresh	75–95
cumuliform cloud	70–90
desert	25–30
savanna, dry season	25–30
tundra	15–20
grassland	15–25
forest (deciduous)	10–20
sea surface (low sun angle)	10–20
sea surface (sun > 25° above horizon)	< 10

Total attenuation

The total depletion of solar radiation in its passage through the atmosphere by the combined effects of scattering, reflection and absorption may be expressed by *Beer's law* in which

$$\ln \frac{Q}{Q_0} = -km \qquad (7.12)$$

or

$$\frac{Q}{Q_0} = e^{-k/m} \qquad (7.13)$$

where Q_0 is the solar radiation incident at the top of the atmosphere, Q is that which reaches the surface, m is the mass of attenuation material in the atmosphere (sometimes called *optical thickness*) and k is the extinction coefficient. The ratio Q/Q_0 is the transmissivity of the atmosphere. The path length of the radiation beam varies during the day from a minimum at solar noon to a maximum when the sun is near the horizon. This is allowed for by introducing the *optical air mass*, which varies as $1/\cos Z$ and is equal to 1 when the sun is vertically overhead. The zenith angle, Z, has been defined in equation (7.1). Equation (7.13) then becomes

$$\frac{Q}{Q_0} = e^{-km/\cos Z} \qquad (7.14)$$

The spectral distribution of solar energy falling on a surface under cloudless conditions is shown in Figure 7.14. The total area under the upper curve (m = 0) represents the energy received at the earth's outer limit and is equal to the solar constant. The selective nature of the attenuation of solar radiation is shown by the curve for an optically thin air mass (m = 1) and for an optically thick air mass (m = 5).

A cloud cover presents a barrier to the transmission of solar radiation through the atmosphere. Reflection occurs from the cloud top and absorption takes place within the cloud. Optical thickness increases with the depth of cloud, as do reflection and absorption relative to transmission (Fig. 7.15). Clearly reflection from clouds produces far greater attenuation of solar radiation than does absorption. The transmission of *diffuse* radiation declines to about 20 per cent with clouds deeper than 100 m.

The effect of latitude on controlling the solar radiation received at the ground is shown in Figure 7.16 for the December solstice. This is the time of maximum solar radiation in the southern hemisphere (as seen from the receipt of insolation at the top of the atmosphere). In high latitudes, the low water-vapour content of the air limits atmospheric absorption. The obliquity of the sun's rays and increase in the

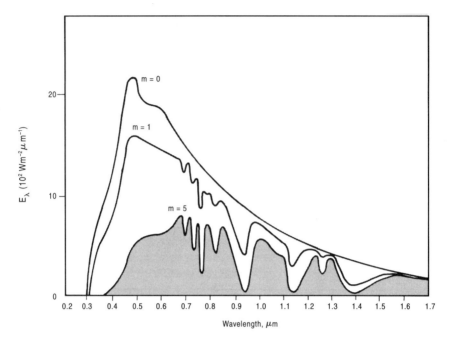

Fig. 7.14 Spectral distribution of solar radiation falling on a flat surface for two optical air masses (after Munn, 1966).

optical thickness of the atmosphere adds to the depletion of solar radiation. However, it is largely in response to the high reflectivity of snow and of ice surfaces, as well as the high average cloudiness of the arctic in summer, that attenuation is as large as is observed.

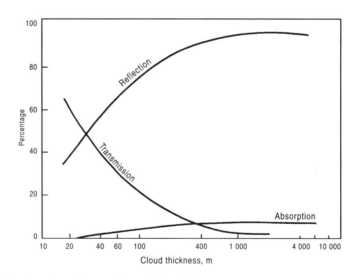

Fig. 7.15 Reflection, absorption and transmission of solar radiation by clouds of different thickness (after Hewson and Longley, 1944).

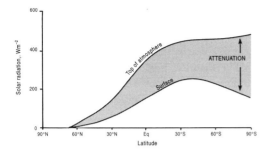

Fig. 7.16 The average receipt of solar radiation with latitude at the top of the atmosphere and at the earth's surface at the December solstice (after Barry and Chorley, 1998).

THE EFFECTIVE TEMPERATURE OF THE EARTH

The earth absorbs solar radiation in the amount $(Q + q)(1 - \alpha)$ where Q is direct beam radiation, q is diffuse radiation and α is surface albedo. The surface is consequently heated, which in turn leads to the emission of long-wave, terrestrial radiation. The amount depends on the surface temperature, T, and emissivity, ϵ, and is given by the Stefan-Boltzmann equation as

$$I \uparrow = \epsilon \sigma T^4 \qquad (7.15)$$

If it is assumed that no long-period change in the average temperature of the atmosphere is occurring, then an equilibrium may be assumed between the streams of incoming solar radiation and outgoing terrestrial radiation. Thus the mean temperature, T, at which the earth's surface and atmosphere radiate to space is given by the balance between the energy absorbed by the earth, which is the product of the irradiance and the area of the disc of the earth in receipt of energy (Fig. 7.17), and the energy emitted from the earth, which is the product of the outgoing radiation and the total area of the earth given by $4\pi r^2$. The balance is thus

$$\pi r^2 S_0 (1 - \alpha) = 4\pi r^2 \, \epsilon \sigma T^4 \qquad (7.16)$$

where r is the radius of the earth and S_0 the solar constant. Assuming an average earth emissivity of 0.90 and an average albedo of 0.36, the *effective temperature* of the earth's surface would be 256 K. The actual mean surface temperature of the earth is about 288 K, a difference of 32 K. The reason for this discrepancy lies in the absorption in infrared wavebands by low cloud, water vapour and carbon dioxide, much of which is returned to the earth's surface to raise the mean surface temperature 32 K above the effective temperature.

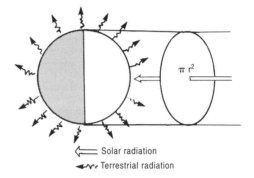

Fig. 7.17 The area of the disc of the earth in receipt of solar energy.

TERRESTRIAL RADIATION

On average, the atmosphere allows 49 per cent of direct, Q, and diffuse, q, solar radiation to reach the ground. By comparison, only 12 per cent of the earth's terrestrial (long-wave) radiation escapes unattenuated to space. The atmosphere readily absorbs infrared terrestrial radiation, primarily owing to absorption by water vapour and carbon dioxide. Water vapour absorbs over several infrared bands, notably between 2.5 μm and 3.5 μm, 5 μm and 7 μm and beyond 13 μm (Fig. 7.18). Carbon dioxide absorbs terrestrial radiation partially in discrete bands centred at 2.7 μm and 4.3 μm and completely in a wide band beginning at 15 μm. Methane, oxides of nitrogen, ozone, chlorofluorocarbons (freons) and other chemical components in the atmosphere likewise absorb terrestrial radiation selectively. The gases absorbing infrared radiation are termed *greenhouse gases*. Many of them are the result of pollution of the atmosphere. The steady increase of pollutants is giving rise to the serious concern that anthropogenic heating of the atmosphere by enhanced absorption of infrared radiation will lead toward long-term climatic change.

Just as important absorption bands are to be observed in the radiation spectrum, so are there bands in the spectrum of terrestrial radiation in which no absorption takes place. These transparent *windows* occur at wavelengths which are observed from 8–9.5 μm and 10–13 μm, the latter being in the region of maximum emission from the earth (Figs. 7.8 and 7.9). The band from 8 μm to 13 μm is generally considered to be transparent throughout, since the absorption

by ozone at around 10 μm is slight. It is through this transparent window that the earth loses heat to space by infrared radiational cooling.

The manner in which infrared heating and cooling of the atmosphere takes place is best considered by examining the interplay of radiation between the surface of the earth and the air above in the following way. Assume an atmosphere that is divided into layers, each containing unit volumes of water vapour. The depth of each layer is controlled by the water-vapour content of the air. Owing to the decrease of water vapour with height, the depth of each layer increases with height above the surface. Consider a simple model. Ignore all absorption from solar radiation and ignore all infrared radiation in the transparent window, thus considering the atmosphere fully opaque to terrestrial radiation. If the surface is to be in a state of

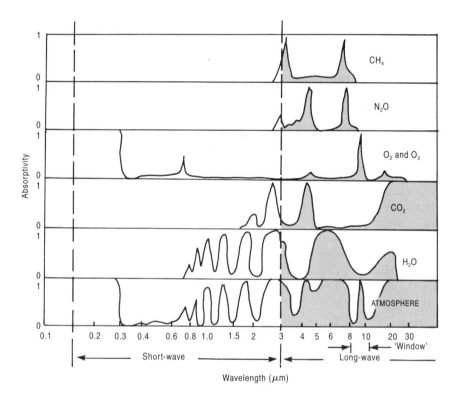

Fig. 7.18 The absorption of long-wave terrestrial radiation by atmospheric constituents (after Fleagle and Businger, 1963).

radiative equilibrium, then the gain in energy by absorption of solar radiation must balance the loss by infrared cooling. If each successive layer is to maintain equilibrium, then it must lose by emission of infrared radiation as much energy as it gains. Finally, the net upward flux of infrared radiation between adjacent layers (including the surface and top of the atmosphere) must be the same. Infrared radiation passes from the surface to the layers above, with each layer absorbing in all possible wavebands radiation from the adjacent layer and re-radiating upward and downward with an intensity determined by the temperature of the air in the layer (Fig. 7.19). In order to sustain a net upward flux of radiation through the opaque model atmosphere to provide at the top of the system the energy loss to balance the surface gain, it is necessary that emissions must decrease with height. By the Stefan-Boltzmann law it follows that temperature must decrease with height as well.

The simplified model fails to take into account the fact that about 20 per cent of the solar beam is absorbed in the atmosphere and that infrared energy is lost from the surface and layers above in the transparent band and gained from the atmosphere in the same band. When these effects are added, the calculated temperature profile shows a greater decrease with height, so much so that were radiative equilibrium to be the only process operating, the lapse rate would be unstable in the lower levels. That it is not unstable is due to the vertical transport of heat by convective (turbulent) and latent heat transfer processes.

Of the earth's infrared energy emission, 20 per cent is absorbed within 80 cm of the surface and 99 per cent within a layer 4 000 m deep. Clouds as well as temperature and moisture variations in the atmosphere complicate the simple model given above. A cloud behaves virtually as a black body by absorbing all infrared radiation incident upon it and re-radiating in all infrared wavelengths with an intensity that depends on its temperature. A cloud is therefore as effective as the ground as a radiating surface. This is particularly noticeable at night when the base of low cloud radiates energy to earth in all wavelengths, thereby inhibiting the normal decrease in surface temperature. Owing to its comparative transparency to short-wave solar radiation and near-opacity to long-wave terrestrial radiation, and thus its ability to raise the surface temperatures to values well above those that would prevail in the absence of an atmosphere, the atmosphere is often said to exert a *greenhouse effect*. This term is a misnomer, since a greenhouse is warmed as much by preventing convective heat loss as by selective absorption of radiation. It is, however, accepted by common usage. Clearly any change in the composition of the atmosphere that will enhance the so-called greenhouse effect will lead to atmospheric warming.

THE RADIATION BALANCE

The balance between incoming and outgoing radiation has been mentioned a number of times above. In such a balance, the incoming direct, Q, and diffuse radiation, q, not reflected at the surface (i.e., the amount absorbed and available for heating) is the same as the

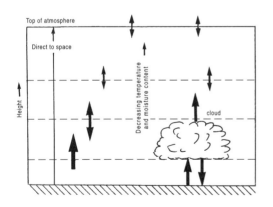

Fig. 7.19 A simplified diagram to illustrate the greenhouse effect and infrared radiational heating of the atmosphere.

difference between the infrared energy lost to space by the surface, $I\uparrow$, and gained from the atmosphere by *counter radiation*, $I\downarrow$. Thus on a cloudless day

$$(Q + q)(1 - \alpha)\kappa = I\uparrow - I\downarrow \qquad (7.17)$$

where κ is the transmission through the atmosphere. From this the *net all-wave radiation* may be defined as

$$R_n = (Q + q)(1 - \alpha)\kappa - I \qquad (7.18)$$

where I is the net infrared radiation $I\uparrow - I\downarrow$.

A number of empirical models have been developed to compute counter radiation and net long-wave radiation at the surface *under cloudless skies*. For example, in the *Brunt equation*

$$I\downarrow = \sigma T^4(a + b\sqrt{e}) \qquad (7.19)$$

and

$$I = \sigma T^4(1 - a - b\sqrt{e}) \qquad (7.20)$$

where the constants a and b have the average values of 0.61 and 0.05 respectively and e is the vapour pressure of the air. In the *Swinbank equation*, for temperatures above 0 °C

$$I\downarrow = 1.20\sigma T^4 - 171 \qquad (7.21)$$

and

$$I = 0.21\sigma T^4 - 171 \qquad (7.22)$$

Alternatively, a useful measure of determining counter and net infrared radiation is the *Brutsaert method* in which

$$I\downarrow = \epsilon_a \sigma T_a^4 \qquad (7.23)$$

and

$$I = \epsilon_s \sigma T_s^4 - \epsilon_a \sigma T_a^4 \qquad (7.24)$$

where $\epsilon_a = 1.24(e_a/T_a)^{1/7}$ and the subscript a refers to the air and s to the surface.

With *cloud cover*, counter radiation and net infrared radiation may be determined using a non-linear cloud factor such that

$$I\downarrow = I\downarrow (1 + an^2) \qquad (7.25)$$

and

$$I = I (1 - bn^2) \qquad (7.26)$$

where the constants a and b vary with cloud height (and therefore the temperature of the cloud base) as shown in Table 7.3 and n is the fraction of the sky covered in tenths.

Using one of the above ways to determine net infrared radiation it is now possible to calculate net all-wave radiation from equation (7.18). An example of its diurnal variation is shown in Figure 7.20. During daylight hours the net short-wave gain exceeds the net long-wave loss causing a radiation surplus at the surface. This is reversed at night and R_n values become negative.

Table 7.3 Value of coefficients used in equations (7.25) and (7.26) to allow for decreasing cloud temperature with height (after Sellers, 1965; Oke, 1978)

Cloud type	Typical cloud height (km)	Coefficients	
		a	b
Cirrus	12.20	0.04	0.16
Cirrostratus	8.39	0.08	0.32
Altocumulus	3.66	0.17	0.66
Altostratus	2.14	0.20	0.80
Stratocumulus	1.22	0.22	0.88
Stratus	0.46	0.24	0.96
Fog	0	0.25	1.00

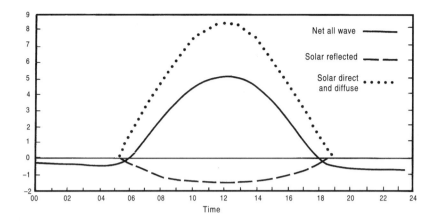

Fig. 7.20 A schematic illustration of the diurnal variation of net all-wave radiation (after Munn, 1966).

THE ENERGY BALANCE

As a closed system the earth–atmosphere system permits the transfer of energy, but not matter, into and out of the system. Solar radiation absorbed at the surface is output in the form of radiant heat as well as sensible and latent heat. This transfer takes place in accordance with the principle of the conservation of energy, which dictates that energy may neither be created nor destroyed, but can be converted from one form to another.

Over time the energy input into a system must equal the energy output. In a generalized form the *energy balance*, therefore, may be written as

$$R_n = H + LE + G \tag{7.27}$$

where R_n is the net radiation, H is the transfer of sensible heat by conduction, convection and advection, LE is the latent heat transfer, and G is the soil heat transfer term (Fig. 7.21). By convention the non-radiative fluxes are positive when directed away from and negative when directed towards the surface. Positive terms in equation (7.27) indicate that the surface loses heat, while a negative term would indicate a

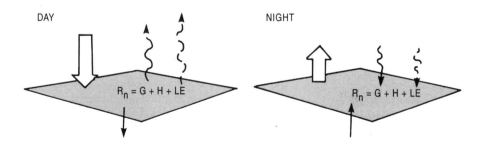

Fig. 7.21 Energy fluxes involved in the energy balance by day and by night (after Oke, 1978).

gain of heat by the surface.

At any one time the energy balance may not approach equality due to the retention or loss of energy stored in the system. Changes in *G*, whether positive (an increase in stored energy) or negative (a decrease in stored energy), are reflected in changes in temperature. When *G* is positive, soil temperature rises. It is necessary, therefore, to add an energy-storage term, *S*, to equation (7.27) to give

$$R_n = H + LE + G + \Delta S \qquad (7.28)$$

Changes in net energy storage, ΔS, are caused by imbalances in R_n, *H*, *LE* or *G*. These changes are reflected by the warming of the system volume when the input of heat energy, *Q*, exceeds the output (*vertical flux convergence*) and its cooling when input is less than output (vertical flux divergence) (Fig. 7.22, *left* and *centre*). When $\Delta S = 0$, an energy balance is achieved, since input equals output.

Horizontal flux convergence or *divergence* occurs when air with specific temperature properties moves from one area to another causing warming or cooling to take place (Fig. 7.22, *right*). This is known as *advection*. The transfer of energy by each of the terms in equation (7.27) will be discussed in the following sections.

Sensible heat flux

Sensible heat, *H*, and latent heat, *LE*, are transferred upward in the atmosphere by the turbulent mixing of air. Turbulent motion is characterized by vortex motions transverse to the mean wind. The convection responsible for the generation of turbulence is usually a mixture of both *forced* and *free* convection. Forced convection depends upon the roughness of the surface and is particularly important near the ground (below 2 m). Free convection involves the mixing of air in a buoyant environment and under unstable conditions is the dominant mechanism for the vertical transport of sensible and latent heat in the lower atmosphere.

The flux of sensible heat in the *laminar* and *turbulent boundary* layers causes considerable change in the shape of the near-surface temperature wave over the course of a calm, clear day. After sunrise a heat surplus begins to accumulate in the soil layer and must be dissipated in order to balance the energy equation. Sensible heat first passes through the lowest few millimetres or centimetres of the atmosphere in the *laminar boundary layer* by conduction such that

$$H = - \rho C_p k_h \ \frac{dT}{dz} \qquad (7.29)$$

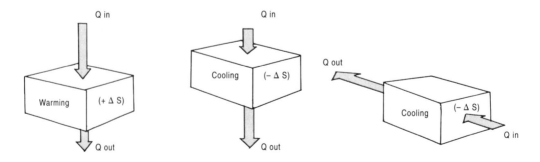

Fig. 7.22 Vertical and horizontal flux divergences of energy and resultant heating and cooling (after Oke, 1978).

where C_p is the specific heat of air at constant pressure, k_h is the molecular diffusion coefficient of heat and dT/dz the lapse rate of temperature.

In the *turbulent boundary layer* extending to 50 m or more into the atmosphere the flux of sensible heat is given by

$$H = -\rho C_p K_h \left(\frac{dT}{dz} + \Gamma \right) \qquad (7.30)$$

where K_h is the eddy diffusivity of heat and Γ the dry adiabatic lapse rate. Since the observed lapse rate in the atmosphere decreases with height (i.e., is negative), dT/dz may be replaced by $-\alpha$ and equation (7.30) becomes

$$H = -\rho C_p K_h (\Gamma - \alpha) \qquad (7.31)$$

Thus when $\alpha > \Gamma$ and the air is unstable, H becomes positive and the flux of heat is upward; when $\alpha < \Gamma$ and the air is stable, H is negative and the flux of heat is downward. Put another way, when the atmosphere is unstable, wind (turbulence) will cause cooling; when the atmosphere is stable, wind will cause warming. This is one of the reasons why a Berg wind has such a warming effect (the air is very stable, the wind is turbulent, and heat is transferred down the temperature gradient towards the surface – see Chapter 12 for Berg winds).

Another consequence following from equation (7.31) is that in an atmosphere that has been well mixed by turbulence the heat gradient will have become vertical; that is, no further heat flux will be possible and $H = 0$. This can only occur if $\alpha = \Gamma$. Thus in a well-mixed atmosphere the observed lapse rate tends to the dry adiabatic.

Sensible heat fluxes are usually determined using measurements of flux gradient, the eddy correlation method or the aerodynamic method. In the *flux gradient method*, temperatures T_1 and T_2 are taken at heights z_1 and z_2 to determine the bulk temperature gradient, $\Delta T/\Delta z$. Sensible heat is then obtained from

$$H = -\rho C_p K_h \frac{\Delta T}{\Delta z} \qquad (7.32)$$

In the *eddy correlation method*, measurements of T (temperature), u (wind velocity in the x direction), w (velocity in the z direction) and q (specific humidity) are made simultaneously at an appropriate height. From these means, deviations, variances and covariances are determined such that $T' = T - T$, $w' = w - w$, and so on. Products of deviations, such as $w'T'$, and the means thereof, $\overline{w'T'}$, are likewise calculated. From these

$$H = \rho C_p \overline{w'T'} \qquad (7.33)$$

Using the *aerodynamic method* and measurements of temperature T_1 and T_2 and wind u_1 and u_2 at heights z_1 and z_2, the sensible heat fluxes are determined from

$$H = -\rho C_p k^2 \frac{(u_2 - u_1)(T_2 - T_1)}{(\ln z_2 / z_1)^2} \qquad (7.34)$$

where ln is the natural logarithm to the base e and k is von Karman's constant (0.4).

The effects of radiative and turbulent processes on the diurnal variation of lapse rates may now be considered. Selected temperature profiles in Figure 7.23 show that at night the upward radiative flux causes the air immediately above the surface to cool below the temperature of the overlying air and a temperature inversion forms. The depth, duration and

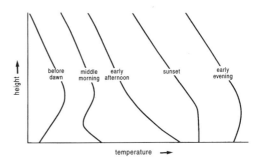

Fig. 7.23 The diurnal variation of lapse rates near the ground.

strength of this inversion is a function of wind speed, cloud cover, the nature of the radiating surface and the moisture content of the air. Should wind speed increase above a critical limit the inversion will be dissipated by the downward flux of sensible heat. Should clouds be present, counter radiation from them will retard cooling. Likewise, if the air has a high moisture content, then counter radiation will increase. Thus calm, clear nights and dry air govern the development of strong surface radiation inversions.

After sunrise, surface heating causes the air in the surface layer to warm and become unstable. The resulting turbulence causes convective mixing as warm buoyant parcels of air rise and are replaced by cooler air aloft. This creates a layer of well-mixed air which begins to erode the inversion from below. This continues until the inversion is eliminated and the mixed layer, with a lapse rate conforming to the dry adiabat, extends throughout the boundary layer. In the late afternoon the near-surface air begins to cool once more and a new cycle begins.

Latent heat flux

The flux of water vapour and latent heat takes place in a manner similar to that of sensible heat. Evapotranspiration, E, from the surface determines the humidity of the atmosphere. The upward flux of water vapour is greatest by day and continues at a reduced rate by night as it passes through the laminar boundary layer into the turbulent boundary layer with the flux given by

$$E = -\rho K_w \frac{dq}{dz} \tag{7.35}$$

where K_w is the eddy diffusivity of water vapour and q is the mean specific humidity. Whereas sensible heat may be returned to the surface by the turbulent mixing of air, water vapour is returned by the process of precipitation.

When a change of state occurs (i.e., from water to water vapour), energy is taken up in the form of latent heat, LE. For example, at 10 °C the *latent heat of vaporization*, L, takes up 2.48 $\times 10^6$ J kg^{-1}. When the vapour condenses back into a liquid state (i.e., with the formation of cloud droplets or dew), this energy is liberated. Using the *flux gradient method* the latent heat flux may be determined from

$$LE = -\rho L K_w \frac{\Delta q}{\Delta z} \tag{7.36}$$

where L is the latent heat of vaporization and $\Delta q/\Delta z$ is the bulk moisture gradient.

In the *eddy correlation method* the latent heat flux is

$$LE = \rho L \overline{w'q'} \tag{7.37}$$

whereas, using the *aerodynamic method* it is determined from

$$LE = -\rho L k^2 \frac{(u_2 - u_1)(q_2 - q_1)}{(\ln z_2/z_1)^2} \tag{7.38}$$

On a large scale latent heat exchange is a major agent of heat transfer in the atmosphere, in both the vertical and the horizontal. In the vertical, heat is transferred in the following way. Evaporation at the surface causes cooling. The heat taken up in the evaporated water vapour is then transported upward by convection until condensation occurs. When this happens the heat is released into the atmosphere and the air warms. Tropical areas of the atmosphere derive much of their heating in this way. Often the heat released is sufficient to sustain the formation and maintenance of tropical storms and, in some instances, severe storms such as hurricanes. Horizontal export of heat out of the tropics takes place if the moisture within the air is carried out of the region to condense in higher latitudes. This happens, for instance, in the formation of extra-tropical cyclones when air is transported in the warm conveyor of mid-latitude depressions (see Chapter 10).

The Bowen ratio

The values of H and LE are determined by the nature and characteristics of heat and water sources and sinks. The relative importance of these fluxes is given by the Bowen ratio, β, as

$$\beta = \frac{H}{LE} \qquad (7.39)$$

When $H > LE$, as occurs over non-vegetated, dry surfaces, $\beta > 1$. When $H < LE$, $\beta < 1$ and cool, moist conditions tend to prevail. Negative β values may also occur at night if the two values have different signs (i.e., downward sensible heat flux and upward latent heat flux). Typical Bowen-ratio values are:

Tropical ocean	< 0.1
Tropical forest	0.1–0.3
Temperate forest	0.4–0.8
Semi-arid	2.0–6.0
Desert	> 10.0

Soil heat flux

After sunrise, a downward (positive) heat flow is initiated into the soil by thermal conduction. This changes to an upward (negative) heat flow after sunset. The rate of flow is determined by the product of the thermal conductivity, k_s, (a measure of the ability of the soil to conduct heat) and the mean temperature gradient, $\Delta T/\Delta z$, and is given by

$$G = -k_s \frac{\Delta T}{\Delta z} \qquad (7.40)$$

The thermal conductivity varies with soil moisture, soil porosity and soil conductivity. The effect of moisture is particularly important. For example, the thermal conductivity of water (at 4 °C) exceeds that of air (at 10 °C) by a factor of 22.8. By expelling air from the soil pore space in a dry, sandy soil, the percolation of water into the soil increases the thermal conductivity of the soil.

The amount of heat necessary to raise the temperature of unit volume of the soil by 1 K defines its *heat capacity*. Since the heat capacity of water (at 4 °C) exceeds that of air (at 10 °C) by a factor of 3 483, considerably more heat must be added to water to effect a temperature change than is the case with air. The addition of water to soil displaces air from soil pore space and thereby reduces the rate of warming. The thermal response of a soil to heating is therefore determined, firstly, by its ability to transmit heat determined by thermal soil conductivity, k_s, and, secondly, by the amount of heat required to effect a temperature change, determined by soil heat capacity, C_s. The ratio of the two properties is the *thermal diffusivity*, K_s, of the soil given by

$$K_s = \frac{k_s}{C_s} \qquad (7.41)$$

Thermal diffusivity determines both the rate at which temperature changes penetrate the soil and the form of the temperature wave. In a homogeneous soil the downward penetration of the temperature wave is rapid if the diffusivity is high and slow if diffusivity is low. With increasing depth the amplitude of the temperature wave decreases and the times of maximum and minimum temperatures occur progressively later (Fig. 7.24, *left*). This time lag explains why at any one time the soil may be cooling near the surface yet warming beneath. The local rate at which heat is conducted into the sub-surface governs the rate of temperature change.
    Thus

$$\frac{\partial T}{\partial t} = K_s \frac{\partial}{\partial z}\left(\frac{\partial T}{\partial z}\right) = K_s \frac{\partial^2 T}{\partial z^2} \qquad (7.42)$$

and the rate of temperature change with time is proportional to the rate of change with the depth ($\partial/\partial z$) of the local temperature profile ($\partial T/\partial z$) in the sub-surface.
    The form of the annual temperature wave is similar to the diurnal wave (Fig. 7.24, *right*).

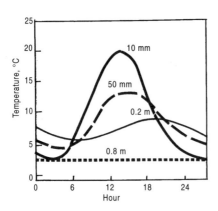
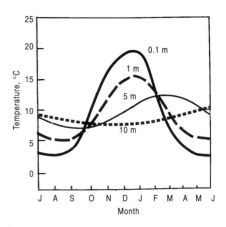

Fig. 7.24 A schematic illustration of the diurnal and seasonal variation of temperature with depth in soil (after Oke,1978).

During summer the predominantly downward flux builds up a heat store in the soil. In autumn the heat flux gradient reverses and the predominance of upward flux begins to deplete the stored heat. This continues until spring when the soil heat gradient reverses once again.

ADVECTIVE HEAT TRANSFER

Advection is the net horizontal transfer of both sensible and latent heat by wind and is often included as an individual extra term in equation (7.27). The importance of this transfer mech-

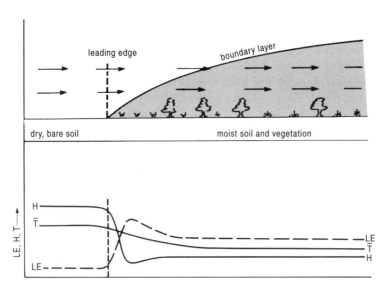

Fig. 7.25 *Upper:* heat advection from a hot, dry unvegetated to a rougher, cooler and more moist vegetated surface and the development of the boundary layer; *lower:* horizontal profiles of sensible and latent heat fluxes and mean temperature (modified after Oke, 1978).

anism lies in the varied nature of earth surfaces, each with different radiative, thermal, moisture and aerodynamic properties. These differing properties impart to each surface specific climatic characteristics in terms of temperature, moisture and wind speed. When wind blows from one distinct surface (such as a ploughed field) to another (such as a moist soil and vegetation surface), the advected air must adjust to new boundary conditions. This adjustment begins at the *leading edge* between the two surfaces with modification to the temperature, moisture and wind-speed properties taking place downwind within a deepening *boundary layer*. The characteristic changes that take place in temperature and sensible heat by day when air from a sun-warmed ploughed field moves over the cooler moist soil and vegetation surface are shown in Figure 7.25. The net result is an adjustment to the energy fluxes over the downwind area. Evaporative cooling would be enhanced by the moist vegetated surface which would reduce the level of both T and H. Under these conditions a surface-based inversion may also form due to the near-surface cooling. This in turn would cause H to reverse direction toward the surface.

The effect of advection becomes visible when advection fog is formed. This occurs with great regularity off the coast of Namibia. When warm air over the Atlantic Ocean moves across the leading edge of the cold Benguela Current flowing northward along the Namibian coast, temperature is depressed to dew-point and fog forms (Fig. 7.26 and Figs. 12.13 and 12.41). The coastline constitutes another leading edge with air moving over a hot, arid desert. Inland movement of the fog is therefore limited by the arid nature of the new surface conditions and the fog thins and evaporates downwind. By day this is hastened by surface heating.

ADIABATIC HEAT TRANSFER

A great deal of heating and cooling occurs in the atmosphere through adiabatic expansion or compression of air. Rising air parcels expand because of the decrease of inward pressure exerted by the ambient air on them with increasing height. The work done by such parcels in expanding is at the expense of their internal energy and their temperature falls accordingly. Conversely, layers of air may be forced to subside in anticyclonic systems and warm adiabatically as a consequence. This is the case with Berg winds. The physics of adiabatic processes need not be considered further here (see Chapter 4).

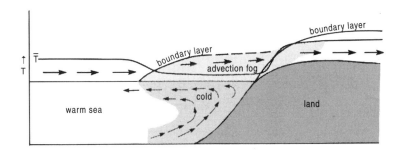

Fig. 7.26 The occurrence of advection fog over the cool waters of the Benguela upwelling system and the adjacent Namib desert.

DIURNAL AND ANNUAL CHANGES IN ENERGY BALANCE

The radiation surplus over a surface on cloud-less days is balanced in equation (7.28) by the combination of *H, LE, G* and *ΔS*. The proportional difference between the components depends on the nature of the surface (crop, grassland, forest, desert, water body) and subsequent variations in cloud cover, albedo, evapotranspiration and heat storage.

In a humid climate, such as occurs in the KwaZulu-Natal Midlands, evapotranspiration from well-vegetated surfaces is likely to cause the latent heat flux to exceed the sensible heat flux by day (Fig. 7.27, *upper*). To maintain the energy balance, since net radiation will not change, the sensible heat flux diminishes accordingly. In contrast, in the case of a semi-arid environment such as the dry western Karoo, the latent heat flux is small and the sensible heat flux becomes the dominant term balancing the energy budget. A similar pattern is apparent in the case of the annual cycle (Fig.

7.27, *lower*). During winter all the terms in the energy balance are small. In summer they increase to a maximum at the time of the solstice. In the case of the KwaZulu-Natal Midlands example, in summer the sensible heat flux exceeds the latent heat flux before the solstice and the opposite occurs after it. Over the dry western Karoo, the sensible heat flux is the dominant term required to balance net radiation throughout the summer. Only in early autumn (the time of the rainfall maximum), does the latent heat flux act to diminish the sensible heat flux to some extent.

In both diurnal examples given above, the excess of energy input over output by day results in the accumulation of an energy surplus which causes temperature to rise. The temperature rise continues beyond the peak solar input at noon until the time of maximum energy input which occurs at the point at which input equals output (Fig. 7.28, *upper*). This occurs a few hours after solar noon. Thereafter the surface loses heat by radiation and the temperature drops to a minimum value just after sunrise, again at the point

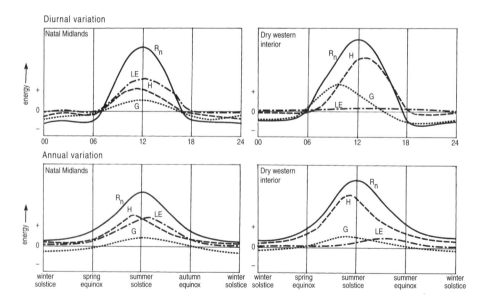

Fig. 7.27 Schematic diurnal and seasonal variations of energy-balance components (*R_n*, net radiation, *H*, sensible heat, *LE*, latent heat, and *G*, soil heat flux) in the humid KwaZulu-Natal Midlands and dry western interior of the Karoo.

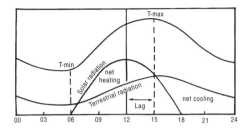

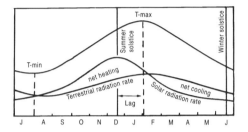

Fig. 7.28 Diurnal and seasonal variations of absorbed solar and net terrestrial radiation and air temperature.

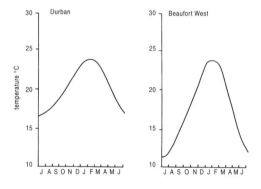

Fig. 7.29 The annual march and seasonal range of temperature at Durban and Beaufort West (Weather Bureau, 1954).

when energy input equals energy output.

The annual variation is similar to the diurnal variation with maximum temperature occurring after the summer solstice and minimum temperature occurring after the winter solstice (Fig. 7.28, *lower*). Annual temperature curves for Durban and Beaufort West illustrate the lag in maximum and minimum temperatures (Fig. 7.29). Instead of being in December and June

they are in January and July. The greater influence of sensible heat in the energy balance at Beaufort West is apparent in the large seasonal range in temperature, whereas at Durban the greater latent heat flux term lowers the sensible heat flux and the temperature range.

THE ENERGY BALANCE OF THE EARTH–ATMOSPHERE SYSTEM

If the solar radiation of 342 W m^{-2} incident at the top of the atmosphere is expressed as 100 units, then, averaged over a long period of time, the sum of short-wave scattering and reflection to space and long-wave radiation from the surface and atmosphere must also total 100 units. This radiative equilibrium is achieved by the complex set of energy flows illustrated in Figure 7.30. The incoming stream of radiation is shown to be attenuated by scattering and reflection by clouds and the earth's surface, all of which return 31 units to space without achieving any heating in the earth–atmosphere system. A further 20 units are absorbed by dust, water vapour and clouds to warm the atmosphere. This leaves 49 units to heat the surface made up of direct solar radiation, diffuse radiation through cloud and scattered radiation from the atmosphere.

The earth disposes of the 49 units absorbed at the earth's surface by the combination of infrared terrestrial radiation and sensible and latent heat fluxes. The net terrestrial radiation is 19 units (114 − 95), of which 7 (102 − 95) are absorbed by greenhouse gases and only 12 reach space through the window in the absorption spectrum. In addition to the radiative transfer, sensible and latent heat fluxes transfer 7 and 23 units of heat respectively into the atmosphere. Within the atmosphere, the net loss of infrared terrestrial energy (made up of the combined losses of 69 units made up of 9 units resulting from emission from clouds, net emission from the atmosphere of 48 units and direct ground radiation to space of 12 units minus the 19 units of net ground infrared radiation; i.e., 69 − 19 = 50 units) must

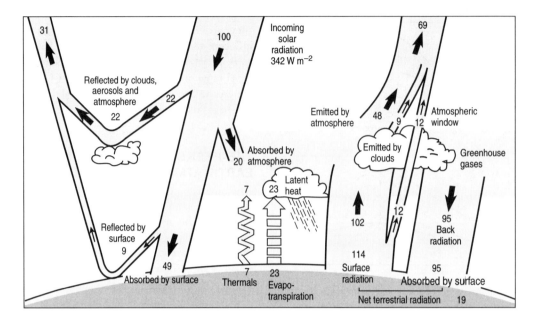

Fig. 7.30 The annual mean global energy balance (modified after IPCC, 1996). The incoming solar radiation of 342 W m^{-2} is taken to be 100 and all other quantities are scaled accordingly.

balance the energy gained (made up of 20 units from the absorption of solar radiation and of 7 units due to sensible and 23 units due to latent heat fluxes; i.e., 20 + 7 + 23 = 50 units). The total emission of the earth–atmosphere system, therefore, equals the total absorption.

Although the earth–atmosphere system *as a whole* is in a state of radiative equilibrium with the sun, this does not extend to all parts of the earth, due to latitudinal seasonal variations in the amount of radiation. The elevation of the sun above the horizon determines the thickness of

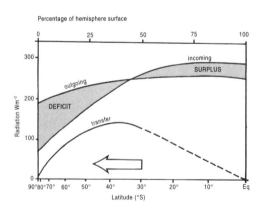

Fig. 7.31 Latitudinal balance between incoming solar radiation and outgoing terrestrial radiation (after Newell, 1964).

Table 7.4 Elevation of the noonday sun above the horizon by season at selected latitudes.

	Latitudes	Elevation
Summer solstice	23° S	90.0°
	30° S	83.5°
	50° S	63.5°
	70° S	43.5°
Winter solstice	23° S	43.0°
	30° S	36.5°
	50° S	16.5°
	70° S	0.0°

the atmosphere through which the radiation must pass before reaching the ground. Attenuation of radiation by reflection, scattering and absorption will clearly be greater when the elevation of the sun is low. The examples of solar elevation by latitude and season given in Table 7.4 illustrate the large differences in the slant of the sun's rays in summer and winter, which, together with the variation in day length, accounts for the large temperature differential between summer and winter in extra-tropical latitudes. The generally high elevation of the sun in the tropics accounts for the mean annual absorption of about 50 per cent of the incoming solar radiation at the ground. This falls to less than 20 per cent near

the poles, due both to the low solar elevation and the high albedo over snow-covered surfaces (see Fig. 7.16). For the combined earth–atmosphere system, the average energy flux is positive between 40° N and 30° S and negative poleward of these latitudes (Fig. 7.31). This means that there exists a surplus of energy in tropical and subtropical latitudes and a deficit of energy in mid-latitude and polar regions. In order to maintain the energy budget of the earth, heat must be transported from the energy-excessive tropics towards the energy-deficient poles. It is this energy imbalance that in large measure drives the atmospheric heat engine and the general circulation of the atmosphere (see Chapter 11).

Seasonal sunshine

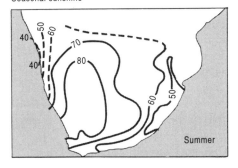 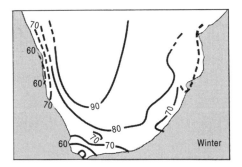

Total incoming radiation

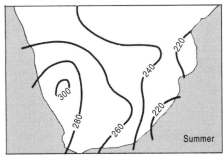 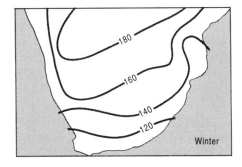

Fig. 7.32 *Upper:* seasonal sunshine duration as a percentage of the total possible (after Schulze and McGee, 1976); *lower:* total incoming radiation ($Q + q$) in summer and winter (10^{-5} J m^{-2} day^{-1}) (after Schulze, 1984).

SUNSHINE AND RADIATION OVER SOUTHERN AFRICA

Sunshine duration shows a marked spatial variation over southern Africa, with the western interior of South Africa receiving more than 80 per cent of the possible annual sunshine, in both summer and winter (Fig. 7.32, *upper*). This percentage decreases to a minimum duration (less than 50 per cent of the possible) in spring and summer over the KwaZulu-Natal Escarpment and coastal margins.

The effect of cloud-producing weather systems on the total radiation (direct + diffuse) is clear. In summer, mean total radiation maxima occur in the least cloudy western areas of southern Africa (Fig. 7.32, *lower*). The increase in cloud-cover frequency from west to east explains the north–south orientation of isolines of incoming radiation over the eastern half of the subcontinent. In winter, however, the generally clear skies explain the east–west orientation of isolines of mean incoming radiation.

ADDITIONAL READING

BARRY, R. G. and CHORLEY, R. J. 1998. *Atmosphere, Weather and Climate*, 7th Edition. London: Methuen.

HOUGHTON, J. T. 1997. *The Physics of Atmospheres*. Cambridge: Cambridge University Press.

SCORER, R. S. 1997. *Dynamics of Meteorology and Climate*. Chichester: Wiley.

STURMAN, A. S. and TAPPER, N. 1996. *The Weather and Climate of Australia and New Zealand*. Melbourne: Oxford University Press.

HARTMANN, D. L. 1994. *Global Physical Climatology*. San Diego: Academic Press.

MCILVEEN, R. 1992. *Fundamentals of Weather and Climate*. London: Chapman and Hall.

PEIXOTO, J. P. and OORT, A. H. 1992. *Physics of Climate*. New York: American Institute of Physics.

OKE, T. R. 1978. *Boundary Layer Climates*. London: Methuen.

MONTEITH, J. L. 1973. *Principles of Environmental Physics*. London: Edward Arnold.

MUNN, R. E. 1966. *Descriptive Micrometeorology*. New York: Academic Press.

SELLERS, W. D. 1965. *Physical Climatology*. Chicago: University of Chicago Press.

8

HORIZONTAL MOTION AND WINDS

It is a matter of common experience that air is seldom calm for any length of time and that winds may range in velocity from the almost imperceptible breeze of a gentle katabatic flow to the gale force of a tornado. Winds may also blow from many different directions at any one location and range in scale from planetary circulations affecting the entire hemisphere, to minuscule eddies set up by imperfections on smooth surfaces.

It is important to distinguish between horizontal *advection* and vertical *convection* in the atmosphere. Wind refers to horizontal air motion, whereas vertical movements are usually

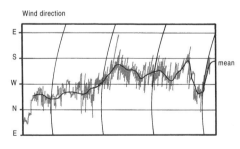

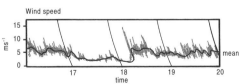

Fig. 8.1 Turbulence superimposed on the mean direction and the speed of air motion. The turbulence manifests itself as deviations from the mean.

termed *updraughts* and *downdraughts* in mesoscale cumulus convection or *uplift* and *subsidence* in large-scale systems. Winds are seldom steady. Instead they constantly fluctuate in both speed and direction as a consequence of short-period accelerations and decelerations (deviations, u') being superimposed on the steady state mean flow, u (Fig. 8.1). The actual wind speed, u, is thus defined as

$$u = u + u' \tag{8.1}$$

It is convenient to distinguish between turbulent fluctuations in wind on the one hand and the mean *steady wind* on which the *turbulence* is supported on the other. Turbulence is the response to rapid accelerations in the wind field; the steady motion is flow in which sudden accelerations are not present.

In a theoretical atmosphere in which no forces are operating, no air movements will occur. Once forces begin to operate and pressure surfaces are tilted, air will begin to move horizontally and in accordance with Newton's laws of motion such that, if U is the initial and V the final velocity, then the velocity acquired in time t is

$$V = U + at \tag{8.2}$$

the distance travelled in time t is

$$s = Ut + \frac{at^2}{2} \tag{8.3}$$

the velocity acquired in distance s is

$$V^2 = U^2 + 2as \tag{8.4}$$

where a is the force of acceleration. Although important in understanding the physics of the atmosphere, these equations find little direct application in meteorology since they do not allow for the immediate prediction of wind velocities from routinely observed meteorological variables.

In the atmosphere, wind is caused by a number of forces, the sum of which (the net force) moves the air in the direction of the resultant force. The relationship between the sum of forces, ΣF, acceleration, a, and mass, m, is stated by Newton's second law of motion in the form

$$a = \frac{\Sigma F}{m} \tag{8.5}$$

The acceleration of air is the rate of change of wind velocity with time and is directly proportional to the magnitude of the net force producing the wind. Wind velocity is a measure of the rate of change of position with time and is therefore a vector which specifies both wind speed (magnitude) and wind direction.

Newton's second law implies that changes in velocity are produced by unbalanced forces acting over time, since for a balance of forces the net force must be zero.

In this chapter it is *steady motion* or *motion under balanced forces* that will be considered; that is, motion where no accelerations are present. Various states of balance may occur, depending on the various forces involved. The balance between pressure-gradient and Coriolis forces defines geostrophic flow; the balance between pressure-gradient, Coriolis and centrifugal forces defines gradient flow; the balance between pressure-gradient and centrifugal forces determines cyclostrophic flow; finally, the balance between the pressure-gradient and frictional forces is responsible for local antitriptic winds. The forces and some of the winds they produce will now be considered.

CONCEPTS AND FORCES

The gravitational force

If the earth were a perfect sphere with a uniform distribution of mass within it, then the true *grav-*

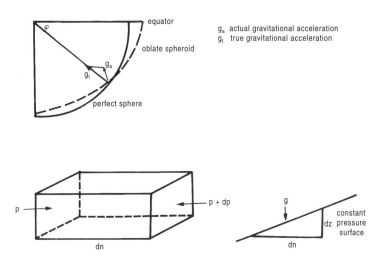

Fig. 8.2 *Upper:* actual and true gravitational forces; *lower left:* horizontal pressure forces acting on a parallelepiped of length *dn*; and *lower right:* on a sloping constant pressure surface.

itational force would act everywhere with uniform strength and directly toward the centre of the earth. Instead, the earth is an oblate spheroid, flattened at the poles, the mass is non-uniform and the actual gravitational force is different from the true force (Fig. 8.2, *upper*). Although for most purposes the gravitational force, *g*, is taken as constant in atmospheric science, it is in fact not so and varies with height and latitude, being slightly less at the equator and slightly greater at the poles. This has consequence in the determination of heights above the surface.

A horizontal surface is one that is everywhere equidistant from mean sea level. A level surface is one to which gravity is always perpendicular and along which gravity is everywhere constant. Because of the variation of *g* over the surface of the earth, horizontal and level surfaces seldom coincide. They are termed *equipotential surfaces* or *geopotential surfaces* (i.e., surfaces of constant potential energy in the earth's gravitational field). In atmospheric science data are usually referred to these surfaces and not to surfaces of constant geometric height. In this way tangential components of gravity may be ignored and computations and modelling are thus simplified. The heights of geopotential surfaces are termed geopotential heights and the common unit is the geopotential metre (gpm). Thus, for example, the heights of the 500 hPa surfaces over South Africa vary from day to day from less than 4 500 gpm to more than 5 800 gpm.

Linear and angular momentum

Linear momentum is defined as the product of velocity, *V*, and mass, *m*, thus

$$\text{momentum} = mV \qquad (8.6)$$

For unit mass of air, velocity (and speed) and momentum are equal. Within the atmosphere momentum is conserved. Thus every increase in speed of airflow in one area must be balanced by a decrease elsewhere to preserve a *momentum balance*.

Where air is rotating it is necessary to consider its *angular velocity of rotation*, Ω, which is related to its linear tangential velocity by

$$V = \Omega r \qquad (8.7)$$

for unit mass of air, where r is the radius of curvature. *Angular momentum* is defined as the product of tangential velocity and radius of curvature. Thus

$$\text{angular momentum} = Vr \qquad (8.8)$$

or

$$\text{angular momentum} = \Omega r^2 \qquad (8.9)$$

In the case of the earth, angular momentum is given by ωr^2, where ω is the *angular velocity of the earth's rotation* (the rate of rotation as measured by the angle through which the earth moves in unit time, 7.29×10^{-5} rads^{-1}).

The pressure-gradient force

Pressure may be defined in two ways in the atmosphere: either as the pressure at a constant height, such as isobars at mean sea level (as over the oceans adjacent to South Africa in this book), or, alternatively, as the height of a constant pressure surface, such as contours of the 850 hPa surface (as over land areas of South Africa in this book). It is therefore necessary to specify the *pressure-gradient force* according to each convention.

At a constant height, consider a volume with unit cross-section and length *dn* (Fig. 8.2, *lower left*). If *p* is the pressure at one end and *p* + *dp* the pressure at the other end, the net force, *dp*, acts to move air from high to low pressure. The mass of the air in the volume is ρdn so that the force per unit mass can be written as

$$-\frac{1}{\rho}\frac{dp}{dn} \qquad (8.10)$$

where the negative sign indicates that the force

acts in the direction of decreasing pressure.

On a constant pressure surface (Fig. 8.2, *lower right*), once the surface is tilted it acquires a slope, dz/dn. Air moves down this slope under the force of gravity, g. The pressure-gradient force per unit mass is given by

$$g \, \frac{dz}{dn} \qquad (8.11)$$

The force likewise acts from high to low and at right angles to the contours of the pressure surface.

Once air begins to move in the atmosphere, it is affected immediately by the earth's rotation and suffers a consequent deflection as a result of the Coriolis force.

Coriolis force

The earth rotates in a west to east direction about the polar axis and completes one rotation in a day. At both poles the angular velocity vector is vertical and represents a rotation in the horizontal plane. However, the velocity vector points into the earth at the south pole and out of the earth at the north pole (Fig. 8.3, *left*). Rotations relative to the earth are opposite in sense, being clockwise from above the south pole and anticlockwise from above the north pole. The truth of this statement can be tested by looking at the end of a pencil while at the same time rotating it clockwise. Without changing the rotation, turn the pencil to observe the other end. It will be rotating in a counterclockwise direction.

The earth's angular velocity is at a maximum at the poles. However, the component of rotation around the local vertical changes with latitude until, at the equator, ω lies in the horizontal plane (i.e., tangential to the earth) with no component of rotation about the local vertical.

It is useful to resolve ω into its horizontal and vertical components. At any place, P, at latitude ϕ these are PB = $\omega\cos\phi$ and PA = $\omega\sin\phi$ respectively in both hemispheres (Fig. 8.3). In each case the horizontal component, $\omega\cos\phi$, acts towards local north. The vertical component, $\omega\sin\phi$, has the opposite sign in each hemisphere

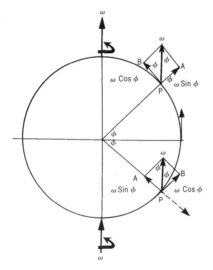
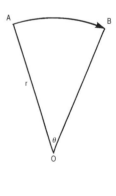

Fig. 8.3 *Left*: horizontal and vertical components of the earth's angular rotation; *right*: motion from A to B in polar coordinates.

(to the right in the northern hemisphere and to the left in the southern hemisphere). The horizontal component of the earth's rotation has to compete with large-scale pressure-gradient and other forces and has negligible effects on winds. In contrast the vertical component, $\omega\sin\phi$, has profound effects.

If air moves from a point 0 at latitude ϕ, where it was initially at rest, to point A with constant velocity V (Fig. 8.3, *right*), then it will move through distance

$$r = Vt \qquad (8.12)$$

in time t. In the meantime the earth has rotated clockwise so that point A is now point B and through the angle AOB or

$$\theta = \omega\sin\phi t \qquad (8.13)$$

The distance AB is given by

$$AB = r\theta = Vt\omega\sin\phi t = \omega\sin\phi Vt^2 \qquad (8.14)$$

From equation (8.3), the distance travelled in time t from A to B is

$$AB = \frac{at^2}{2} \qquad (8.15)$$

since the initial transverse velocity, U, is zero. Thus

$$\frac{at^2}{2} = \omega\sin\phi Vt^2 \qquad (8.16)$$

or

$$a = 2\omega\sin\phi V \qquad (8.17)$$

The acceleration given by equation (8.17) is normally referred to as the *Coriolis acceleration* and the quantity $2\omega\sin\phi$ as the *Coriolis parameter*. The force is apparent and not real, and acts only to cause deflection and alter the direction of motion. The Coriolis force:

- acts at right angles to the direction of motion;
- produces a deviation to the left of the direction of motion in the southern hemisphere, maintains a deviation which is proportional to the wind speed and alters only the direction of the wind and not its speed; and
- declines from a maximum value at the poles ($\sin 90° = 1$) to zero at the equator ($\sin 0 = 0$).

The centrifugal force

When isobars or contours on isobaric surfaces are curved and motion is along a curved path, a third force must be considered. If the radius of curvature of the motion is r and the velocity is V, then the *centrifugal force* is V^2/r for unit mass. The force is directed radially outward at right angles to the axis of rotation and is equal and opposite to the *centripetal acceleration* which is directed radially inward.

THE GEOSTROPHIC WIND

Throughout the middle and high latitudes wind direction tends to parallel closely isobars on synoptic weather maps due to a close balance between the horizontal pressure-gradient force and the Coriolis force. The air motion that results from this balanced flow is called the *geostrophic wind*.

If a parcel of air, initially at rest, is subjected to a pressure-gradient force, it will begin to move towards the region of lower pressure. Once in motion, however, it will be deflected to the left in the southern hemisphere and to the right in the northern hemisphere (Fig. 8.4). When the system reaches equilibrium and the forces eventually balance, geostrophic airflow occurs parallel to the isobars. Until balance is reached the flow is *ageostrophic*.

In geostrophic flow the

pressure gradient force = Coriolis force   (8.18)

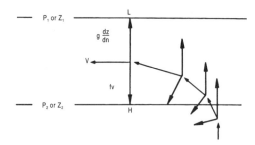

Fig. 8.4 The establishment of geostrophic equilibrium in the southern hemisphere.

Thus on an isobaric chart

$$\frac{1}{\rho}\frac{dp}{dn} = 2\omega\sin\phi V \qquad (8.19)$$

or

$$\frac{1}{\rho}\frac{dp}{dn} = fV \qquad (8.20)$$

where f is the Coriolis parameter $2\omega\sin\phi$.

On a constant pressure surface

$$g\frac{dz}{dn} = fV \qquad (8.21)$$

From equations (8.20) and (8.21) it follows that the geostrophic wind speeds, V, are given by

$$V = \frac{1}{\rho f}\frac{dp}{dn} \qquad (8.22)$$

for isobars at a constant height and

$$V = \frac{g}{f}\frac{dz}{dn} \qquad (8.23)$$

for contours on a constant pressure surface.

The geostrophic-wind relationship illustrates the geodynamic paradox that on the rotating earth air subject to a constant pressure-gradient force does not move parallel to the force (i.e., in the direction of the force) with a constant acceleration, but ultimately moves normally (i.e., at right angles) to the force with a constant speed.

Despite its simplicity, the geostrophic wind equation is one of the most useful in meteorology and finds numerous applications both in forecasting and modelling. From equations (8.20) and (8.21) it is evident that:

• when isobars are drawn at the standard 2 hPa or 10 gpm intervals, as on South African weather maps, dp or dz is constant and the pressure gradient, and hence V, varies inversely with the width of the isobar or contour spacing;

• since $\sin 90° = 1$ and $\sin 30° = 0.5$, for the same isobar or contour spacing, the wind will be twice as strong at latitude 30° S as at the pole; and

• Coriolis force weakens towards the equator as ϕ tends to 0. At the same time V tends to infinity and the physical admissibility of the geostrophic relationship collapses.

In practice the geostrophic relationship is not used for latitudes in the tropics and equatorial regions. As the equator is approached, winds tend to blow at increasing angles across isobars toward lower pressure.

A further consequence of the geostrophic relationship is the confirmation of *Buys Ballot's law* that if observers stand with their backs to the wind in the southern hemisphere, the lower pressure will be on the right-hand side (and on the left in the northern hemisphere).

THE EFFECT OF CHANGING PRESSURE GRADIENTS

Changes in space

According to *Le Chatelier's principle*, if a system is in equilibrium and one of the stresses affecting it is altered, then the state of the system will change so as to relieve the changing stress. So it is with geostrophic flow if the pressure gradient undergoes spatial changes (Fig. 8.5). With nar-

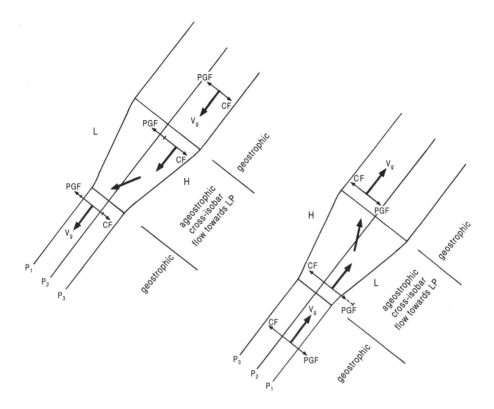

Fig. 8.5 The effect of changing pressure gradients *in space* on geostrophic motion. Pressure-gradient force is designated by PGF and Coriolis force by CF.

rowing of isobars or contours, the pressure gradient and hence pressure-gradient force increases. In the region of narrowing the force constantly increases. While this is happening the Coriolis force begins to increase to maintain the balance of forces and steady flow. Until the isobars or contours cease to converge and until they again become parallel, a state of imbalance exists. Since the pressure-gradient force exceeds the still-changing Coriolis force, an ageostrophic wind blows obliquely across the isobars towards lower pressure. Conversely, where the isobars or contours diverge, the pressure-gradient force decreases faster than the commensurately adjusting Coriolis force and ageostrophic flow occurs obliquely towards higher pressure.

Changes in time

When pressure systems deepen with time, pressure gradients steepen and an ageostrophic component of flow, the *isallobaric* component, develops. The term *isallobar* refers to the lines on charts that join points of equal pressure tendency. Pressure tendency is the change of pressure over a given time period, for example over six hours. While the isallobaric gradient is increasing (usually owing to falling pressure with the approach of a depression), ageostrophic flow develops obliquely across the isobars or contours towards the region of lower pressure (Fig. 8.6). Conversely, while pressure systems are weakening, the ageostrophic flow due to the development of the isallobaric component is obliquely across the isobars or contours towards higher pressure.

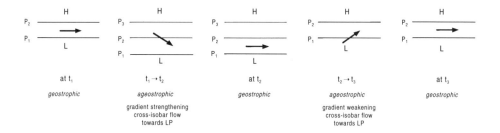

Fig. 8.6 The effect of changing pressure gradients *in time* on geostrophic motion.

The isallobaric component of wind that produces the ageostrophic flow is directed at right angles to the isallobars, acts down the isallobaric gradient and is given, approximately, by

$$V_i = \frac{1}{\rho f} \frac{d}{dn} \left(\frac{dp}{dt} \right) \qquad (8.24)$$

where d/dn (dp/dt) is the isallobaric gradient. Since this component of wind blows towards the lowest isallobars, it follows that convergence and vertical ascent will occur in isallobaric lows and divergence and descent in isallobaric highs.

THE EFFECT OF FRICTION AT THE SURFACE

Departures from geostrophic flow also occur near the earth's surface owing to the effect of frictional forces. Frictional retardation of airflow occurs below 1 000 m and causes the wind in this layer to decelerate. Since the Coriolis force is proportional to the wind speed, a decrease in speed causes this force to diminish. The relatively stronger pressure gradient force then causes the flow to become ageostrophic and to be deflected across the isobars at an angle, α, in the direction in which this force acts, towards

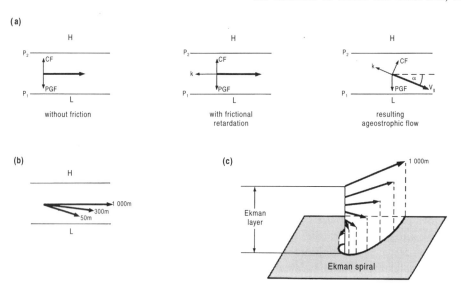

Fig. 8.7 (a) The effect of friction on geostrophic flow; (b) ageostrophic near-surface flow with geostrophic motion at 1 000 m; and (c) the Ekman spiral.

lower pressure (Fig. 8.7a). The angle, α, depends on the magnitude of the coefficient of friction, k, such that

$$\tan\alpha = \frac{k}{f} \qquad (8.25)$$

where f is the Coriolis parameter. Over land the deflection near the ground is about 30° (with a speed reduction of about two thirds). Over the sea the deflection is only about 10° (with a speed reduction of about one third). As the frictional effect diminishes with height, so the flow becomes less ageostrophic, until by about 1 000 m geostrophic flow is established (Fig. 8.7b). The projection of wind vectors onto the horizontal gives a hodograph in the form of the *Ekman spiral* (Fig. 8.7c). The *friction layer* is often termed the *Ekman layer* and within it the geostrophic relationship does not hold. For this reason surface winds seldom blow parallel to isobars or contours.

THE EFFECT OF HORIZONTAL TEMPERATURE GRADIENTS: THE THERMAL WIND

Just as spatial variations in pressure gradients affect the geostrophic wind, so do horizontal variations in temperature gradients. Such temperature variations exert a particular effect on the vertical wind shear (i.e., on the variation of wind velocity with height) through the effect of the thermal wind.

From the pressure–height equation

$$\Delta z = \frac{R\overline{T}}{g} (\ln p_0 - \ln p) \qquad (8.26)$$

it is clear that the thickness between pressure surfaces, for example the 850 and 500 hPa surfaces, remains the same provided the mean temperature of any two adjacent columns of air A and B remain the same (Fig. 8.8a). If, however, column A is cooled and column B is warmed,

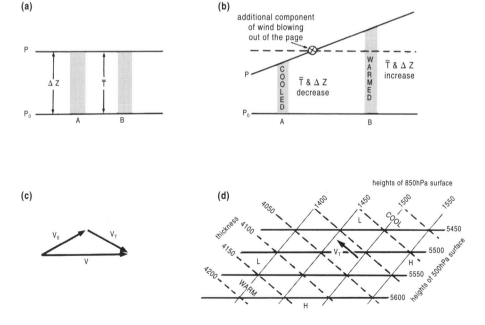

Fig. 8.8   (a) Thickness and mean temperature in a uniform temperature field (i.e., with the same lapse rates in columns A and B); (b) the effect of cooling column A and warming column B on the inclination of a constant pressure surface, P; and (c and d) the vectorial difference between geostrophic winds V_0 and V to give the thermal wind velocity V_T.

then the thickness between p_0 and p at A will decrease because the mean temperature of the column will have diminished. Likewise the thickness in column B will increase with heating. The consequence is that the slope of the pressure surface, p, will increase (Fig. 8.8b) and an additional geostrophic component will have been added to the wind by an amount directly proportional to the change in thickness between the pressure surfaces.

This thermal wind component, V_T, is given by

$$V_T = \frac{g}{f} \frac{dz}{dn} \qquad (8.27)$$

where dz/dn is the gradient of the thickness surface. The equation is identical in form to that of the geostophic wind. In effect the thermal wind is the vectorial difference between the geostrophic winds V_0 and V at the two levels p_0 and p (Fig. 8.8c), or

$$V_T = V - V_0 \qquad (8.28)$$

In practice a thickness map (or surface) is prepared in the manner illustrated in Figure 8.8d. Over South Africa thickness is often determined for the layer 850 to 500 hPa. In such a case the 500 hPa contours are overlaid on those for the 850 hPa surface and for each intersection thickness is determined by the difference in heights at that point. Lines of equal thickness are then drawn. Lowest thickness, and therefore lowest mean temperature, is located to the north-east in Figure 8.8d. Thickness increases to the south-west as does the mean temperature between the 850 hPa and 500 hPa surfaces. Isotherms will be parallel to the thickness lines.

Since the thermal wind bears the same relationship to the thickness lines (isotherms) as does the geostrophic wind to contours of a constant pressure surface, it can be seen that the thermal wind will blow parallel to the thickness lines (and isotherms) keeping low temperatures (decreased thickness) on the right-hand side in the southern hemisphere. The magnitude of the thermal wind components is given by

$$V_T = \frac{g}{f\bar{T}} \frac{dT}{dn} (z - z_0) \qquad (8.29)$$

where dT/dn is the horizontal mean temperature gradient and z_0 and z the heights of the two surfaces between which the thickness has been determined.

Some effects of the thermal wind

Thermal winds are important wherever steep temperature gradients occur in the atmosphere. They find their most regular expressions in the bands of strong westerly winds that occur in the jet streams of the middle and upper troposphere as a consequence of strong horizontal pole-to-equator temperature gradients at these levels. The *polar night westerlies* and the *mesospheric westerlies* (Chapter 11) are further examples of thermal winds.

Since pole-to-equator temperature gradients produce westerly thermal winds in both hemispheres, it follows that in general westerly winds will increase in velocity with height in the troposphere. Conversely, easterly winds, having to overcome the hemispheric thermal wind component, will be weaker than their westerly counterparts and will weaken with height.

In any large rotating system the geostrophic wind relationship will hold over short distances where the isobars are approximately straight. Given cold- and warm-cored systems, the thermal wind component will augment or counteract the geostrophic flow depending on the temperature gradient between the centre and periphery of the rotating system (Fig. 8.9).

Thus it may be seen that
- warm-cored cyclones will weaken with height and be shallow; and
- cold-cored anticyclones will weaken with height and be shallow,

whereas
- cold-cored cyclones will intensify with height

and be deep; and

- warm-cored anticyclones will intensify with height and be deep.

Over South Africa cold-cored, cut-off lows and warm-cored, subtropical highs conform to this expectation. Their oppositely cored counterparts are unusual and insignificant features of the weather.

Thermal winds may also have local consequences such as causing a diurnal cycle in wind direction over the central interior plateau and producing a backing of the wind from north-east to north-west during the day. (Backing is changing direction in an anticlockwise sense, i.e., decreasing azimuth; veering is changing direction in a clockwise sense, i.e., increasing azimuth.)

THE GRADIENT WIND

The existence of cells of high and low pressure in the atmosphere means that isobars are generally curved rather than straight. Nevertheless, wind direction remains approximately parallel to the isobars. This indicates the presence of an additional force to maintain the balanced flow. The extra force is the centrifugal force produced by the rotation about the centre of the cells.

The three-way balance between the pressure-gradient force, Coriolis force and centrifugal force for cyclonic and anticyclonic air motion results in gradient flow. Gradient wind equations may be derived for flow in both types of pressure fields: on constant pressure surfaces (e.g., at 850 hPa) or at constant height (e.g., at mean sea level). Only the equations for winds blowing along contours of constant pressure surfaces will be given here.

In the *cyclonic case* (Fig. 8.10) the pressure-gradient force acts inwards from high to low pressure, the Coriolis force acts to the left of the motion (i.e., outward) and the centrifugal force acts radially outward. The forces balance for clockwise cyclonic rotation in the southern hemisphere when the pressure-gradient force is equalled by the Coriolis and centrifugal forces. Thus

$$g \frac{dz}{dr} = fV + \frac{V^2}{r} \tag{8.30}$$

where V is the gradient wind speed.

In the *anticyclonic case* the rotation is anticlockwise and the Coriolis force must be balanced by the sum of the pressure-gradient and centrifugal forces. Thus

$$fV = g \frac{dz}{dr} = fV + \frac{V^2}{r} \tag{8.31}$$

or

$$g \frac{dz}{dr} = fV - \frac{V^2}{r} \tag{8.32}$$

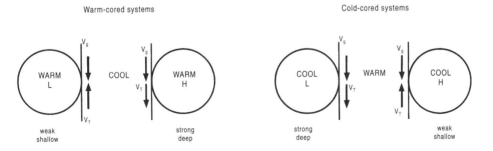

Fig. 8.9 The relative strengths of warm- and cold-cored cyclones and anticyclones as a consequence of the interactive effect of geostrophic (V_g) and thermal winds (V_T). The cyclones and anticyclones are assumed to be sufficiently large for the tangential wind to be approximated by the geostrophic wind.

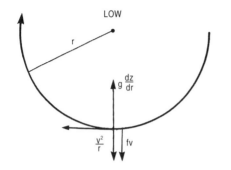
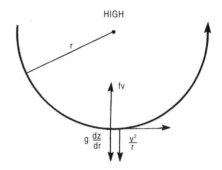

Fig. 8.10 The gradient wind balance of forces in rotating systems in the southern hemisphere.

The only difference between equations (8.30) and (8.32) is in the sign before the centrifugal-force term. In the northern hemisphere the equations are identical, but cyclones rotate in the opposite direction (i.e., anticlockwise) and anticyclones rotate clockwise.

Both gradient-wind equations are quadratic equations of the form

$$y = ax^2 + bx + c \qquad (8.33)$$

where the solution for x is given by

$$x = \frac{-b \pm \sqrt{b^2 - 4ac}}{2a} \qquad (8.34)$$

In the case of cyclonic curvature and rotation $a = 1/r$, $b = f$ and $c = g\, dz/dr$. Thus *for a cyclone*

$$V = -\frac{fr}{2} + \sqrt{\frac{f^2 r^2}{4} + rg\,\frac{dz}{dr}} \qquad (8.35)$$

In equation (8.35) it is apparent that two solutions are possible, one with a positive sign before the radical and the other with a negative sign. The latter must be ignored since it produces the physically indeterminate solution that when the pressure gradient tends to zero the velocity does not.

For an *anticyclone* the solution to the gradient wind equation is

$$V = -\frac{fr}{2} - \sqrt{\frac{f^2 r^2}{4} - rg\,\frac{dz}{dr}} \qquad (8.36)$$

The relative sizes of cyclones and anticyclones

For the velocity to be a maximum in equation (8.36) the radical must be zero:

$$\sqrt{\frac{f^2 r^2}{4} - rg\,\frac{dz}{dr}} = 0 \qquad (8.37)$$

From this it follows that

$$\frac{dz}{dr} = \frac{f^2 r}{4g} \qquad (8.38)$$

Since a maximum, by definition, cannot be exceeded, the pressure gradient must always be below (or equal to) this value. That is,

$$\frac{dz}{dr} \leq \frac{f^2 r}{4g} \qquad (8.39)$$

From this result it may be seen that, in anticyclones, as the radius of the cell decreases, so does the pressure gradient. In other words, for winds to be strong in an anticyclone, the radius must be large. A second consequence that follows from equation (8.39) is that near the centre of anticyclones, where pressure gradients are slack, winds will be light and variable. No such constraints hold for cyclones where strongest winds are associated with small deep systems (Fig. 8.11).

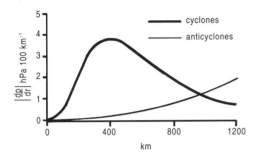

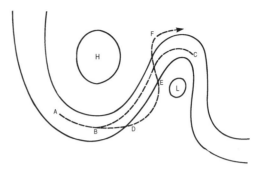

Fig. 8.12 Gradient (ABC) and agradient (ABDEF) flow in a perturbed westerly airstream in which the radius of curvature changes from anticyclonic to cyclonic in wave motion.

Fig. 8.11 The variation of pressure gradient with radial distance from the centre in cyclones and anticyclones (after Hewson and Longley, 1944).

A final consequence that follows from the relationship given in equation (8.39) is that should the pressure gradient, dz/dr, become too great to maintain the relationship for a given radius, r, then *agradient* flow will immediately result. Instead of maintaining the gradient wind path ABC in Figure 8.12, the greater-than-equilibrium value of the pressure gradient causes the air to accelerate and overshoot the contours along the path ABD toward the lower pressure at D. Meanwhile, Coriolis force has increased to compensate for the increased speed and deflection to the left of the wind increases so that the air moves towards E and higher pressure, again across the isobars as agradient flow. Deceleration occurs, the Coriolis deflection diminishes and, until gradient flow is re-established, the pressure gradient is again larger than required to maintain the flow. Consequently the air swings to the right, towards F, where ultimately gradient flow is again achieved. The net result of the loss of mass from D through E is to deepen the cyclonic trough or disturbance ahead of the ridge of high pressure. With this deepening may come a deterioration of the weather.

COMPARISON OF GRADIENT AND GEOSTROPHIC WIND VELOCITIES

If the radius of curvature in wave motion or closed circulations is large ($r > 500$ km), then the centrifugal-force term becomes small enough to ignore (e.g., if $V = 10$ m s^{-1} and $r = 500$ km, then $V^2/r = 0.002$). The gradient wind then tends to the geostrophic and the geostrophic wind becomes a good approximation of its gradient counterpart.

In small systems ($r < 500$ km), provided the pressure-gradient force and Coriolis force are the same (i.e., the pressure gradient is the same), the geostrophic wind will exceed the gradient wind or vice versa, depending on the curvature of the flow.

For *cyclonic curvature* on a constant pressure surface the gradient wind equation is

$$g\ \frac{dz}{dr} = fV_{gr} + \frac{V_{gr}^{\ 2}}{r} \qquad (8.40)$$

and the geostrophic wind equation is

$$g\ \frac{dz}{dr} = fV_g \qquad (8.41)$$

Since gdz/dr is the same in both cases

$$fV_g = fV_{gr} + \frac{V_{gr}^2}{r} \qquad (8.42)$$

or

$$V_g = V_{gr} + \frac{V_{gr}^2}{fr} \qquad (8.43)$$

and

$$V_g > V_{gr} \qquad (8.44)$$

That is, the geostrophic wind exceeds the gradient wind.

For *anticyclonic curvature* the opposite is the case and

$$V_g = V_{gr} - \frac{V_{gr}^2}{fr} \qquad (8.45)$$

and

$$V_g < V_{gr} \qquad (8.46)$$

Here the gradient wind exceeds the geostrophic wind.

CYCLOSTROPHIC FLOW

A special type of balanced motion is provided by airflow around a hurricane. The flow around these storms is sufficiently curved for the centrifugal force to greatly exceed the Coriolis force to the extent that the latter becomes insignificant and may be ignored. Balanced motion occurs, therefore, between the pressure-gradient force and the centrifugal force. In such a *cyclostrophic flow*

$$g \frac{dz}{dr} = \frac{V^2}{r} \qquad (8.47)$$

It is likely that in smaller systems such as tornadoes, which are even more intense systems, cyclostrophic motion also occurs. A prime requirement for such flow is that the circulations be small in diameter.

INERTIAL FLOW

When the pressure gradient is zero and the Coriolis force balances the centrifugal force,

$$fV = \frac{V^2}{r} \qquad (8.48)$$

and the flow is said to be *inertial*. The path of such flow is circular, in the so-called circle of inertia, since no external force is present to counteract the Coriolis force as in geostrophic motion. In practice pure inertial motion rarely occurs since a small pressure gradient is almost always present in the atmosphere.

ADDITIONAL READING

BARRY, R. G. and CHORLEY, R. J. 1998. *Atmosphere, Weather and Climate,* 7th Edition. London: Methuen.

GORDON, A., GRACE, W., SCHWERDTFEGER, P. and BYRON-SCOTT, R. 1998. *Dynamic Meteorology.* London: Arnold.

HOUGHTON, J. T. 1997. *The Physics of Atmospheres.* Cambridge: Cambridge University Press.

SCORER, R. S. 1997. *Dynamics of Meteorology and Climate.* Chichester: Wiley.

SCHNEIDER, S. H. (ed.). 1996. *Encyclopaedia of Weather and Climate,* Vols. 1 and 2. Oxford: Oxford University Press.

STURMAN, A. S. and TAPPER, N. 1996. *The Weather and Climate of Australia and New Zealand.* Melbourne: Oxford.

JAMES, I. N. 1994. *Introduction to Circulating Atmospheres.* Cambridge: Cambridge University Press.

MCILVEEN, R. 1992. *Fundamentals of Weather and Climate.* London: Chapman and Hall.

PEIXOTO, J. P. and OORT, A. H. 1992. *Physics of Climate.* New York: American Institute of Physics.

MCINTOSH, D. H. and THOM, A. S. 1978.
Essentials of Meteorology. London:
Wykeham Publications.
WALLACE, J. M. and HOBBS, P. V. 1977.
Atmospheric Science, an Introductory Survey.
New York: Academic Press.

BYERS, H. R. 1974. *General Meteorology,* 4th
Edition. New York: McGraw-Hill.
GORDON, A. H. 1962. *Elements of Dynamical
Meteorology.* London: English Universities
Press.

9

VERTICAL MOTION AND
CUMULUS CONVECTION

Without the vertical uplift of air no weather would occur. Uplift may occur in a variety of ways and on a variety of scales, the latter ranging from the turbulence produced by convective shimmer above a hot surface, through uplift in Kelvin and gravity waves, to cumulus convection and on to large-scale eddy overturnings associated with planetary waves, fronts, cyclones and the like. Each of these systems is associated with *kinetic energy* (i.e., energy of motion) that may have a horizontal and vertical component. The spectrum of kinetic energies generated in the atmosphere ranges from the energy generated by small systems with wavelengths the order of 1 metre and periods the order of 1 second, to the energy generated by large, planetary-scale

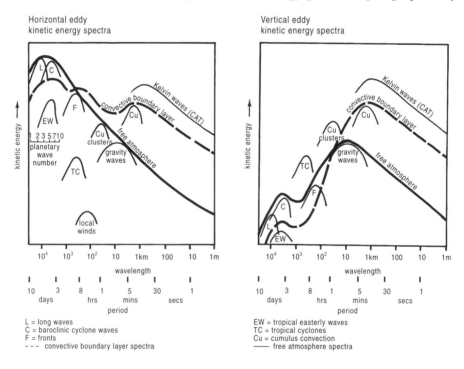

Fig. 9.1 Schematic spectra of horizontal and vertical eddy kinetic energy in the free atmosphere and within the convective boundary layer. Contributions by different-scale systems are also shown (modified after Ludlam, 1980).

disturbances with wavelengths in excess of 10 000 kilometres and periods of up to 30 days (Fig. 9.1).

The overwhelming concentration of eddy kinetic energy resides in the horizontal motion of large-scale, long-wave trough and ridge systems (Fig. 9.1, *left*) of which between two and four span the hemisphere at any one time. Cyclone waves (both westerly and easterly) and cyclonic disturbances (and fronts) likewise make a substantial contribution to the eddy kinetic energy of the free atmosphere, which shows a peak at about 10 days. The spectrum for the convective boundary layer shows a second peak at a scale of a few kilometres due to cumulus convection in the boundary layer. Within the free atmosphere, energy is intermittently and locally introduced on scales the order of 100 metres to one kilometre as a result of transfer from the large-scale flow by the development of Kelvin waves in shallow, stable layers of pronounced wind shear, often at the top of the boundary layer or on the upper surface of an elevated inversion layer. Such waves produce *clear air turbulence* and are the predominant cause of aircraft bumpiness. The spectra of vertical eddy kinetic energy (Fig. 9.1, *right*) are significantly different from their horizontal counterparts. Little energy resides in large-scale systems and major peaks in both free atmosphere and boundary layer spectra are associated with cumulus convective systems, which typically produce uplift in the region of 20–25 m s^{-1} by comparison to that of 1–2 m s^{-1} or less in syn-optic and planetary-scale disturbances.

In this chapter cumulus convective systems will be considered; in Chapter 10 the manner in which uplift may occur in synoptic and larger-scale systems will be examined.

CLASSIFICATION OF CUMULUS CONVECTIVE ACTIVITY

Vertical motion in the atmosphere is the key to most characteristics of weather. Upward motion results in expansion, cooling and the eventual condensation of water vapour. The release of latent heat in condensation often accelerates convection by increasing the buoyancy and instability of air. Downward motion produces compression, warming and an increase in the capacity of the air to hold water vapour. In general, maximum vertical velocities tend to be inversely related to the size of the circulation. Large-scale circulations are associated with weak (but steady) uplift, whereas smaller systems such as cumulonimbus storms may have vertical motion that is similar in scale to horizontal velocities.

Cumuliform cloud types are closely related to the strength of the vertical motions sustaining them and usually evolve, sometimes very quickly, through a series of stages from cumulus to cumulonimbus (Figs. 9.2 and 9.3).

The convective ascent of air is initiated by the development of bursts of bubbles of warm air that initially have the dimensions of tens to

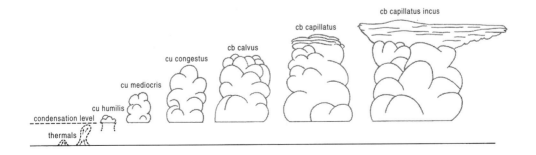

Fig. 9.2 Scale differences and cloud types associated with increasing convective activity.

Fig. 9.3 Cloud photographs to illustrate the stages in the development of cumulus convection. *Upper:* (from left to right) a thermal, cumulus humilis, mediocris and congestus; *lower:* cumulonimbus, capillatus and incus (photos: P. D. Tyson).

thousands of metres. As each bubble rises, expands and cools, it mixes with its surrounding air and is eroded. Successive bubbles penetrate the wakes of previous ones, ascend to greater heights and organized thermals develop. Once these reach condensation level, small puffs of *cumulus humilis* (low or shallow cumulus) form. As bubbles and thermals rise, compensating downward motions develop to preserve continuity. After condensation, latent heat release enhances instability, and with a continued inflow of warm, moist air from below, the circulation becomes more continuous and the cumulus cloud may grow rapidly through the *mediocris* (Latin meaning average) and *congestus* (heaped) stages to become a cumulonimbus cloud advancing through the *calvus* (Latin meaning bald), *capillatus* (meaning hairy) and *incus* (meaning anvil-shaped) types before decaying. As this happens, distinctive updraught and downdraught regions develop within the cloud and storm cells.

Storms may be of a variety of types ranging from single-cell to multi-cell and supercell storms which may occur as isolated, scattered or line storms. Both types of division reflect the processes maintaining the storms. Isolated storms tend to be of the single-cell variety in which surface heating in a fairly homogeneous air mass is the major source of uplift. Scattered storms tend to be stronger, more multi-cellular and clustered in nature and to occur in response to a combination of heating effects, mesoscale forcing and larger-scale synoptic controls. With extended line storms, synoptic forcing becomes a predominant control.

RADAR REFLECTIVITY PATTERNS

In observing cloud and storm structure using radar systems, pulses of electromagnetic radiation are emitted by the radar. These are reflected and scattered by water drops in the cloud and the returned radiation (in dBz units) is recorded by the radar, which may be operated in two modes. The first mode records the horizontal spatial pattern of cloud and rainfall on a *plan position indicator* (PPI). Examples of different types of storms are given in Figure 9.4. PPIs may be taken at different heights (the radar being set at different elevations) and stacked to give a three-dimensional picture of a cloud cell (Fig. 9.5, *left*). The greater the concentration of large drops, the stronger is the radar echo.

Alternatively, a vertical cross-section may be taken through a cloud on a *range height indicator* (RHI). An example of a RHI section is given in Figure 9.5 (*right*). Such sections are useful

since they allow the ready identification of updraughts (where large raindrops are not observed and hence radar echoes are weak) and downdraughts (where rain is falling and echoes are strong). Thus in Figure 9.5 (*right*) the updraught and weak echo region were ahead of the storm (i.e., to the right), while strongest precipitation within the cloud was recorded where echoes exceeded 40 dBz.

SINGLE-CELL STORMS

In the absence of wind shear, single-cell storms are usually 5–10 km in horizontal extent, are short-lived (less than 60 minutes) and change markedly with time. The life cycle of a typical cell is illustrated schematically in Figure 9.6. In the early cumulus stages of growth, updraughts occur throughout the cloud and as radar echoes begin to form they acquire an inverted saucer-like structure (10- and 15-minute stages in Fig. 9.6). As precipitation begins to form in the upper part of the mature cloud, so a strong echo region forms. The main updraught is characterized by weak echo regions (15- and 20-minute stages). The strongest echoes occur in the upper

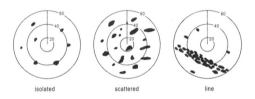

isolated   scattered   line

Fig. 9.4 Simplified radar echoes in the horizontal *plan position indicator* (PPI) mode to show isolated, scattered and line thunderstorms. Range rings are in kilometres.

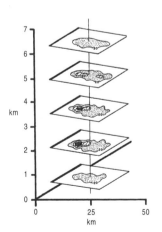

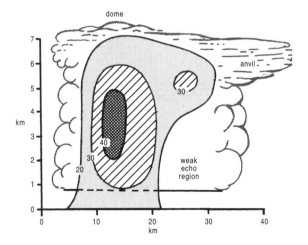

Fig. 9.5 *Left*: stacked PPI radar echo patterns to show the vertical structure of a single-cell thunderstorm; *right*: a vertical radar echo section in *range height indicator* (RHI) mode of the same schematic cell. Echo contours are at 10 dBz intervals after 20 dBz.

regions of the cloud. With further growth in raindrop size, the precipitation begins to fall, a downdraught is initiated and the regions of maximum radar echo and rainfall subside towards the ground (25-minute stage). At this time updraughts cannot sustain the precipitation, fallout is rapid, downdraughts begin to occur uniformly throughout the cell and dissipation proceeds rapidly (30-minute stage). The times given for the different stages are approximate only and vary from cell to cell. Rainfall usually occurs about 20 minutes after the first radar echoes are recorded and the highest rainfall rates are usually recorded when the region of maximum precipitation has sunk to the melting (0 °C) level in the cloud. Throughout their lifetimes isolated single-cell storms retain their symmetry and vertical stance owing to their occurrence in generally light ambient winds with little vertical shear. Isotherms passing through the cells usually bend upward in the warmer updraught air and downward in the cooler downdraught air. Temperature differences of up to 5 °C may be observed between updraught and environmental air.

Over the Highveld of South Africa the life cycle of single-cell storms is usually 18–30 minutes. Cells move at speeds of about 25 km h⁻¹ from south-west to north-east. Hail seldom occurs in such storms since the mechanism for its recirculation within the cell (to acquire a layered structure) is lacking with a nearly vertical updraught. Only with a sloping updraught can such recirculation occur. Such is the case in some multi-cellular storms and in all supercells.

The thunderstorm downdraught

Thunderstorms form in unstable air. As precipitation particles grow in the upper parts of clouds, they reach a size beyond which they cannot be supported by the updraught. At this point they begin to fall. In so doing they exert a drag on the surrounding air which begins to move downward as a consequence. Since the air is unstable, once displaced downward it begins to move in that direction without further impetus (see Fig. 4.1). The evaporation of precipitation causes cooling and the chilled air begins to accelerate towards the ground, often reaching the surface as an explosive and highly turbulent gust front that characterizes the lateral spreading of the downdraught and cold outflow dome (Figs. 9.7 and 9.8). The boundary between the spreading chilled downdraught air and the surrounding warmer air is termed a *line squall*, which occurs along a *pressure surge line*. The higher pressure results from the chilled air and produces a localized *thunderstorm high* beneath the storm. The pressure may rise by a few hectopascals and the surface flow becomes strongly

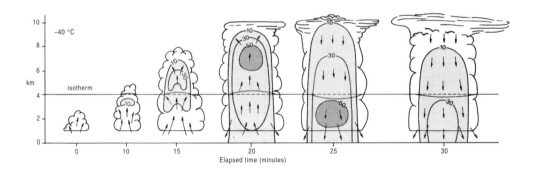

Fig. 9.6 Schematic vertical cross-sections to show the life cycle of a single-cell storm. Contours of echo intensity are given in dBz (after Chisholm and Renick, 1972).

divergent. To the rear of the local high a *wake low* may form. This occurs in a manner analogous to the low forming behind an obstacle, which in this case is the storm as it moves faster through the atmosphere than the ambient air.

The changes that occur in the pressure, temperature and wind fields with the passage of a mature storm cell and its gust-front squall are characteristic. As surface convergence and uplift occur during the early stages of cumulus cloud formation, so the surface pressure drops slightly and steadily. With the onset of the thunderstorm, downdraught pressure rises, temperature drops suddenly, rainfall is initiated abruptly and then diminishes steadily, wind direction shifts suddenly, speed increases markedly and turbulence becomes pronounced (Fig. 9.9). The squalls are localized and never large enough to

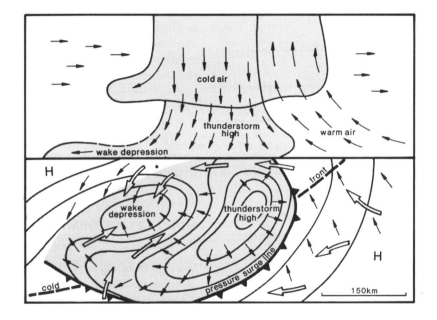

Fig. 9.7 *Upper:* the updraught, downdraught and development of the gust front, thunderstorm high and wake depression; *lower:* the mesoscale, sub-synoptic pressure field associated with a storm cell and the development of the wedge of cold air behind the pressure surge line (after Fujita, 1955).

Fig. 9.8 Thunderstorm downdraughts and gust fronts near Kimberley and Bloemfontein (photos: P. D. Tyson).

be evident on synoptic charts. The wedges of cold air produced by downdraughts often act in a manner analogous to cold fronts and serve to invigorate storms by enhancing slope convection and ascent along their upper surfaces.

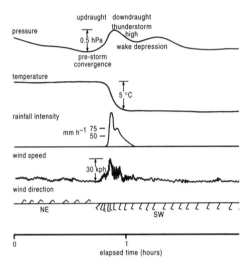

Fig. 9.9 Schematic autographic records to show the passage of a mature single-cell storm.

Thunderstorm downdraughts and line squalls are usually violent. Vertical velocities exceeding 30 m s^{-1} may be associated with downdraughts. The highest instantaneous gust recorded in South Africa, 153 km h^{-1} at Bloemfontein, was probably associated with a thunderstorm downburst.

Mesoscale considerations

Isolated, single-cell storms are usually very localized. As mesoscale effects become significant, so the convection becomes more organized. Thus sea breezes and their fronts, valley and plain–mountain winds blowing over heated slopes and localized topographic modifications to the surface divergence field (see Chapter 10 for a full consideration of divergence) may act to enhance (or suppress) convective activity in the boundary layer. In Figure 9.10 divergence is determined from a mesoscale wind-observation network covering an area of approximately 56 × 33 km extending from the Lowveld of the Kruger National Park to the Highveld near Belfast. The divergence that is shown is an average for the network. Over the heated Escarpment slopes of the Mpumalanga Drakensberg, unstable, up-slope plain–mountain winds by day produce a generally convergent wind field; by night the stable down-slope mountain–plain winds are divergent (Fig. 9.10). The nocturnal regime suppresses convective activity in contrast to the daytime regime, which provides an environment conducive to its enhancement. The divergence minima (i.e., peaks in convergence) at P, Q, R and S represent updraughts into convective clouds or clusters of such clouds. The divergence peaks A, B, C and D represent the effects of downdraughts throughout the day.

MULTI-CELL STORMS

Usually a multi-cell storm consists of a sequence of evolving cells, each of which goes through the life cycle of the single-cell type. A number of cells may develop within an individual cumulonimbus cloud mass. Alternatively, separate, independently formed cells may merge into a multi-cell storm. Multi-cell storms are typically 30–50 km in horizontal extent (Fig. 9.11). In southern Africa new cells form preferentially (but not exclusively) on the left front flank of the storms, while dissipation tends to occur preferentially to the right rear of the multi-cells. In contrast, in the northern hemisphere preferential formation and dissipation is to the front right and left rear respectively. The local pattern of preferential replenishment and dissipation to the left and right leads southern African storms to propagate to the left of the wind steering the storm. The most important determinants of the individual storm characteristics will be the degree of instability and vertical wind shear in the cloud layer.

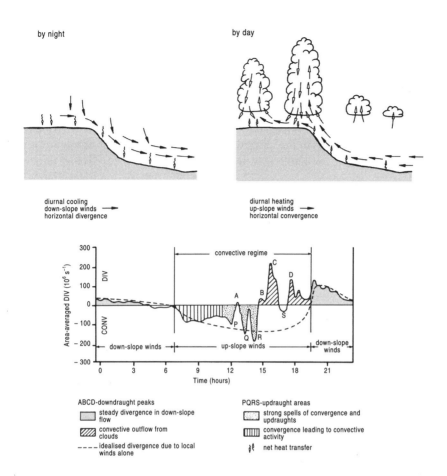

Fig. 9.10 *Upper:* the nature of local topographically induced airflow over the Mpumalanga Escarpment by night and day and the integrated effects of thermal convection and convergence enhanced by up-slope winds; *lower:* area-averaged surface divergence over the mesocale wind-observing network extending from the Kruger National Park in the Lowveld to near Belfast on the Highveld (after Garstang et al., 1985). Troughs P, Q, R and S indicate convergence into cumulus cells developing over the area during the day; peaks A, B, C and D indicate the effects of downdraughts from the cells.

A schematic section along the section line AB in Figure 9.11 is given in Figure 9.12. The four-celled structure with new growth ahead and dissipation to the rear of the complex is clearly evident. The upper parts of the cumulonimbus anvil are glaciated and composed of ice crystals; the middle regions are made up of ice and supercooled water; below the freezing level only water drops occur. Raindrops play an important part in invigorating and maintaining the storms and are usually associated with rainfall, which is heavy in the mature stage of the storm and light

and steady during the dissipating stage. The extent to which the idealized multi-cell storm structure given in Figures 9.11 and 9.12 is representative of the Highveld storms of southern Africa may be assessed in Figures 9.13 and 9.14. Multi-celled storms are common over the Pretoria-Johannesburg region. One such summer, a four-cell cloud cluster and storm showed a typical radar echo structure in both the horizontal and the vertical (Fig. 9.13). The storm moved towards the north-east. Such tracks from the south-west quadrant are common and indicate the extent to which synoptic-scale forcing exerted by middle- and upper-level airflow (and particularly the divergence within such airflow) influences the climatology of storms. The storms often retain their integrity for several hours; one example is given of a storm over the Pretoria-Witwatersrand area that was observed to move from the south-west over a distance exceeding 180 km in four hours (Fig. 9.14, *upper left*). On any one day different storms are likely to follow similar tracks, though significant differences may be apparent (Fig. 9.14, *lower right*).

As a result of the tendency to cluster, the lifespan of multi-cell storms is typically longer than the lives of the individual cells, which typically form every 5–10 minutes and last 30–45 minutes. Up to 30 cells may form in the life of a South African Highveld storm. The climatologies of storms vary from locality to locality. Over the eastern Free State, in summer, around 30 storms, reaching a mean height of ~16 km, may occur per day. Most track from west to east or south-west to north-east. Three quarters of these tracks are less than 3 km and average propagation speed is 11 km h^{-1}. Some storm clusters are highly organized, may extend laterally over distances exceeding 200 km and may track over large distances.

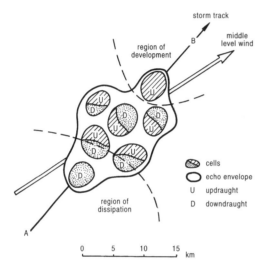

Fig. 9.11   A schematic plan view of a multi-cell Highveld storm.

SUPERCELL STORMS

Although not nearly as frequent in their occurrence as scattered multi-cell storms, supercell storms do occur in Africa from time to time and usually cause havoc and devastation. They are the most powerful of the thunderstorms known to occur over the subcontinent.

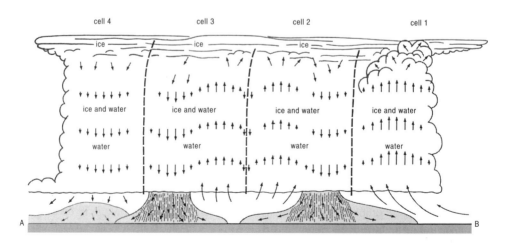

Fig. 9.12   A section along AB in Figure 9.11 to show the vertical structure of a mature multi-cell Highveld storm.

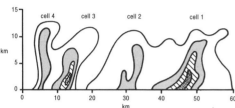

The occurrence of severe local storms is dependent upon the development of strong potential and conditional instability, most often as a result of large-scale differential advection of air in upper and lower layers, with a pronounced wind shear between the layers. Such is the case over Europe and the United States, as well as over southern Africa (Fig. 9.15). Although useful, the above notion is oversimplified and small differences in the dynamics of upper-level airflow over the subcontinent may have profound consequences for the nature of convective rainfall. Over the eastern Highveld variations in the wave structure of airflow at about 300 hPa control the occurrence of isolated, scattered or general convection and rainfall (Fig. 9.15, *lower right*). Supercells are likewise sensitive to general airflow patterns and only occur with an infrequently occurring set of conditions. Such storms are larger, more persistent and produce more severe weather than normal mature cumulonimbus multi-cell storms. In addition, they appear to have a highly organized internal circulation,

Fig. 9.13 *Upper:* radar echo outline of a January storm over Johannesburg; *lower:* vertical section along AB with contours at 25, 40, 50 and >60 dBz to show a four-cell multi-cellular storm moving north-north-east (adapted from Held and Carte, 1979).

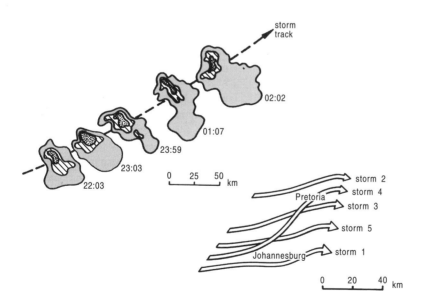

Fig. 9.14 *Upper left:* echo outlines with echo contours at 20, 40 and >60 dBz to illustrate the movement and south-west to north-east track of an October storm over Johannesburg over a 4-hour period; *lower right:* tracks of centroids of echo maxima (i.e., heaviest rainfall in each storm) for five December storms on one day over the Johannesburg–Pretoria area (adapted from Carte, 1979, 1981).

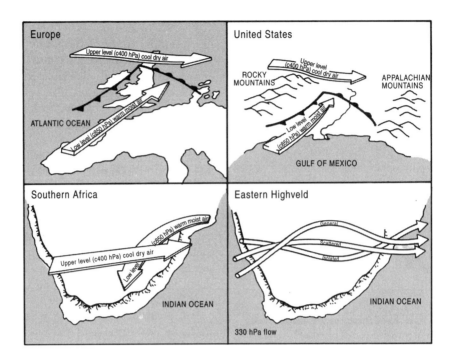

Fig. 9.15 Conditions of differential advection conducive to the development of severe local storms in Europe, the United States (after Atkinson, 1981) and southern Africa. Also shown are 300 hPa flow patterns likely to be associated with different convective regimes over the eastern Highveld (adapted from Steyn, 1984).

which is maintained in a nearly steady state for long periods. They are closely related to pronounced vertical shear in the ambient winds and appear to propagate continuously rather than by the development of discrete cells as is the case with multi-cell storms.

Supercells typically occur in an atmosphere characterized by widespread conditional instability with a greater than 4 °C difference in parcel and ambient air temperatures at 500 hPa; with strong (10 m s⁻¹) mean sub-cloud winds and with strong wind shear within the cloud layer. They usually exhibit a massive single-cell structure, circular to elliptical in shape, and extending 20–30 km in horizontal and 12–15 km in vertical dimensions. In plan view they usually show a weak echo region with dimensions of 5–12 km to the left of the direction of motion (in the southern hemisphere) and strongest echoes to the right and behind the weak echo region. Most hail occurs between the

maximum and weak echo regions. A shield of anvil cirrus extends ahead of the storm by up to 150 km. A typical vertical radar echo structure is shown in Figure 9.16. Warm, moist and conditionally unstable air is fed in from the front (left) of the storm and ascends over the cold outflow air in the area of the weak echo region beneath the echo overhang. Some of the updraught air is swept into the anvil outflow region; most ascends into the upper reaches of the cloud before cooling and subsiding into the downdraught in the manner described earlier. An additional and major influx of cool, dry downdraught air enters the cloud from behind at middle levels and enhances evaporation and hence the downdraught itself. The updraught slopes towards the rear and may be very strong. Speeds of 25–40 m s⁻¹ are not uncommon. The cumulonimbus cells that constitute the storms are embedded in a flow pattern that is maintained by the condensation of water vapour in one region (the updraught) and the partial evaporation of precipitation in another (the downdraught). Hail forming in the updraught is carried into the upper ice regions of the cloud, falls into the supercooled layer below, is carried aloft again and so recirculates, acquiring alternating layers of rime and clear ice before eventually falling out of the cloud, often as very large stones (Fig. 9.17).

The model given in Figure 9.16 is two-dimensional in the along-wind and vertical planes. It is necessary to add the cross-wind structure as well. This is done schematically for the southern hemisphere in Figure 9.18 (*upper left*). The relationship between the radar echo structure and airflow is clear but still incomplete, since no rotation is shown. In reality it appears that the middle-level cool, dry air descends anticyclonically into the storm, sweeping around the cold outflow dome in the manner illustrated in Figure 9.18 (*lower right*), before passing behind the storm in a reverse airstream at the surface. The low-level warm, moist air ascends cyclonically over the cold outflow dome and in so doing begins to spiral upward in a cyclonic sense before being expelled

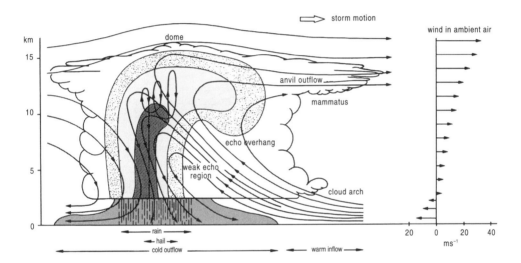

Fig. 9.16 A two-dimensional model of airflow and radar echo patterns in a supercell storm (adapted from Browning and Ludlam, 1962; Browning and Foote, 1976).

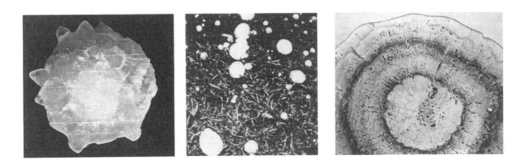

Fig. 9.17 *Left to right:* a 6-cm diameter hailstone from a Highveld storm (photo: D. vd S. Roos); large hailstones from a storm over Johannesburg; the layered structure of a hailstone (photo: D. vd S. Roos).

anticyclonically at the top of the system in the anvil outflow.

Combining the ideas developed thus far, it is possible to postulate a model of southern African supercells in the manner suggested in Figure 9.19. Ideally, warm and moist surface air will tend to be fed into the system from the front left flank of the storm; that is, from the north-east. A second source of air, cool and dry, will be fed into the system from the south-west at about the 500 hPa level. It will descend and contribute to the formation of the down-draught, before exiting from the storm from the right rear. The middle-level winds will steer the storm from a direction in the south-west quadrant; the storm itself will tend to move to the left of the steering wind, which passes around the storm as if around an obstacle. Anvil cirrus will normally advance ahead of the storm and from the south-west quadrant. Supercell storms will exhibit quasi-steady-state characteristics, will be intense and will exhibit longevity and storm tracks greater and longer than those of their multi-cell counter-parts. They are usually associated with severe hail. Stones with diameters exceeding 6 cm are common and occasionally exceed 10 cm. The terminal velocity of a stone of 8 cm diameter may reach 45 m s^{-1}. In heavy hailfalls the kinetic energy of hailstones may reach

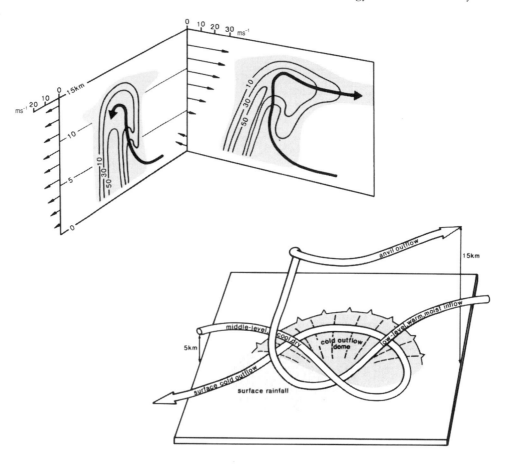

Fig. 9.18 *Upper:* schematic across-wind (*left*) and along-wind (*right*) reflectivity patterns and relative airflow in a Highveld supercell storm (adapted from Chisholm and Renick, 1972); *lower:* three-dimensional airflow in a Highveld supercell travelling to the left of middle-troposphere winds (adapted from Browning, 1964).

1 000 J m⁻¹, and in violent storms it exceeds this value considerably. Hail from a supercell storm over Pretoria once produced energy values in excess of 3 000 J m⁻¹.

Massive supercell storms have been observed over the eastern Free State extending 20–30 km in the horizontal and up to 16 km in the vertical. They may occur in isolation as the only storm in the area. On one occasion one such storm lasted for more than 4 hours and

was tracked for nearly 100 km before it passed out of the field of the tracking radar.

LINE STORMS

To a greater extent than is the case with scattered, multi-cell storms, line storms owe their alignments to either major mesoscale forcing or more usually to synoptic-scale processes. On

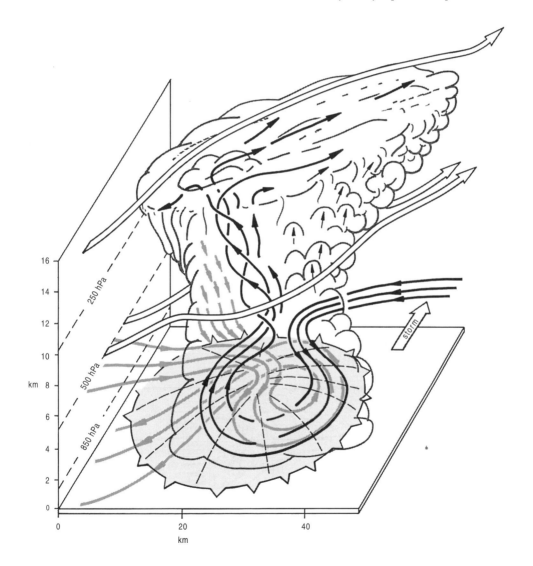

Fig. 9.19   A three-dimensional representation of a Highveld supercell storm.

many occasions in southern Africa, storms tend to build up in an organized fashion along lines extending in a south-east to north-west direction from the east coast to the northern interior of the subcontinent. They tend to form preferentially ahead of a moisture discontinuity (or front) with the same orientation as the line of storms and separating a warm, moist anticyclonic airstream to the north-east from a cooler, drier airstream (often also anticyclonic) to the south-west (Fig. 9.20, *upper*).

Cumulonimbus clouds and thunderstorms appear to develop preferentially parallel to the moisture front within a distance of about 300 km to the north-east. Warming of the surface air by thermal convection probably initiates the instability, while convergence of the surface air over the subsiding south-westerly air realizes the instability and leads to heavy precipitation in rain and often hail. Strong downdraughts in the storm cells act to invigorate the storms by providing wedges of cool air over which the warm unstable air may rise in slope convection. At the same time the downdraughts give rise to the characteristic storm collar of a line squall and the high wind speeds of variable direction and considerable turbulence that are usually experienced.

In 1958 a model was proposed for such line storms (Taljaard, 1958) (Fig. 9.20, *lower*). It is remarkably like that offered in Figure 9.16 for a steady-state supercell, and predated the pion-

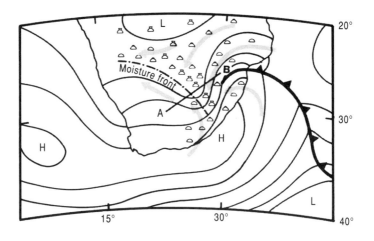

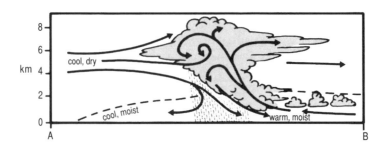

Fig. 9.20 *Upper:* line storms and showers in relation to surface pressure and wind fields and a moisture front or discontinuity; *lower:* a vertical section through the thunderstorm belt to show the structure of a single storm within the line of storms (after Taljaard, 1958).

eering work of Ludlam and Browning by several years. Taljaard's conceptual work was of great prescience and at the time received too little recognition. Hopefully the record now stands corrected.

Today it is recognized that only a small number of line storms are supercells; most are multicellular. Almost all line storms originate in the manner first suggested by Taljaard. Such storms contribute substantially to South African rainfall since they occur over several hundreds of kilometres and during their north-easterly movement they sweep over large tracts of country. Thus storms which develop in the afternoon over the northern Free State and Gauteng spread northward to the southern Northern Province during the evening and further to the north later in the night. Maximum development of the storms takes place in the late afternoon and early evening. On one occasion, consolidation of individual cells into a line took over four hours over a 180 km front and the storms lasted for six hours. Typically line storms propagate to the north-east at 30–70 km h^{-1}.

With all types of storm, the prevailing synoptic circulation fields either promote or inhibit the development of storm cells. Large-scale fields that are convergent in the lowest layers and divergent aloft promote effective thunderstorm growth. Those fields in which deep anticyclonic flow is present produce widespread subsidence and increase the atmospheric stability over the whole of southern Africa. This effectively inhibits deep cumulus convection and storm development.

Storms have been considered in general in this chapter. They are considered more specifically in relation to southern Africa in Chapter 12.

INTER-ANNUAL VARIABILITY

Much of southern Africa is affected by a quasi-periodic 18-year oscillation in mean annual rainfall (see Chapter 16). This is particularly so in the summer rainfall areas of South Africa most affected by mesoscale convective activity. The rainfall oscillation manifests itself in runs of about nine years during which rainfall is either predominantly above or below normal (see Fig. 16.17). An example of consecutive wet and dry spells is given by the generally dry 1960s and wet 1970s that were observed over eastern South Africa (Fig. 9.21). The dry spells are characterized by increased thunderstorm activity and hailfall frequencies. The inter-annual variability of mean annual hail-day frequency is the inverse of that of mean annual rainfall. During wet spells general rainfall resulting from widespread convergence in the windfield results from stratiform cloud decks, unlike the more cumuliform convective processes producing the more spatially variable rainfall of the dry spells.

TORNADOES

Tornadoes are associated with the severest of local storms and are vortices of small horizontal extent and great intensity. They extend downward from the bases of cumulonimbus clouds in a cloud form known as *tuba*, which has an inverted cone shape with the point on or above the ground. In their localized consequences, tornadoes are the most destructive and feared of all atmospheric phenomena. Associated wind speeds may exceed 800 kph. The force of the wind is sufficient to transfix grass straws through wooden planks 5 cm thick. The pressure in a tornado may drop suddenly by the order of 25 hPa. Non-robust buildings simply explode outward as a consequence of differential inside and outside pressures.

Tornadoes occur relatively rarely in southern Africa. Between 1984 and 1990, 19 tornadoes were reported in South Africa. Of these 12 were seaward of the Escarpment over KwaZulu-Natal and the Eastern Cape; 7 were over the Highveld. Tornadoes are associated with intense thermal instability and usually rotate cyclonically at great speed, sucking up objects before throwing them centrifugally outward from the

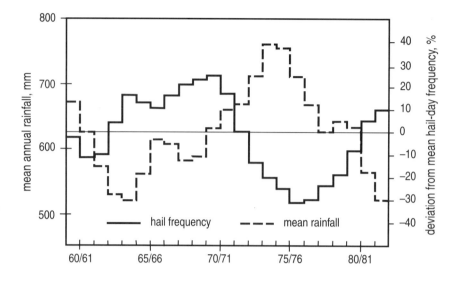

Fig. 9.21 Inter-annual variation of area-averaged rainfall and hail frequency over the central plateau of South Africa during the dry spell of the sixties and wet spell of the seventies. Data are smoothed with a five-term binomial filter; hail-day frequencies after Olivier (1990).

vortex funnel in a devastating rain of debris (branches of trees, roofs, timber, metal, etc.). Where exactly tornadoes form in relation to the structure of southern African storms is not known. If the northern-hemisphere experiences can be transposed to the southern hemisphere, then they are likely to occur in the left rear quadrant of storms. The average length of a tornado damage swath is 5–10 km, though much longer ones have been reported.

LIGHTNING AND THUNDER

Within thunderclouds precipitation particles acquire different electrical charges. Small cloud drops and ice crystals become positively charged. Because these particles are swept upward and concentrate in the upper parts of cumuliform clouds, those parts become positively charged. Larger particles and drops acquire a negative charge. Such drops are found in the lower parts of clouds, which thus increasingly become negatively charged. As the electrical potential gradient between the upper and lower parts of the cloud, or between the lower parts and the ground, increases beyond a critical limit, a discharge takes place in the form of lightning.

Lightning may be of various kinds (Fig. 9.22, *upper*). *Forked lightning* and *sheet lightning* are the most common, with the latter probably being the reflection off a neighbouring cloud of an obscure flash, or far-distant flashes seen through intervening clouds. *Cloud flashes* that occur within clouds to neutralize main centres of positive and negative charge are far more numerous than flashes to the ground (in the ratio of six or more to one). *Ribbon lightning* is caused by a strong wind separating the component charges of a flash. When a discharge channel is broken into fragments *beaded lightning* may occur, usually in heavy rain. *Ball lightning* is rare in southern Africa and is a luminous sphere about 20 cm in diameter. *Air discharges* are flashes emerging from the cloud and not reaching the ground. Cloud-to-cloud discharges are common.

The cloud-to-ground lightning flash begins in the form of an invisible discharge originating just above cloud base as a stream of electrons. This moves quickly towards the ground in a series of discrete steps and is termed the *stepped leader* (Fig. 9.22, *middle left*). It reaches the ground in about ten milliseconds. Just before it does so, a spark from the ground completes the connection. A rapid (50 micro-second) massive upward return surge of electrons then takes place in a *return stroke* in the ionized channel left by the leader stroke (Fig. 9.22, *middle right*). The current surge, in a channel only a few cen-timetres wide, produces an incandescent flash and an explosive expansion of air heated to temperatures in excess of 30 000 °C in an instant. The shock waves from this sudden expansion reverberate as *thunder* and may rattle loose objects far from the lightning. The initial light-ning discharge is seldom complete and several

further strokes may occur. These are initiated by a downward stream of electrons moving along the main path of the first-stroke flash in a *dart leader* (Fig. 9.22, *lower*). This in turn is followed by a second and highly visible return stroke. Whereas the stepped leader and first return stroke have a typical forked path, the dart leader and subsequent return strokes follow only the main channel and thus are not forked. A com-posite lightning flash is made up of about four strokes lasting about 50 milliseconds, just per-ceptible as a flicker in the discharge. Only then is discharge between cloud and ground com-plete. Recharging begins again prior to the next flash tens of seconds or minutes later. Some examples of lightning flashes are given in Figure 9.23.

Thunder travels at the speed of sound,

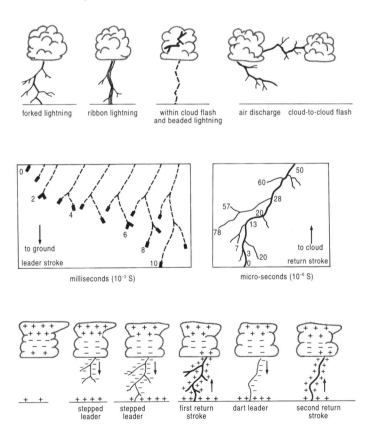

Fig. 9.22 Types of lightning and lightning strokes (after Malan, 1963).

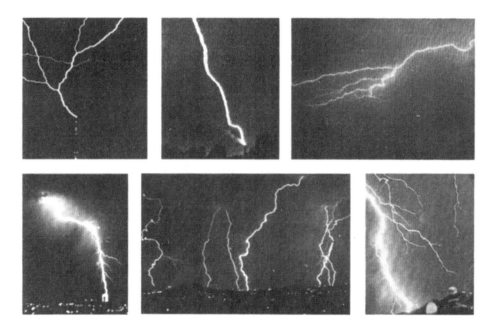

Fig. 9.23 *Upper* (left to right): forked lightning striking the Brixton Tower, Johannesburg (photo: D. Briscoe), ribbon lightning striking a chimney stack (photo: F. W. Lane), a within-cloud discharge (photo: CSIR); *lower* (left to right): a return stroke from Sandton City, Sandton (photo: D. Hartnagel), multiple discharges over Pretoria (photo: CSIR), a return stroke over Arizona (photo: © Gary Ladd, 1972).

whereas the visible lightning flash is virtually instantaneous. A distant flash produces thunder after a time delay of about 3 seconds per kilometre.

Brief and intense downpours of rain often fall a minute or so after a lightning flash. In fact as long ago as 580 BC Lucretius wrote that gushes of rain were caused by thunder. Today it is thought that the agglomeration of droplets is due in part to electrostatic forces. Lightning discharges leave droplets in their channels with a large opposite polarity to those in the vicinity. Attraction occurs, drops coalesce, grow rapidly and precipitate in a sudden downpour.

ADDITIONAL READING

PIELKE, R. A., SR., and PIELKE, R. A., JR. (eds.). 2000. *Storms, Natural Hazards and Disasters.* London: Routledge. In press.

SCORER, R. S. 1997. *Dynamics of Meteorology and Climate.* Chichester: Wiley.

SCHNEIDER, S. H. (ed.). 1996. *Encyclopaedia of Weather and Climate,* Vols. 1 and 2. Oxford: Oxford University Press.

ATKINSON, B. W. 1981. *Meso-scale Atmospheric Circulations.* London: Academic Press.

EAGLEMAN, J. R. 1980. *Meteorology, the Atmosphere in Action.* New York: Van Nostrand Company.

LUDLAM, F. H. 1980. *Clouds and Storms.* University Park: Pennsylvania State University Press.

WALLACE, J. M. and HOBBS, P. V. 1977. *Atmospheric Science, an Introductory Survey.* New York: Academic Press.

BYERS, H. R. 1974. *General Meteorology*, 4th Edition. New York: McGraw-Hill.

10

LARGE-SCALE WEATHER-PRODUCING PROCESSES AND SYSTEMS

Convective and mesoscale vertical motion resulting from cumulus-cloud and thunderstorm formation play a key role in producing weather, particularly in the tropics. In higher latitudes, large synoptic-scale disturbances and the processes governing the uplift and subsidence of air in these disturbances assume a more important, and indeed dominant, role in governing weather. In this chapter the processes responsible for large-scale, sustained vertical motion and the systems in which such motion occurs will be considered.

THE NATURE OF CONVERGENCE AND DIVERGENCE

Given the occurrence of planetary-scale turbulence in the form of waves, cyclones and anticyclones, airflow is seldom uniform, straight and parallel in the atmosphere. Instead the atmosphere is in a state of continuous perturbation in which mass convergence and divergence of air is always taking place. This convergence and divergence may take many forms, such as motion directly to and from a point, or as a result of cross-isobar ageostrophic or gradient flow, or from inward or outward spiralling of air (Fig. 10.1). Deceleration or acceleration of the wind causes kinematic (speed) convergence or divergence, as does confluence and diffluence of air motion. Streamlines of airflow may converge or diverge. Air movement up slopes produces slope convergence, either orographic across topographic barriers or frontal over cold and warm fronts. All these processes produce the vertically upward or vertically downward air motion so essential in weather-forming systems. Sometimes the processes act together to produce strong convergence or divergence; sometimes they act in opposition to cancel each other or to produce indeterminate states. Thus

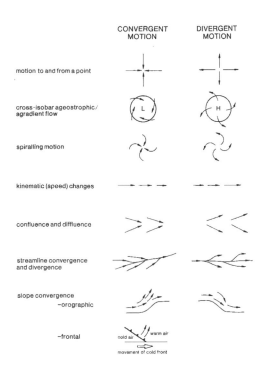

Fig. 10.1 Various types of convergent and divergent motion in the atmosphere.

kinematic convergence and confluence acting together produce strong convergence, whereas kinematic convergence and diffluence produce an indeterminate situation in which the actual occurrence of convergence will depend on the individual magnitudes of the two processes (Fig. 10.2).

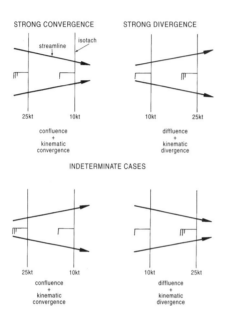

Fig. 10.2 The effect of combining various sources of convergence and divergence (the feathers indicate 10 kt increments in wind speed).

When an incompressible fluid (e.g., water) is squeezed horizontally, it expands vertically. Though it is somewhat compressible, air behaves in much the same way and the horizontal squeezing of a volume of air is to a large extent complemented by its vertical expansion. Thus the convergence of air into an initial volume of depth D_1 will result in a shrinking of the horizontal cross-sectional area with time (Fig. 10.3). Since the total mass of air must remain constant, it follows that the depth must increase to D_2 and that air must move upward within the volume. In so doing it acquires cyclonic rotation (vorticity). Conversely, divergence within a col-

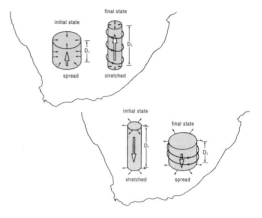

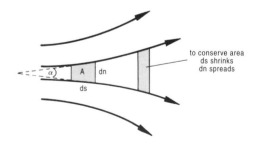

Fig. 10.3 The effect of spreading and stretching and the resulting patterns of convergence and divergence.

umn of vertical depth D_1 will result in spreading, downward motion and anticyclonic rotation.

In the horizontal, divergence may be determined using the notion of shrinking and spreading. Consider different streamlines as shown in Figure 10.4. As the element of air defined by distance ds downstream and dn across the flow is deformed, ds decreases (shrinks) and dn increases (stretches) as the area A is conserved. Horizontal divergence is dependent on the rate of change of area with time and is

$$\text{div}_h = \frac{1}{A} \frac{dA}{dt} \qquad (10.1)$$

Fig. 10.4 Diffluent flow and the deformation of an element of area. In order to preserve area A ($dn \times ds$), dn must increase (spread) as ds decreases (shrinks).

This is the same as saying that the horizontal divergence is the sum of the spreading and the stretching effects, or expressed differently, is the sum of the effect of acceleration or deceleration and the effect of confluence or diffluence. Expressed more quantitatively

$$\text{div}_h = \frac{dV}{ds} + V\frac{d\alpha}{dn} \qquad (10.2)$$

where V is velocity, s is distance along the streamline, n is distance orthogonal (at right angles) to the streamlines, measured to the left of the wind, and α the angle between the tangents bounding area A.

THE EQUATION OF CONTINUITY

By assuming that the initial mass of a body of air is conserved no matter how it may subsequently deform, expand, divide or combine and that nowhere in the system is fluid created or destroyed, it is possible to develop the equation of continuity. Consider Figure 10.5. The initial mass of air in the box is given by the product of its density, ρ, and volume; that is, by $\rho\ dxdydz$. The net flux of mass into or out of the box depends on the amount of air entering and leaving. In the x direction with velocity component u,

Net mass flux = influx at x_0 − outflow at x_1

$$= \rho u dy dz dt - \left(\rho u + \frac{\partial \rho u}{\partial x}\right) dx\ dy dz dt \qquad (10.3)$$

Similar expressions may be developed for the net fluxes through the other faces of the box by the v and w components of wind in the y and z direction respectively. The local rate of change of air density with time, $\partial\rho/\partial t$, is given by

$$\frac{\partial \rho}{\partial t} = -\left(\frac{\partial \rho u}{\partial x} + \frac{\partial \rho v}{\partial y} + \frac{\partial \rho w}{\partial z}\right) \qquad (10.4)$$

(Most meteorological equations deal with changes of some property with distance or time, such as temperature with height, pressure with horizontal distance, or pressure with time. This is designated in mathematical terms by the use of the *total derivative*, such as dT/dz, which gives the change of temperature with height *following the parcel*. The *local change* (i.e., change at a fixed point) is indicated by a *partial derivative*, such as $\partial T/\partial t$, which gives the local change of temperature with time.)

The *equation of continuity* follows from equation (10.4) and is usually stated as

$$\frac{\partial \rho}{\partial t} + \frac{\partial \rho u}{\partial x} + \frac{\partial \rho v}{\partial y} + \frac{\partial \rho w}{\partial z} = 0 \qquad (10.5)$$

If the air is considered to be incompressible, and thus its density to be constant, the equation of continuity then simplifies to

$$\frac{\partial u}{\partial x} + \frac{\partial v}{\partial y} + \frac{\partial w}{\partial z} = 0 \qquad (10.6)$$

The terms on the left-hand side of the equation constitute the divergence of the mass flux. Equation (10.6) is the most usual form of the equation of continuity and states that the local changes in horizontal u and v components of wind in the x and y directions must be balanced by an equal and opposite local change in the vertical w component in the z direction:

$$\frac{\partial u}{\partial x} + \frac{\partial v}{\partial y} = -\frac{\partial w}{\partial z} \qquad (10.7)$$

Put another way, equation (10.7) shows that the horizontal divergence, which comprises the left-hand terms of the equation, must be accompanied by vertical shrinking, vertical convergence (i.e., negative divergence) and negative (i.e., downward) vertical velocities. In contrast, horizontal convergence must be accompanied by vertical stretching, upward velocities and vertical divergence. For continuity to be preserved, surface horizontal convergence must be accompanied by horizontal divergence aloft and vice

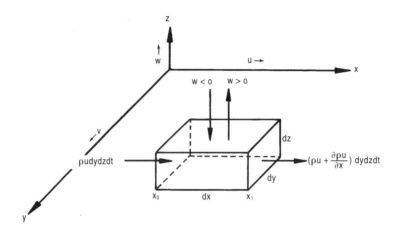

Fig. 10.5 Continuity and its derivation.

versa (Fig. 10.6). The resulting vertical motions explain the general weather associated with surface convergence and divergence. Thus large systems of clouds and precipitation are invariably associated with uplift in regions of low-level convergence and upper-level divergence. Similarly, cloud-free air is usually found in regions of subsiding air associated with low-level divergence and upper-level convergence. It is also apparent that between the surface and upper levels, the sign of the divergence must change at the level of non-divergence. This most often occurs at about the 500 hPa level, which happens to be the level at which large-scale steering of storms and weather systems occurs.

THE EFFECT OF LATITUDE

Provided the pressure gradient remains constant in time and space, the change in the Coriolis parameter with latitude causes distinctive patterns of convergence and divergence depending on whether the air is moving towards or away from the equator (Fig. 10.7).

Consider a situation over the west coast of South Africa in which air is moving *equator-*

ward with velocity V_1 from an initial position at a higher latitude ϕ_2 to a lower latitude ϕ_1, where it has velocity V_2. From the geostrophic assumption, at ϕ_2

$$V_1 = g \ \frac{dz}{dn} \ \frac{1}{2\omega\sin\phi_2} \qquad (10.8)$$

Similarly

$$V_2 = g \ \frac{dz}{dn} \ \frac{1}{2\omega\sin\phi_1} \qquad (10.9)$$

The difference between the velocities is

$$V_1 - V_2 = g \ \frac{dz}{dn} \ \frac{1}{2\omega} \left(\frac{1}{\sin\phi_2} - \frac{1}{\sin\phi_1} \right) \quad (10.10)$$

and since $\phi_2 > \phi_1$, it follows that $V_1 - V_2 < 0$ or $V_1 < V_2$. The air is thus accelerating and divergence will occur between the two latitudes. Similarly, air moving *poleward* over the east coast of South Africa will be convergent. These effects are not strong, but may help to explain the relative ease with which vertical motion is enhanced on the western side of the South Indian anticyclone over eastern South Africa and inhibited on the eastern side of the South

in cells

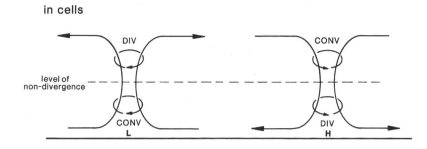

in horizontal airflow

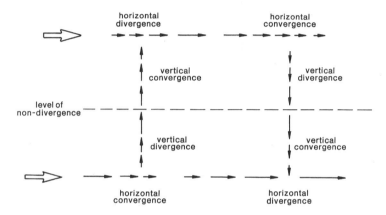

Fig. 10.6 Large-scale patterns of convergence and divergence and consequent vertical air motion. *Upper*: associated with rotating systems in the southern hemisphere; *lower*: in horizontal linear airflow.

Atlantic anticyclone over the west coast. This process has climatological implications and contributes to the fact that the eastern half of South Africa is wetter than the western half.

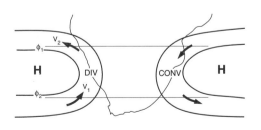

Fig. 10.7 The convergence and divergence associated with variations in geostrophic wind with latitude when the pressure gradient remains constant in time and space.

THE EFFECT OF CHANGING PRESSURE GRADIENTS

Changes in space

In Chapter 8 it was shown that as pressure gradients steepen, ageostrophic flow occurs toward lower pressure (see Fig. 8.5). Under such circumstances a component of surface convergence will be generated. However, since kinematic divergence is also likely to occur because of the acceleration of the wind, the net result will depend on the stronger of the two effects. With slackening pressure gradients, ageostrophic flow towards higher pressure will occur, as too will kinematic convergence.

Changes in time

The *pressure tendency* at any height, z, is given by the horizontal mass divergence above z in a column of air and by the vertical transport of air into or out of the column through level z. The local change of pressure with time may be expressed as

$$\frac{\partial p}{\partial t} = - g \int_{z}^{\infty} \text{div}(\rho V)dz + g\rho w \qquad (10.11)$$

where ρ is air density, V is horizontal velocity along a streamline, g is the force of gravity and w the vertical velocity. If more air moves out of the column than into it, either by horizontal divergence (first term on the right-hand side) or by downflow through the bottom (second term), the pressure at z will fall. By itself the *tendency equation* is not operationally useful. However, it does illustrate how pressure tendency and divergence are linked. In regions of falling pressure isallobaric convergence will occur (Fig. 10.8) and weather will deteriorate; in regions of rising pressure isallobaric divergence will be present and the weather will clear. (Definitions of isallobars and isallobaric changes are given in Chapter 8.) Where the pressure gradient is constant in time, geostrophic (or gradient) flow will occur. The most important consequence arising out of the tendency equation is that *in conditions of geostrophic flow, and with pressure gradients remaining constant in both time and space, no convergence or divergence will occur and no weather will be produced.*

THE EFFECT OF CHANGING RADIUS OF CURVATURE

The changes from anticyclonic to cyclonic curvature in a continuous wave in the horizontal have profound consequences for changing

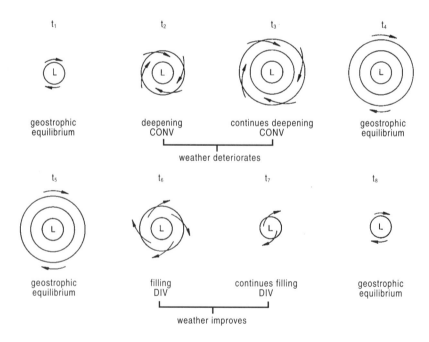

Fig. 10.8 Changes in pressure tendency (change of pressure with time, t) and the occurrence of isallobaric convergence and divergence.

weather. In Chapter 8 it was shown that, for anticyclonic curvature, the gradient wind velocity, V_{gr}, exceeds the geostrophic velocity, V_g, for the same pressure gradient, whereas for cyclonic curvature the opposite holds and the geostrophic velocity exceeds the gradient flow. Thus in Figure 10.9 on the ridge line

$$\text{at A} \quad V_{gr} > V_g \tag{10.12}$$

whereas on the trough line

$$\text{at B} \quad V_{gr} < V_g \tag{10.13}$$

Since V_g is the same at both A and B, and because the pressure gradient is everywhere the same, it follows that

$$V_{gr} \text{ at A} > V_{gr} \text{ at B} \tag{10.14}$$

and hence convergence will occur between A and B. Similarly

$$V_{gr} \text{ at B} < V_{gr} \text{ at C} \tag{10.15}$$

and divergence will occur between B and C. Since mass will be transferred from A to B, the pressure at B will rise while it is being lowered at A. Similarly, the pressure at C will fall with the removal of mass towards the east. Hence the ridge R_1 will change to R_2 and trough T_1 will become trough T_2. The wave will thus propagate towards the east. Such waves in the horizontal are known as *Rossby waves* and at any time their number in the westerlies varies from about 4 to 8 around the hemisphere. Wavelengths are typically the order of 60° to 120° of longitude. It is important to note that owing to changes in radius of curvature in Rossby waves *convergence occurs behind the trough*

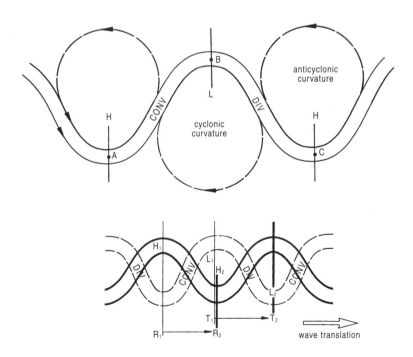

Fig. 10.9 *Upper:* changes in radius of curvature and the occurrence of convergence and divergence; *lower:* westerly wave propagation. Light broken lines give the wave at time t_1, heavy lines the wave at time t_2. Trough and ridge positions are given by T_1 and T_2, R_1 and R_2 respectively.

line and divergence occurs ahead of it. This is the exact opposite of the pattern produced by variations in geostrophic wind with latitude (and illustrated in Fig. 10.7). Of the two effects, the convergence and divergence patterns produced by changes in the radius of curvature are the stronger and more important in the dynamics of waves.

VORTICITY

Closed-vortex perturbations in the circulation of the atmosphere are common, and at any one time many such vortices can be observed in the southern hemisphere. They frequently sweep across southern Africa from the west and may be quite long lived.

Consider a circular vortex with a perimeter length L and velocity V along the perimeter. The *circulation* along the closed curve of the perimeter is defined by

$$C = VL \qquad (10.16)$$

If the radius of the vortex is r and it rotates with an angular velocity Ω, then the linear velocity along the perimeter is $V = \Omega r$ and the length of the perimeter – the circumference – is $2\pi r$. Hence

$$C = \Omega r 2\pi r \qquad (10.17)$$
$$= 2\Omega \pi r^2 \qquad (10.18)$$

where πr^2 is the area of the vortex. The quantity 2Ω is known as the *relative vorticity* (i.e., relative to the rotating earth). Effectively, it is twice the 'spin' of the vortex. From equation (10.18) it is clear that circulation is given by vorticity multiplied by area. The relationship is general and holds for all shapes.

Since the angular velocity of the earth's rotation is ω, the earth's vorticity is 2ω at the poles. Elsewhere at any latitude ϕ it is $2\omega\sin\phi$; that is, the Coriolis parameter, f. The *absolute vorticity* of a rotating vortex perturbation in the earth's atmosphere, which is itself rotating, is thus made up of the vorticity relative to the earth, the relative vorticity, ζ, and the earth's vorticity, f. Absolute vorticity is consequently specified as $\zeta + f$.

Local rotation of fluid particles (and hence vorticity) is produced by both shear (variation of wind speed with height or distance) and *curvature effects* (Fig. 10.10). In the case of curvature, the *curvature vorticity* is defined by

$$\text{curvature vorticity} = \frac{V}{r} \qquad (10.19)$$

In the case of shear, cyclonic vorticity is generated if a particle moves from a slowly moving westerly air current into a faster one on its equatorward side (Fig. 10.10). Conversely anticyclonic shear and vorticity is generated if a particle moves into a faster airstream on its poleward side. The *shear vorticity* generated is specified by

$$\text{shear vorticity} = \frac{dV}{dn} \qquad (10.20)$$

The total relative vorticity, ζ, is thus given by the sum of the curvature and shear effects, namely

$$\zeta = \frac{V}{r} + \frac{dV}{dn} \qquad (10.21)$$

Expressed in u and v velocity components, relative vorticity is given by

$$\zeta = \frac{dv}{dx} - \frac{du}{dy} \qquad (10.22)$$

which is the difference between the rate of change of the v (meridional) wind component in the x (east-west) direction and the u (zonal) component in the y (north-south) direction.

The vorticity or spin of a vortex or particle can only change if there is a net tangential force acting on its surface (i.e., a torque), as happens with particles between the airstreams illustrated in Figure 10.10 (*lower*). First, particles acquire cyclonic or anticyclonic vorticity with cyclonic or anticyclonic shear respectively. Secondly, vorticity changes if the distribution of mass

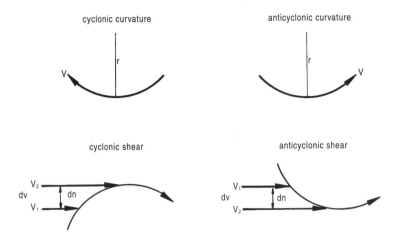

Fig. 10.10 Cyclonic and anticyclonic curvature and shear.

changes. For example, in Figure 10.3, if a column of air is rotating clockwise (i.e., cyclonically) around a vertical axis, convergence and concentration of mass near the axis of rotation will increase the rotation (in a fashion analogous to skaters who increase their spin by pulling their hands to their bodies) and hence the cyclonic vorticity will increase. The sense of vorticity is defined as follows: *cyclonic vorticity is positive or increasing; anticyclonic vorticity is negative or decreasing.*

The relationship between divergence and vorticity in the linearized, geostrophic case is given by

$$\text{div} = -\frac{1}{f}\frac{d\zeta}{dt} \tag{10.23}$$

Thus cyclonic vorticity is associated with negative divergence (i.e., convergence) and anticyclonic vorticity with divergence.

The conservation of absolute vorticity

The shrinking and stretching of columns of air are directly linked to the divergence in the wind field and to the generation of vorticity. Horizontal divergence implies vertical shrinking of a column of air (see Fig. 10.3) and a decrease in its vorticity. On the other hand, horizontal convergence is associated with vertical stretching of a column of air and an increase in its vorticity.

Just as matter and energy are conserved in the atmosphere, so it is useful to consider the conservation of absolute vorticity. In such a case

$$\frac{\zeta + f}{D} = \text{constant} \tag{10.24}$$

where D is the depth of the system (and hence defines the shrinking or stretching taking place). *If the depth of the system does not change* (i.e., D remains constant) then the conservation of absolute vorticity becomes

$$\zeta + f = \text{constant} \tag{10.25}$$

with D being subsumed into the new constant. Consider the movement in a Rossby wave in which anticyclonic curvature and vorticity is at a maximum at the ridge lines (R_1 and R_2) and cyclonic curvature and vorticity is at a maximum at the trough line (T_1) in Figure 10.11. As air moves equatorward from point A at latitude ϕ_2 to point B at ϕ_1 the latitude decreases. The earth's vorticity, given by the Coriolis parameter,

f, will thus diminish accordingly. To preserve the constancy of absolute vorticity, the relative vorticity, ζ, must increase, becoming more cyclonic. From equation (10.25), increasing ζ means that negative divergence (i.e., convergence) will occur between A and B to the rear of the trough T_1. Similarly, poleward movement from C to D will generate anticyclonic vorticity and divergence ahead of the trough. Put another way, convergence (divergence) will occur upwind (downwind) from a vorticity maximum at a trough line. These characteristic patterns of convergence and divergence are the same as those produced by changes in the radius of curvature (see Fig. 10.9) and are strong enough to overcome the effect of changing geostrophic flow with latitude (see Fig. 10.7). They are of profound importance in influencing the weather everywhere, not least in the southern African sector of the southern hemisphere.

If zonal flow is considered, then latitude, ϕ, remains constant. Under such circumstances, consider air approaching a longitudinal mountain barrier in the southern hemisphere (Fig. 10.12). As the airstream ascends the mountain, its depth D decreases. To maintain constancy in equation (10.24) the relative vorticity, ζ, must decrease commensurately and in so doing becomes more anticyclonic. The airflow is thus deflected northward and acquires an anticyclonic curvature. As the air descends the mountain on the lee side, D increases, ζ increases and the airstream acquires cyclonic vorticity and curvature. The formation of a trough is thus facilitated. Such lee troughs are common features of the circulation to the east of the Andes and the New Zealand Alps (Fig. 10.13, *upper left*).

In southern Africa as air descends over the Escarpment from the high interior plateau to the coast, cyclonic vorticity is generated and coastal lows form. They form at any locality along the coast between Namibia and Mozambique, provided that strong offshore gradient winds prevail. Without the Escarpment, which forms the edge of the elevated inland plateau of southern Africa, they would not come into being. They owe their existence to topography and the local generation of cyclonic vorticity and are regular features of the circulation of coastal regions, particularly on the east coast (Fig. 10.13, *upper right*).

Conservation of absolute vorticity in both westerly and easterly waves over southern Africa produces convergence, uplift, cloud and often precipitation to the rear of the trough, and divergence, subsidence and clear skies ahead of the trough (Fig. 10.13, *lower*).

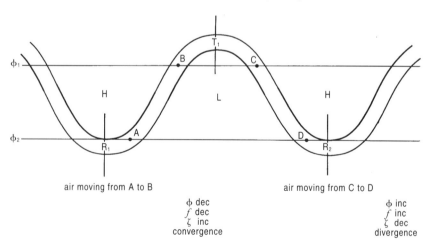

Fig. 10.11 The conservation of absolute vorticity and the occurrence of convergence and divergence.

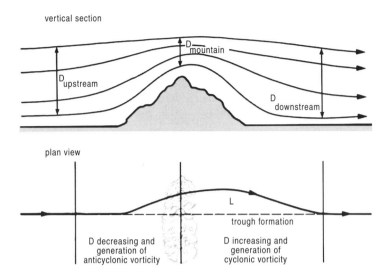

Fig. 10.12   The formation of lee troughs and orographic cyclogenesis through the conservation of absolute vorticity in the southern hemisphere.

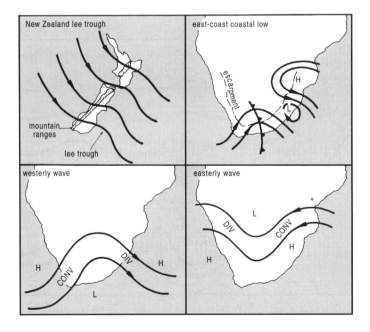

Fig. 10.13   *Upper*: the formation of a lee trough over New Zealand and a coastal low over South Africa; *lower*: characteristic patterns of convergence and divergence associated with westerly and easterly waves over southern Africa.

WAVE PERTURBATIONS

Wave motion is common in the atmosphere, both in the vertical and in the horizontal. Common waves include gravity, lee and Kelvin waves, which occur in the vertical, and Rossby and semi-stationary long planetary waves, which occur in the horizontal.

Gravity waves

Gravity waves form because of vertical displacements of air which are then restored by gravity. A special case of gravity wave is to be found in the formation of *lee waves* downwind of mountains in stably stratified air (see Fig. 4.2). These waves may maintain their positions for many hours at a time. More usually gravity waves are carried by the current of air in which they form. Two conditions govern their speed of translation. If the depth, h, of a discontinuity between adjacent layers exceeds 40 per cent of the wave length (i.e., the system is deep) then

$$c = V \pm \left(\frac{g\lambda}{2\pi}\right)^{1/2} \tag{10.26}$$

where c is the speed (celerity) of the wave, V the wind speed, g gravity, and λ the wavelength. When the waves form in shallow flows (i.e., $h < \lambda/40$) then

$$c = V \pm (gh)^{1/2} \tag{10.27}$$

Gravity waves are often associated with wave clouds, the exact form of which depends on the variations of wind speed and temperature with height (Fig. 10.14). Many different forms are possible; most produce clouds of striking morphology and great beauty. Examples of lenticular, orographically induced clouds and gravity waves are given in Figure 10.15.

Kelvin waves

Kelvin waves are vertical waves that result from the shear instability between two air masses of

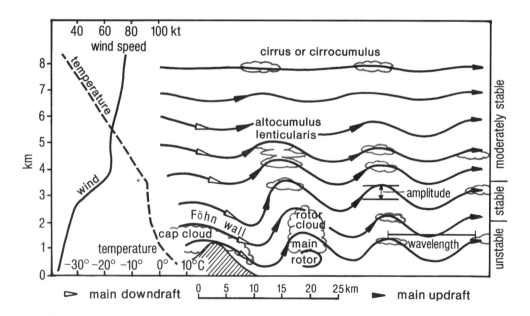

Fig. 10.14   Types of lenticular clouds produced by airflow across mountains with the occurrence of appropriate wind and temperature profiles (after Ernst, 1976).

different density, particularly if the interface between the masses is tilted. The wavelike motion can develop within a stably stratified atmosphere, provided the vertical wind shear exceeds some critical value. The resulting motions take the form of *billow clouds*, particularly the undulations in *altocumulus* or *cirrocumulus undulatus* cloud. The rippling effect is smoothed by the thermal stability, but when the Richardson number (Chapter 14) falls below 0.25 the shear waves grow to the point where the waves develop crests and break down in turbulence. These shear waves are much shorter in wavelength than gravity waves and are common features of the nocturnal boundary layer. They are frequently found on the tops of inversions (Fig. 10.16, and see Fig. 14.19).

Fig. 10.15 Some examples of lenticular wave clouds forming over KwaZulu-Natal (*upper left* and *right*), over the Sierra Nevada, California (*centre left*), over the Southern Alps, New Zealand (*centre right*), over the Magaliesberg (*lower right*) and over Table Mountain, Cape Town (*lower left*) (photos: P. D. Tyson, except *upper right*, D. Tunnington).

Rossby waves

Rossby waves are horizontal waves in which air particles move north and south, while the waves propagate generally along circles of latitude. The wave speed is given by

$$c = V - \frac{\beta \lambda^2}{4\pi^2} \qquad (10.28)$$

where $\beta = \partial f / \partial y$; that is, the local change of the Coriolis parameter with latitude. The waves propagate downwind (see Fig. 10.9) and at speeds dependent on their wavelength. The shorter the wavelength, the faster the wave propagation. Clearly wave speed cannot exceed the speed of the wind. Shorter waves travel with speeds close to those of the zonal current and tend to be steered by the quasi-stationary long waves.

Long waves tend to remain stationary. In the case where $c = 0$, the wavelength is given by

$$\lambda = 2\pi \left(\frac{V}{\beta}\right)^{1/2} \qquad (10.29)$$

At 45° S for a wind of 4 m s^{-1} the stationary wavelength is about 3 000 km. With a zonal current of 12 m s^{-1} the wavelength increases to over 5 000 km.

Whether Rossby waves will be stable (i.e., will not increase in amplitude with time) or not depends on the degree of vertical wind shear present in baroclinic conditions (see later in this chapter for a definition of baroclinicity). In general the extra-tropical westerlies are *baroclinically unstable* because of the strong shear present in the westerlies and short waves increase two- to three-fold in amplitude in about two to three days.

Wave perturbations undergo irregular oscillations in intensity and form according to the *index cycle*. The atmosphere often settles into periods in which disturbances are at a minimum and strong zonal flow patterns are evident (Fig. 10.17). Thereafter perturbations increase and wave-like motion becomes more evident. This may continue until the waves become sufficiently unstable to form vortices in the form of depressions, cut-off lows and cells of high pressure. The anticyclones frequently become blocking highs which, as the name implies, exert a blocking effect on the zonal passage of storms. The weather consequences are immediate, with the blocking area locally experiencing hot, dry conditions, whereas in the upstream region the blocked storms may precipitate copiously.

THE RELATIONSHIP BETWEEN LOW- AND UPPER-LEVEL AIRFLOW

Since pressure decreases most rapidly with height in cold air and least rapidly in warm air (equation 3.26), it follows that the axes of troughs and ridges are seldom vertical in extra-tropical latitudes. In troughs, the coldest air lies

Fig. 10.16 Kelvin waves; *left:* on stratocumulus over the Pilansberg; *right:* on the top of the boundary layer and in air pollution at that height over Johannesburg (photo: P. D. Tyson).

behind the cold front or trough line, hence the axis slopes backward, to the west, with height. With anticyclones or ridges in the southern hemisphere, the warmest air lies to the west of the features and the axes of the highs or ridges will likewise slope to the west. Consequently, in wave motion in middle latitudes and in the subtropics surface waves are displaced to the west with height (Fig. 10.18). The most important effect of this is to produce, within the same tilted system, opposite divergence fields in the low- and upper-level flow. Thus surface convergence is overlain by upper-level divergence, and surface divergence by upper-level convergence; strong upward and downward motion will occur respectively with direct effects on the weather. In wave sequences large synoptic-scale convective cells are set up in the manner shown in Figure 10.18. Should surface convergence not be accompanied by upper-level divergence, then the condition for large-scale uplift will not occur. It is for this reason that seemingly favourable conditions for rainfall on surface synoptic charts may in fact not be accompanied by the anticipated weather. Only if the upper-level airflow is also conducive to rainfall will the necessary uplift, and hence precipitation, occur. An example of the divergence fields, opposite at the 850 and 500 hPa levels, associated with a deep Rossby wave over southern Africa is given in Figure 10.19. From the satellite photograph it is clear that the cloud band occurred in the region where the 850 hPa convergence was overlain by 500 hPa divergence.

JET STREAMS

Regions of locally concentrated, fast-moving streams of air are referred to as jet streams. They occur in regions of strong horizontal temperature gradients and hence exhibit a substantial thermal wind component; they occur near the top of the troposphere and equatorward of frontal discontinuities. Often two jets may be identified climatologically. The first is the *subtropical jet*, the second is the more southerly *polar (Antarctic) jet*. Of the two, the subtropical jet has an immediate consequential effect on southern African weather. In winter its mean position moves equatorward to near 30° S (Fig. 10.20). Mainstream mean wind speeds exceed 30 m s^{-1} at a mean height of about 200 hPa. In summer the jet core is more diffuse, speeds are on average between 20 and 30 m s^{-1} and the jet is located poleward of 40° S.

The dynamics of upper-air jet streams are such that they exert an important influence on surface weather. Consider a jet of the kind illustrated in Figure 10.21 over South Africa. If the jet is associated with a strongly curving trough, then the effects of both the curvature of the streamlines and the confluence and diffluence will be significant. Whichever of the effects is greater will determine whether or not convergence or divergence occurs. Usually it is found that on the poleward, right front, exit side of the jet (the so-called delta region) strong diffluence occurs. Inertial (momentum) effects ensure that the velocity of the air is greater than the spacing

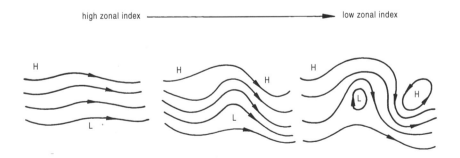

Fig. 10.17 Changes in the zonal index with the formation of waves, cut-off lows and blocking highs.

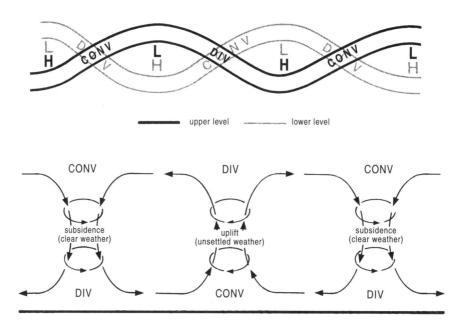

Fig. 10.18 The relationship between surface and upper-level patterns of convergence and divergence in deep tilted waves.

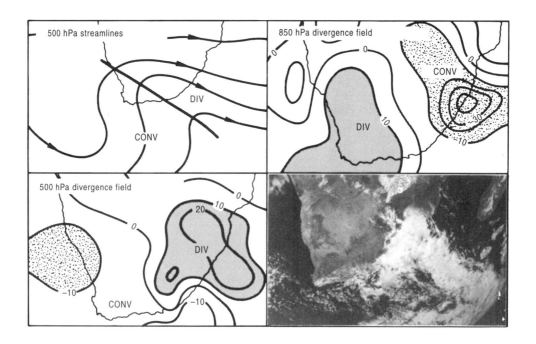

Fig. 10.19 A westerly wave at 500 hPa together with derived divergence fields at the 850 hPa and 500 hPa levels and the Meteosat image for the same day (after Poolman, 1986). Units of divergence are 10^{-7} s^{-1}.

of the contours of the pressure surface would dictate; that is, the Coriolis force is greater than is required for balance (see Fig. 8.5). The motion becomes ageostrophic and deflected to the left towards higher pressure so increasing the diffluence, which is now dominant. Divergence is the net result. On the equatorward left front side of the exit region convergence is predominant. In the jet entry region the opposite patterns of convergence and divergence prevail. If a vertical section is taken through the jet along AB in the exit region, then it can be seen (Fig. 10.21, *lower*) that the upper-level divergence may be sufficient to produce uplift on the poleward side of the jet, so promoting surface convergence and initiating or sustaining surface cyclogenesis.

AIR MASSES

The general circulation of the atmosphere is such that vast quantities of air over particular regions may acquire reasonably uniform properties while the air resides or is passing only slowly over those regions. Such bodies of air are termed air masses and their quasi-homogeneous properties are described by the properties of the air in the source regions. Once useful in day-to-day forecasting, air-mass analysis has lost its importance. However, the notion of air masses retains some conceptual usefulness and often goes a long way to explain the relationships between airstreams of different directions and the large-scale temperature fields likely to occur over southern Africa. The principal airstreams affecting the subcontinent are shown in Figure 10.22 (*upper*) and include tropical continental air with a northerly component originating over central Africa, more easterly tropical maritime air from the Indian Ocean, recurved subtropical maritime air from the Atlantic Ocean and two major cold airstreams. The first is loosely described as a south-westerly subpolar maritime airstream originating from high latitudes in the South Atlantic; the

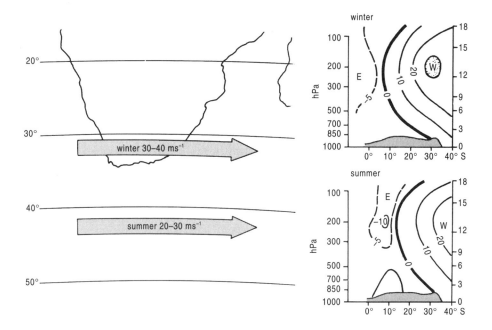

Fig. 10.20 Mean summer (December–February) and winter (January–August) positions and structure of the jet stream over South Africa (modified after Newell et al., 1972). Wind speeds given in m s⁻¹.

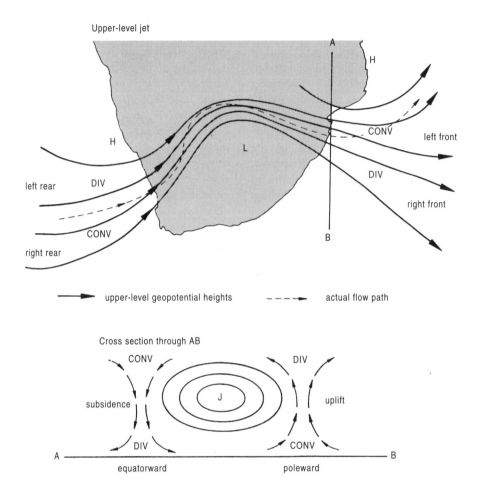

Fig. 10.21 Characteristic patterns of convergence and divergence associated with an upper-level jet stream over South Africa and resulting vertical motions on the poleward and equatorial sides of the jet stream.

second is a southerly polar maritime airstream forming off the Antarctic coast. As these airstreams blow over South Africa, departures from the mean temperature field are characteristic as a consequence of the advection taking place (Fig. 10.22, *lower*). Thus with a winter anticyclonic airstream over the subcontinent, temperature departures are positive (due also to dynamic warming effects), except where the air is directly onshore over north-eastern South Africa and some advective cooling is experienced. In contrast, southerly airflow, even in summer, produces negative departures until the air has penetrated far enough north over the land to become sufficiently heated and modified to result in a warming effect.

FRONTS, FRONTOGENESIS AND FRONTOLYSIS

Wherever two air masses with substantially different temperature characteristics come into contact, a zone of strong discontinuity and tem-

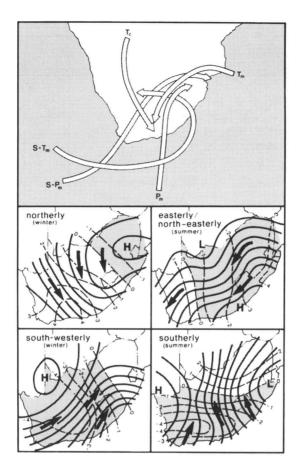

Fig. 10.22 Some characteristic air masses affecting southern Africa (*upper*) and temperature deviations from the means for airflow patterns of the type and direction indicated (modified after Longley, 1976, by Lengoasa, 1987). Tropical maritime air is denoted by T_m, tropical continental by T_c, subtropical maritime by $S–T_m$, subpolar maritime by $S–P_m$ and polar maritime by P_m. Shaded areas indicate cooling by advection.

perature gradient will exist between them. Such zones are termed fronts and may be of two basic types. *Cold fronts* occur when, with the passage of time, cold air replaces warm; *warm fronts* occur when warm air replaces cold. A sloping cold front separating cold south-westerly subpolar air from warm north-westerly air over southern Africa is illustrated in Figure 10.23. The characteristic inversion of temperature separating the two air masses in the vertical is also shown.

The equilibrium boundary of separation between neighbouring masses of cold and warm air is never vertical, but always slopes to a degree dependent on both the temperature and the velocity differences in the air masses. If T_w and T_c represent the temperature and V_w and V_c the velocity components of the wind parallel to the front respectively (Fig. 10.24), then the slope, θ (which is equal to dz/dn), of the front is given by

$$\tan\ \theta\ =\ \frac{f}{g}\left(\frac{T_c V_w - T_w V_c}{T_w - T_c}\right) \qquad (10.30)$$

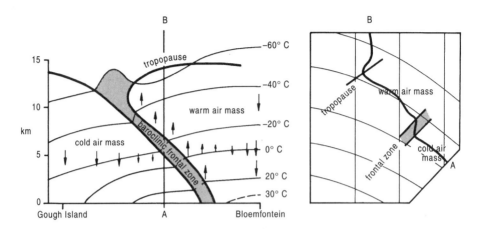

Fig. 10.23 A schematic section through a cold front showing vertical air motion (*left*) and a typical temperature profile through such a front along AB (*right*).

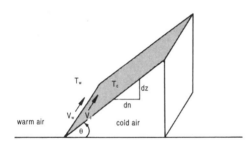

Fig. 10.24 The slope ($\phi = dz/dn$) of a frontal surface separating warm air with temperature T_w and velocity parallel to the front V_w from cold air with T_c and V_c.

Certain immediate consequences follow from this equation. Obviously T_w must be greater than T_c. In addition, since $f = 2\omega\sin\phi$, fronts are not possible at the equator where $\phi = 0$. In the case of cyclonic shear, $V_w > V_c$ and the slope of the front will be positive and physically admissible. On the other hand, with anticyclonic shear $V_c > V_w$, the slope of the surface will be negative and physically inadmissible. From these results it follows that:

- wind change through a front will be such that the component of wind parallel to the front, taken as positive with cold air to the left, increases to the left;
- vorticity is positive (cyclonic) in a frontal zone;

- fronts cannot occur in regions of negative (anticyclonic) vorticity - that is, in anticyclones or ridges; and
- surface fronts are generally accompanied by maximum upper-level winds (the westerly jet) near the tropopause to the poleward side of the surface front.

As will be discussed later, fronts are associated with distinctive assemblages of cloud and sequences of weather. They are fundamental features of atmospheric structure and circulation and, except in equatorial regions, are major determinants of weather variability.

Fronts are zones of *baroclinicity* in which the pattern of isobars and isotherms (or isopycnals, which are lines of constant density), either in the vertical or the horizontal, is such that the pressure and temperature (or density) lines intersect to form *solenoidal fields* in which circulations are readily generated. Figure 10.25 (*centre*) gives a schematic representation of a *baroclinic* field of the type that frequently occurs in middle latitudes. In such a case, isobars and isotherms (or isopycnals) are not parallel, but intersect. In contrast, a *barotropic* field is one in which isobars and isotherms (or isopycnals) are parallel (Fig. 10.25, *left* and *right*). In such a field there can be no thermal wind and the geostrophic wind will not increase with height.

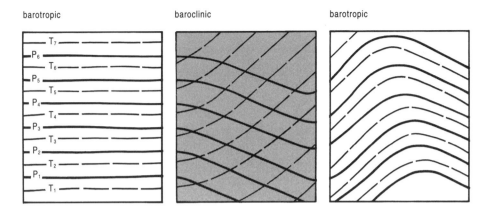

barotropic   baroclinic   barotropic

Fig. 10.25 Temperature (*T*) and pressure (*P*) distributions in barotropic and baroclinic stratifications.

Between the tropics of Cancer and Capricorn the atmosphere is close to barotropic. The near barotropicity of the tropics and the strong baro- clinicity of higher latitudes is basic to the quite different behaviour in the weather and climate in the two regions.

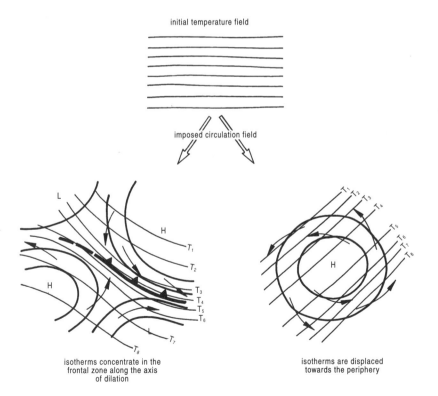

initial temperature field

imposed circulation field

isotherms concentrate in the frontal zone along the axis of dilation

isotherms are displaced towards the periphery

Fig. 10.26 Deformation of an initially uniform temperature field to produce frontogenesis (*left*) and frontolysis (*right*).

It is also necessary to distinguish the more general case of an *equivalent barotropic* atmosphere. This is one in which horizontal temperature gradients exist, but with the constraint that thickness contours are everywhere parallel to height contours and not to isotherms. In such an atmosphere the geostrophic wind speed may in fact vary with height, although there will be no change in the wind direction. Centres of high and low pressure will also be centres of positive and negative temperature anomalies and the speed of the geostrophic wind will increase with height in such systems; in cold highs and warm lows the opposite is the case and the wind may even reverse with height.

Given an initial uniform field of temperature, whether or not *frontogenesis* will occur is dependent on the nature of the *deformation field*. If the circulation is such as to concentrate the isotherms by bringing together air masses of different temperature, as in Figure 10.26 (*lower left*), then a front will form along the

axis of dilation or outflow. On the other hand, a circulation such as shown in Figure 10.26 (*lower right*) would act to concentrate isotherms along the periphery of the anticyclone and no front would form within the system. If the deformation field is such that the deformation has the effect of separating contiguous air masses, existing fronts will weaken and decay in the process of *frontolysis*.

CYCLONES AND CYCLOGENESIS

According to the classical Bjerknes or Norwegian model, the development of extratropical depressions or cyclones occurs along a front separating warm air (of tropical origin) from cold air (of polar or subpolar origin) in a recognizable sequence of events illustrated in Figure 10.27. In stage (*b*) a wave begins to form in both the surface front and in the upper-level jet lying on its poleward side. As low-level

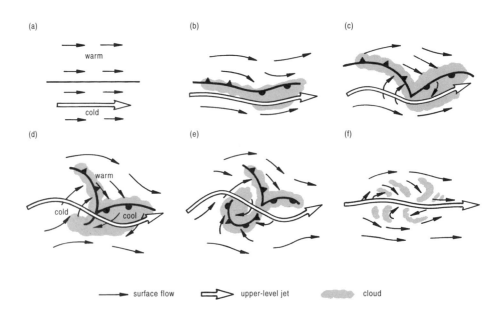

Fig. 10.27   The life cycle of a mid-latitude cyclonic depression in the southern hemisphere according to the classical Norwegian model (after McIntosh and Thom, 1969; Flohn, 1969, and transposed and modified for the southern hemisphere).

north-westerly air within the warm sector converges to replace air rising in frontal uplift in the middle and upper atmosphere, the planetary vorticity given by the Coriolis parameter is concentrated and the whole mass of air begins to rotate cyclonically with the generation of local relative vorticity. The wave is baroclinically unstable and grows steadily in amplitude so that by stage (c) these developments have been enhanced. Striking patterns of high- and middle-level cloud and precipitation form about the rapidly deforming trailing cold front and leading warm front. The mature stage of the cyclone has been reached. The frontal zones slope upward on the cold side of their surface positions, so that in the upper troposphere they lie several hundred kilometres poleward of their surface expressions. A temperature gradient exceeding 5 °C per 100 km may be observed across a local cold front making a sharp thermal boundary between the warm-sector air and that in the cold sector behind (and to the west of) the front. Ahead of the warm sector of the depression the warm front is a weaker, more diffuse boundary. Warm fronts rarely, if ever, affect southern Africa and, given the general north-west to south-east alignment of mature depressions, affect only regions to the south of Africa, for example Gough and Marion Islands. Across the cold front, in addition to the temperature dropping suddenly, pressure reaches its minimum before rising sharply. The wind swings from north-west to south-west, often with the passage of a gust front, and atmospheric stability drops. Storms may accompany the passage of the front, but not necessarily over South Africa. Often the front may be associated only with stratiform cloud.

Two types of cold front need to be recognized. *Ana-fronts* occur where the warm air in the warm sector is rising relative to the frontal zone and the fronts are consequently very active. *Kata-fronts* are accompanied by subsidence of air relative to both the cold air mass and frontal zone, are less intense and are usually associated with a subsidence inversion (Fig. 10.28). Cold fronts tend to travel more rapidly than their warm-front counterparts and thus overtake them to merge in an *occluded front*. In stage (d) of Figure 10.27 this front begins to bend back on itself and the characteristic *comma cloud* formation begins to appear and develop in stage (e). At this time the jet stream has attained its most prominent wave-like configuration. Thereafter decay sets in and by stage (f) the jet has straightened out, so discontinuing the upper-level convergence and divergence necessary to complete the large-scale convective forcing (see Fig. 10.18). The cyclonic storm collapses quickly and all that remains is a series of inward-spiralling cloud bands. Although precipitation tends to be associated with the fronts, it is more often than not related to a greater degree with the position of the jet stream and the upper-level forcing.

The lifetime of an individual frontal low is usually three to five days, during which time it may travel up to 10 000 km. The mature stage is often attained within 24 hours. The comma-cloud stage tends to be much more persistent. The schematic relationship between low- and upper-level flow at this stage is given in Figure 10.28. The upper-level divergence ahead of the trough axis plays a vital role in the dynamics of the system. A cross-section along AB and through the warm sector shows the relative slopes of the fronts, the clouds associated with the uplift produced by their movement (in the case of cold fronts) and by air movement over them (in the case of warm fronts), and the regions of subsidence and suppressed cloud development for both ana- and kata-fronts.

Satellite photographs allow the cloud patterns associated with the various stages of development of southern hemisphere extra-tropical cyclones to be readily identified (Fig. 10.29). The Norwegian model for surface conditions and its subsequent modification to incorporate upper-level jets is a persuasive descriptive model. It fails, however, to explain fully all the characteristics of mid-latitude depressions, notably the mesoscale details of mid-latitude precipitation systems within the simple archetypes of warm, cold, and occluded fronts, and secondly, the flow dynamics associated with comma clouds. The stages involved in the development of these cloud features include the early-stage formation of a high-level cloud canopy, a westward extension of a large cloud shield into the characteristic comma form and then the final development of low-level stratiform cloud to complete the accentuated comma

(Fig. 10.30).

It is possible to model horizontal deformation in atmospheric flow in such a way as to replicate the streaks of cloud that form approximately along the direction of flow at cloud level. The typical streaks are due mainly to the strong deformation of flow in baroclinic zones. Figure 10.31 shows an example of the model-

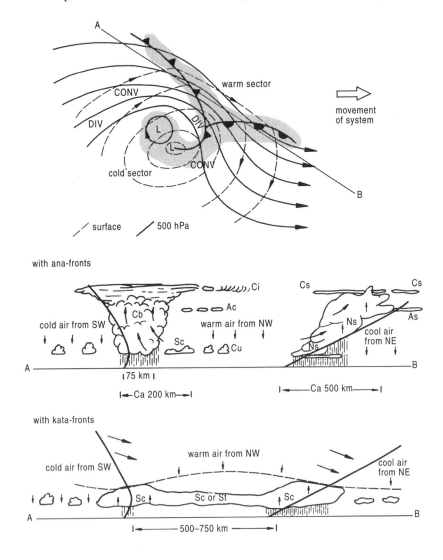

Fig. 10.28 Surface isobars (broken lines) and upper-level geopotential height contours (solid lines) in the late-mature stage of a depression (*upper*), together with schematic cross-sections through the warm sector along AB (*centre*) with ana-fronts (warm air rising relative to cold air masses) and (*lower*) kata-fronts (warm air sinking relative to cold air masses) (sections modified after Pedgley, 1962).

ling of the deformation of a grid of squares of size 300 km on an initial streamline pattern with the deformation taking place at successive 6-hour intervals. After 24 hours, the resulting drawing out of the gridded sheet of fluid by the large-scale flow into long filaments bears a striking resemblance to the typical comma-cloud form illustrated in Figure 10.31 (*lower right*).

By subtracting the velocity of movement of the baroclinic disturbance that constitutes the depression from the overall measured field of horizontal motion, the flow relative to the disturbance may be determined. A useful concept to emerge from relative flow analysis of this kind is that of the *conveyor belt*, a broad stream of air originating in the lower layers of the atmosphere before undergoing baroclinic

Fig. 10.29 Satellite images to illustrate, *upper*, the early, open-wave stage in the development of a depression, and the beginning of the comma-cloud stage, and *lower*, the mature comma-cloud stage and the dissipating stage of the depression (after Troup and Streten, 1972).

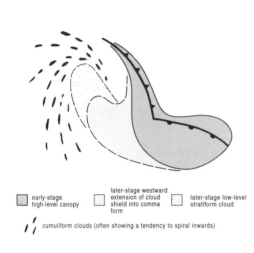

early-stage
high-level canopy

later-stage westward
extension of cloud
shield into comma
form

later-stage low-level
stratiform cloud

cumuliform clouds (often showing a tendency to spiral inwards)

Fig. 10.30 Schematic stages in the development of a comma-cloud pattern in a mid-latitude wave disturbance (after Carlson, 1980).

slantwise (or *sloping*) *ascent* over the cold air in the depression and thence rotating to end up as a current streaming ahead of the disturbance in the upper troposphere. Both *warm* and *cold conveyor belts* may be distinguished.

In the warm conveyor, warm air is drawn into the cloud belt of the depression from the convective boundary layer in subtropical or tropical latitudes. In so doing it conveys large quantities of heat, moisture and westerly momentum poleward and upward. The air rises into the middle atmosphere as it travels within the cloud belt and eventually, after rotation, produces a deck of upper tropospheric cirrus which decays ahead of the frontal system. In low levels the air flows within the warm sector and along the length of the cold front as a low-level boundary-layer jet. The warmest air originates farthest north and is normally found immediately ahead of the surface cold front. As the airstream begins to rise it may undergo

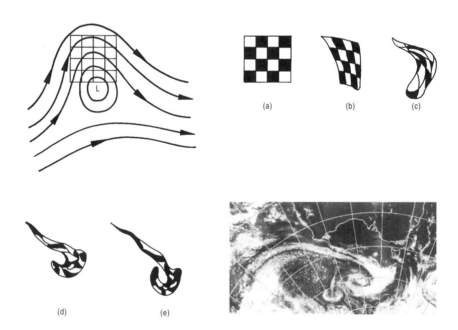

Fig. 10.31 Deformation, as computed by a barotropic model, of motion at the 500 hPa surface. The diagrams show the initial streamline pattern with a grid of squares of side 300 km superimposed and the deformation of that grid thereafter at successive 6-hour intervals (modified after Ludlam, 1980). The satellite photo shows a cloud pattern similar to the deformation produced in stage (*e*).

rearward and forward sloping ascent.

With *rearward sloping ascent* some or all of the warm airstream rises with a rearward component above a clearly defined advancing front of cold air and corresponds to the classical ana-cold front (Fig. 10.32, *upper*). The warm air in the boundary layer is lifted abruptly at up to several metres per second within a narrow belt adjacent to a surface front. The region is one of intense cyclonic shear on the western (cold front) side of the pre-frontal, low-level jet. The air only rises some 2–3 km during its abrupt oblique ascent across the front. It then undergoes less rapid slantwise ascent at rates of only tens of centimetres per second. The two regions of ascent produce a narrow band of heavy rain at the surface front and a broad belt of much less heavy rain behind (and sometimes even ahead of) the front. Subsidence of cold air from middle levels both within and down the frontal zone acts to enhance uplift generally.

With *forward sloping ascent* the situation is more complicated and corresponds to a kata-cold front condition (Fig. 10.32, *lower*). The main ascent in the warm conveyor occurs ahead of the cold front. The warm conveyor is the dominant cloud- and precipitation-producing flow in mid-latitude depressions. Since the air is warm and quite moist it does not have to ascend much before it reaches saturation and fills with stratiform cloud. As air in the warm conveyor ascends, not vertically, but extensively sloping upward, eventually the flow turns anticyclonically (i.e., to the left and eastward) above the air in the cold conveyor.

There is a secondary, cold-conveyor, cloud-producing airstream originating in the anticyclonic, low-level flow to the east of the cyclone in the cool air ahead of the warm front. The air is dry and evaporates the precipitation falling from the cloud in the warm conveyor, thereby moistening the cold air as it undercuts the warm. As the cold conveyor travels westward toward the cyclonic centre, it begins to ascend, reaching into the middle troposphere near the apex of the warm sector where it continues to ascend and begins to curve anticyclonically.

Thereafter the airstream may *either* descend cyclonically around the cyclone centre producing the inward-circling edge to the comma cloud before resulting in cloud-free air with further descent (see Fig. 10.29, *lower right*), or

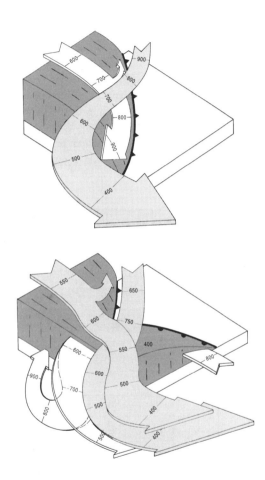

Fig. 10.32 *Upper:* the warm conveyor belt with rearward, slantwise, sloping ascent over a cold front corresponding to the classical ana-front situation; *lower:* the warm and cold conveyors with forward-sloping ascent corresponding to a kata-front situation (after Browning, 1985, but modified and transposed for the southern hemisphere). The broken cold conveyor shows the alternative cyclonic downward sweep of the airstream that produces the comma-cloud swirl. Approximate pressure levels (hPa) are indicated along the conveyors.

the airstream may merge with the warm conveyor belt, as indicated in Figure 10.32 (*lower*). The great anticyclonic sweep of the warm conveyor produces the pronounced flange of cloud that contributes to the comma form. The typical sharp, right-hand cloud edge, leading to the comma in the lower- and middle-level warm conveyors (see Fig. 10.29, *upper right* and *lower left*) is usually on the immediate poleward side of the low-level jet of warm moist air.

Overrunning both belts is middle-level air, recently descended and with a low wet-bulb potential temperature. Part of this air wedges beneath the warm conveyor; part overruns it in the middle troposphere leading to the generation of potential instability, which is realized as convection once sufficient further lifting over the warm conveyor has occurred. Flow patterns

are complicated and have been referred to by some as the 'tangled skein'.

ANTICYCLONES AND THEIR EFFECT ON WEATHER

Continuity requires that on every scale of motion each type of circulation has a counterpart. The counterpart to low-pressure cyclones is the high-pressure anticyclone. As has been shown earlier, anticyclones are large systems with centres characterized by weak pressure gradients and no frontal activity. They are features associated with subsidence of air, compression and adiabatic warming of the local atmosphere. Consequently, they are, in general, fine-weather systems (Fig. 10.33). When located over central

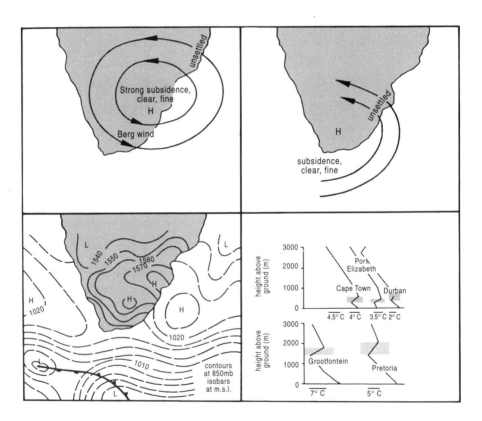

Fig. 10.33 *Upper*: the dependence of anticyclonic weather over southern Africa on the position of the anticyclone with respect to the land mass; *lower*: subsidence inversions at 13:30 produced by subsidence in a winter (July) anticyclone.

South Africa, subsidence and settled fine and dry weather prevail over large areas. Berg winds are likely to occur along the southern and south-eastern coasts. In some instances when the centre of the anticyclone is such that airflow is directly onshore, localized unsettled conditions may result. If the high is centred to the south or south-west, then ridging of the system over the land may cause widespread unsettled weather and rainfall, particularly in coastal areas (Fig. 10.34, *left*).

Warm-cored anticyclones increase in intensity with height, just as cold-cored highs are weak and shallow. However, an upper-level, warm-cored anticyclone diminishes quickly towards the ground, whereas the upper cold-cored high may increase in intensity toward the ground. The areas of high-pressure centres always slope upward in the direction of the warmest air, which is to the west over southern Africa. Upper-level anticyclonic curvature in airflow, or the occurrence of closed upper-level anticyclones, is important in providing upper-level divergence to compensate for low-level convergence and to provide the upper-level forcing for uplift. This is seen in upper-level anticyclonic curvature of the warm and cold conveyor belts in extra-tropical cyclones, and is even more evident in the case of tropical cyclones.

Anticyclones may be semi-permanent features of the subtropics due to large-scale planetary subsidence of air on the poleward side of the Hadley cell (see Chapter 11), or they may be travelling systems intruding into the westerlies. Often the travelling anticyclones stagnate and cause blocking. In so doing they exercise a major control of weather. However, it is their production of regularly-occurring subsiding air that is one of their main features over southern Africa. Almost all prolonged heat waves are caused by anticyclonic weather. All droughts, whether on the scale of days, seasons or years, may be linked in one way or another to a predominance of anticyclonic conditions. They are a dominant control of southern African weather and climate and exert this control through the subsidence they produce. This subsidence is responsible for high atmospheric stability and frequent subsidence inversions.

The subsidence inversions shown for a specific day and time in Figure 10.33 are characteristic. Such inversions are common, even in summer when their base heights are somewhat more elevated than in winter (Fig. 10.35). Typical winter subsidence inversion bases rise from 750 m over the Namib to more than 1 750 m over the central plateau areas. Over the oceans, similarly forming trade-wind inversions

Fig. 10.34 Characteristic cloud patterns to show; *left*, a coastal cloud mass aligned parallel to the coast and Escarpment that results from onshore anticyclonic airflow behind a retreating cold front; and *right*, suppressed cloud growth over the Atlantic Ocean as a result of large-scale subsidence.

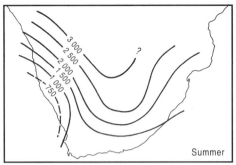
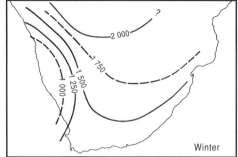

Fig. 10.35 The height of the base of the subsidence inversion over southern Africa in summer and winter.

occur on average throughout the year at heights of 500 m over the Benguela Current rising to 1 500 m over the middle of the subtropical South Atlantic Ocean. The subsidence inversion typically caps the surface boundary layer and greatly inhibits convection and dispersion of pollutants (see Fig. 14.6). Clouds forming in regions of anticyclonic subsidence are suppressed vertically and the broken *stratocumulus* or *altocumulus lacunosus* clouds frequently observed in areas of such subsidence often have a characteristic appearance on satellite photographs (Fig. 10.34, *right*).

TROPICAL CYCLONES

Tropical revolving storms (*cyclones* in the Indian Ocean, *hurricanes* in the Atlantic Ocean and *typhoons* in the Pacific Ocean) are notoriously destructive because of the extremely strong winds they produce and the huge waves and abnormally high tides they generate. These may cause massive inundation in low-lying areas near the coast. The storms last for several days and produce copious rainfall. When the systems pass over the land, the rainfall they deposit may be the order of mean annual rainfall. Such was the case when cyclone Domoina swept over KwaZulu-Natal in January 1984 (Fig. 10.36).

Fig. 10.36 Tropical cyclones. *Left:* over the South Pacific Ocean; *right:* cyclone Domoina over northern KwaZulu-Natal in January 1984.

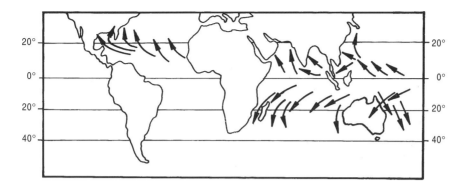

Fig. 10.37 Principal tropical-cyclone tracks in the Indian and other oceans (after Bergeron, 1954).

Typical tropical cyclones are much smaller than their extra-tropical counterparts and have a diameter of about 650 km. Sea-level pressures may drop to 900 hPa and below. Hurricane-force winds are defined arbitrarily as those above 33 m s⁻¹; in many storms the wind speed may exceed 50 m s⁻¹. Within cyclones cumulonimbus clouds are organized into inward spiralling bands and tower to heights in excess of 12 km. Immense convective activity is concentrated in the storms. Because of their destructive potential and since they are almost unique among tropical weather systems in having a sharply defined structure, tropical cyclones have been studied extensively. Despite this, their exact method of formation remains incompletely understood. A number of conditions are necessary, but not necessarily sufficient, for their initiation. One such is an extensive ocean area with sea-surface temperature in excess of 27 °C. That there is also a definite link between the seasonal position of the equatorial trough and the zone of cyclone formation is clear from the fact that no hurricanes form in the tropical South Atlantic Ocean (where the trough never lies south of 5°) or typhoons in the eastern South Pacific Ocean (where the trough remains north of the equator). Most tropical cyclones form near the trough line in a zone 5°–10° latitude poleward of it. Nearer the equator, where the Coriolis parameter tends to zero, and further poleward, where zones of strong vertical

wind shear occur beneath jet streams, the development of intense organized vortices is inhibited. In the South Indian Ocean, cyclones form to the west of Australia and are carried westward in the tropical easterlies before recurving southward in the South Indian Anticyclone as they approach Africa (Fig. 10.37). In the Australian/ Indian/Pacific region, 97 per cent of tropical cyclones form in the Monsoon Trough between the equatorial westerlies and tropical easterlies overlain by a 200 hPa ridge. Cyclones are mainly a summer and autumn phenomenon.

Tropical disturbances mostly take the form of easterly waves, some of which develop into cold-cored tropical storms (equivalent to weak mid-latitude depressions). Only a few of these evolve into fully-fledged warm-cored tropical cyclones (Fig. 10.38). Tropical cyclones almost always appear to develop from pre-existing disturbances, possibly due to the leaking into lower latitudes of some baroclinicity in an otherwise barotropic atmosphere. A key to the development of the systems appears to be the presence of upper-level anticyclonic divergence to produce the high-level outflow (see Fig. 10.6), which in turn allows the development of very low pressure and high winds near the surface (Fig. 10.39). The centre of the storm is characterized by the eye, a region of quiet in otherwise tempestuous fury. The eye is 30 to 50 km in diameter and is a region of pronounced subsidence of warm, dry air in an often cloud-free

Fig. 10.38 Stages in the horizontal development of a tropical cyclone.

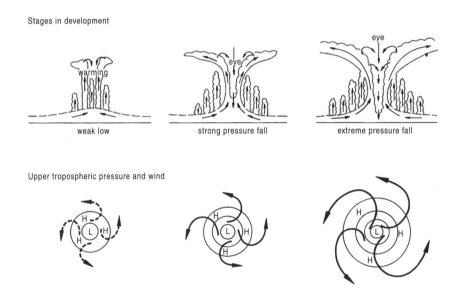

Fig. 10.39 Schematic stages in the vertical development of a tropical cyclone (after Palmén and Newton, 1969).

sky. The warm core is caused not only by the subsidence, but also by the massive release of latent heat of condensation in up to 200 cumulonimbus towers within and around the storm (Fig. 10.40). The warm core is vital to the growth of cyclones because it acts to intensify the upper-air anticyclone, which in turn produces a feedback effect by stimulating a low-level influx of heat and moisture, which fuels and intensifies the convection, latent-heat release and so on. Heat is converted into potential energy by a thermally direct circulation and only a small fraction, about 3 per cent, is converted into kinetic energy. The remaining ener-

gy is transported away in the upper-level anticyclonic circulation at about 200 hPa (12 km).

If the rotating air near the surface were to conserve its angular momentum, wind speeds would become infinite at the centre. Clearly this does not happen. Instead, the strong winds in the region of the cloud wall surrounding the eye are in cyclostrophic balance (see Chapter 8), with the small radius providing a strong centrifugal force. The air rises when the pressure gradient can no longer force the air inward (Fig. 10.40). Essential conditions for the maintenance of tropical cyclones include the low-level supply of heat and moisture combined with a

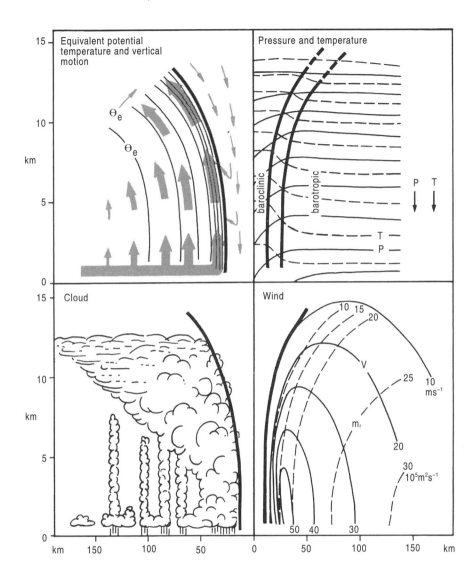

Fig. 10.40 A radial section through a mature tropical cyclone to show: *upper left*, equivalent potential temperature, θ_e, and vertical motion; *upper right*, pressure, P, and temperature, T; *lower left*, cloud structures; *lower right*, tangential velocity, V, (solid lines) and angular momentum, M_r (dashed lines) (after Palmén and Newton, 1969).

low frictional drag at the surface, the massive release of latent heat through condensation and the eviction of air aloft. Remove any of these prerequisites and the storm will decay rapidly. Such is the case if the cyclonic vortex moves over a region of low sea surface temperature or over land. In recurving over land the vastly increased friction causes the system to fill and at the same time the moisture supply is cut off. Without the latent heat derived from the condensation of moisture, the storm declines. However, it may still produce copious rains, even though its total destructive capability will have diminished substantially.

ADDITIONAL READING

GORDON, A., GRACE, W., SCHWERDTFEGER, P. and BYRON-SCOTT, R. 1998. *Dynamic Meteorology.* London: Arnold.

HOUGHTON, J. T. 1997. *The Physics of Atmospheres.* Cambridge: Cambridge University Press.

SCORER, R. S. 1997. *Dynamics of Meteorology and Climate.* Chichester: Wiley.

BARRY, R. G. and CHORLEY, R. J. 1996. *Atmosphere, Weather and Climate.* 7th Edition. London: Methuen.

SCHNEIDER, S. H. (ed.). 1996. *Encyclopaedia of Weather and Climate.* Vols. 1 and 2. Oxford: Oxford University Press.

STURMAN, A. S. and TAPPER, N. 1996. *The Weather and Climate of Australia and New Zealand.* Melbourne: Oxford.

JAMES, I. N. 1994. *Introduction to Circulating Atmospheres.* Cambridge: Cambridge University Press.

MCILVEEN, R. 1992. *Fundamentals of Weather and Climate.* London: Chapman and Hall.

PEIXOTO, J. P. and OORT, A. H. 1992. *Physics of Climate.* New York: American Institute of Physics.

ATKINSON, B. W. (ed.). 1981. *Dynamical Meteorology, an Introductory Selection.* London: Methuen.

LUDLAM, F. H. 1980. *Clouds and Storms.* University Park: Pennsylvania State University Press.

MCINTOSH, D. H. and THOM, A. S. 1978. *Essentials of Meteorology.* London: Wykeham Publications.

WALLACE, J. M. and HOBBS, P. V. 1977. *Atmospheric Science, an Introductory Survey.* New York: Academic Press.

BYERS, H. R. 1974. *General Meteorology.* 4th Edition. New York: McGraw-Hill.

PALMÉN, E. and NEWTON, C. W. 1969. *Atmospheric Circulation Systems.* New York: Academic Press.

11

THE GENERAL CIRCULATION OF THE SOUTHERN HEMISPHERE

SCALES OF MOTION

Weather-producing systems are distinguished by their characteristic scales and circulation patterns. These patterns are caused by energy imbalances which produce temperature and, therefore, pressure variations. Pressure gradients generate winds, which in turn are influenced by such factors as the earth's rotation, frictional drag, temperature variations, moisture variations and mountain barriers. It is most convenient to think of these wind systems as occurring at a variety of scales, all of which are interdependent. At the *microscale*, airflow is characterized by small, short-lived, high-frequency turbulent eddies (Fig. 11.1). These range in size from a few centimetres to a kilometre and, near the surface in the boundary layer, tend to be generated by variations in surface roughness and heating. At the *mesoscale*, which ranges from a kilometre to a few hundred kilometres, local wind systems such as land–sea breezes, mountain–valley winds and convective phenomena such as tornadoes and thunderstorms may persist for a number of hours in the day. *Macroscale* circulations extend over a few hundred to several thousand kilometres in the forms of extra-tropical cyclones and anticyclones and may persist for a week or more. At the *large macroscale* the atmospheric circulation is determined by long waves, which extend from a few thousand to more than ten thousand kilometres and persist over a period of weeks to months. At the *global scale* the wind systems, which in their simplest form are commonly described by the strength and location of the easterly wind belt in tropical latitudes, the westerly wind belt in middle latitudes and the polar easterlies in polar latitudes, are called the general circulation. The nature of the general circulation in the southern hemisphere, particularly in the southern African sector, is the subject of this chapter.

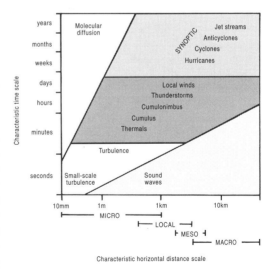

Fig. 11.1 Time and space scales of atmospheric disturbances (after Smagorinsky, 1974).

ENERGY AND MOMENTUM BALANCE CONSIDERATIONS

The energy imbalance between tropical and higher latitudes is such that an energy surplus exists between the equator and about 39° S, whereas at higher latitudes a deficit occurs (Fig. 11.2, *upper left*). Since the balance between incoming and outgoing radiant energy is almost constant over long periods of time, mechanisms must exist to transfer energy between equatorial and polar regions. The total energy transfer is greatest between 30° and 40° latitude in both hemispheres and is made up of various components within the atmosphere–ocean system (Fig. 11.2, *upper right*). Sensible heat transport is largest when meridional temperature gradients are strongest; that is, in middle latitudes. The transport has a double maximum with high-level transfer of sensible heat being particularly significant over the subtropics, whereas

the slightly larger maximum at 50°–60° is related to travelling low-pressure disturbances (eddies). The transport of energy in latent heat form (taken up in evaporation and released in condensation) is highest in tropical latitudes, particularly in the southern hemisphere. Ocean currents account for an important poleward transfer of energy, particularly at low latitudes.

The energy imbalance in the atmosphere causes the air to circulate on a global scale. The mean circulation breaks down into standing and transient disturbances. Together the mean flow and superimposed eddies transport and redistribute energy and momentum between the poles and the equator. In Chapter 8 it was shown that angular momentum was given by $m\omega r^2$, where m is the mass of rotating air, ω is the angular velocity of rotation and r is the radius of curvature of the rotating mass. If no torques are present (i.e., no forces acting to produce rotation), the angular momentum will be

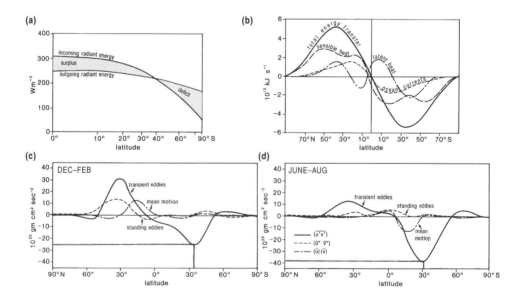

Fig. 11.2 (a) Zonally averaged fluxes of solar and terrestrial radiation absorbed and emitted by the earth–atmosphere system (after London and Sasamori,1971); (b) average annual latitudinal distribution of the components of the poleward energy flux (after Sellers, 1965); (c) and (d) vertically integrated components of the horizontal momentum transport in December–February and June–August (after Newell et al., 1972).

conserved so that, per unit mass of air,

$$\omega r^2 = \text{constant} \qquad (11.1)$$

or

$$Vr = \text{constant} \qquad (11.2)$$

where V is the linear (tangential) velocity. As in the case of the ice skater rotating faster with arms vertical (small radius), when air moves poleward over the earth the radius of rotation about the earth's axis, r, decreases and the velocity of rotation increases, at the same time becoming more westerly as a consequence of the Coriolis deflection. Relative westerly momentum is thus transferred to high latitudes. Similarly, with equatorward motion, winds become more easterly and slow down as r increases. Thus relative easterly momentum is transferred to low latitudes.

In the upper poleward-moving air in the Hadley cells, the conservation of angular momentum does much to maintain the subtropical jet streams. Were the principle of conservation to be acting alone, air moving from zero velocity at the equator would acquire a speed of about 130 m s^{-1} by latitude 30°. That this does not happen is due to the collapse of a purely meridional-type circulation in the tropics into smaller circulations, which take the form of large-scale eddies and waves in middle latitudes. As the upper branch of the Hadley cell transports angular momentum poleward, it becomes impeded more and more as it reaches into the subtropics by air moving equatorward from higher latitudes. There is thus a transition from a smooth poleward flux of zonal angular momentum to an eddy flux.

Actual transport of momentum in the atmosphere may be broken down into various components: that due to the mean circulation; that due to standing eddies (and waves); and that due to transient (moving) eddies. The relative roles of these components at different times of year is apparent in Figure 11.2 (*lower left* and *right*). At all times of the year, except in the tropics, the mean circulation term is less significant than the eddy terms. In the tropics the mean circulation dominates through the persistence and steadiness of the tropical easterly trade winds. Standing eddies likewise exert their greatest influence in the near-equatorial regions. Transient eddies, taking the form of perturbations in the westerlies, are the main agent of horizontal momentum transfer. During summer (December–February) their efficacy is much less than during winter (June–August). In summer the maximum transient eddy transport of momentum occurs at about 35° S; in winter the latitude of greatest transport is nearer to 30° S.

Within the atmosphere a clear picture emerges. In the barotropic tropics, the mean circulation and standing eddies superimposed thereon transport heat out of the region mainly in the form of latent energy. Within the baroclinic middle latitudes, the global energy balance is maintained by the transport of sensible heat in synoptic-scale transient eddies. Consideration will now be given to the mean circulation, the standing eddy and the transient eddy components of the general circulation of the southern hemisphere.

MEAN TEMPERATURE AND MOISTURE CONDITIONS

In the southern hemisphere summer (December–February), the highest near-surface, zonally averaged temperatures occur in the tropics when the thermal equator (line of highest surface temperature) is in that hemisphere. The lowest temperatures (less than −80 °C) occur at the 100 hPa level between about 0°–10° S (Fig. 11.3). The temperature gradient between equator and pole is pronounced. In contrast, in winter (June–August) the highest near-surface temperatures occur in the northern hemisphere, while upper-air polar temperatures at the 50 hPa level drop to below −85 °C and the temperature gradient between the equator and pole reverses at this level. The summer to winter reversal in the meridional upper-atmospheric temperature gradient produces corresponding wind reversals through the thermal wind effect.

The tropics are the main source of atmospheric moisture (Fig. 11.3). Most of this is confined to levels below 500 hPa. In summer, zonally averaged specific humidities exceeding 1 g kg^{-1} occur below a height of 9 km at 10° S and below 3 km at 70° S.

THE PRESSURE FIELD

In both hemispheres, and in all seasons, the dominant feature of the general circulation of the atmosphere is the large tropospheric circumpolar vortex of westerly winds, which reaches its maximum in the upper troposphere. At any one time a proliferation of perturbations

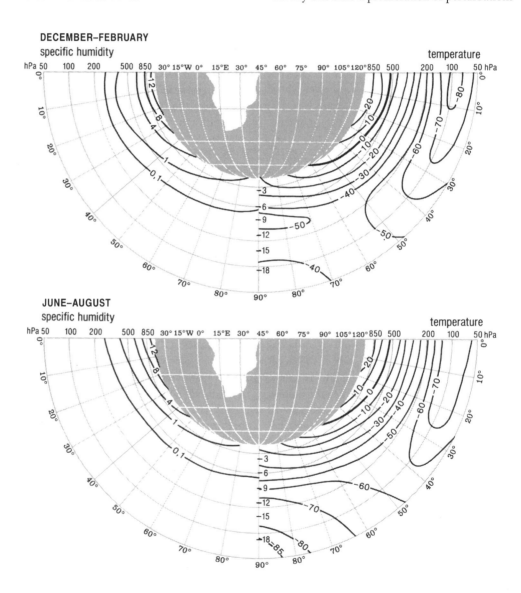

Fig. 11.3 Meridional sections of zonally averaged specific humidity, g kg^{-1} (data from Oort, 1983), and temperature, °C (data from Newell et al., 1972), in December–February and June–August.

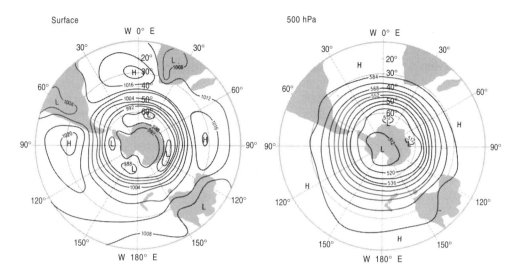

Fig. 11.4 Mean pressure fields in January. *Left:* isobars (hPa) at sea level; *right:* contours of the 500 hPa surface (gpdam) (after Taljaard et al., 1969).

is embedded within the vortex, taking the form of the waves and eddies that control daily weather. Mean pressure fields seldom reflect these disturbances. The mean January pressure field for the southern hemisphere is given in

Figure 11.4, that for July in Figure 11.5. The differences between the southern-hemisphere mean January and July conditions are not pronounced. At all times the surface pressure field is characterized by large, semi-permanent high-

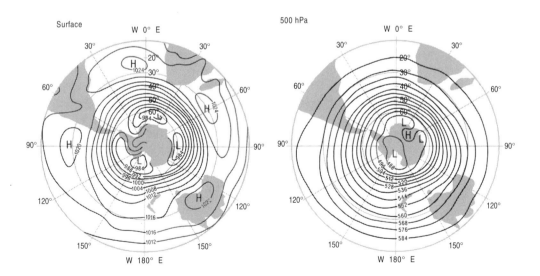

Fig. 11.5 Mean pressure fields in July. *Left:* isobars (hPa) at sea level; *right:* contours of the 500 hPa surface (gpdam) (after Taljaard et al., 1969).

pressure cells in the subtropics: the South Atlantic Anticyclone, the South Indian Anticyclone and the East Pacific Anticyclone. In the lower troposphere, over land at these latitudes, pressures tend to be lower in summer than in winter as a result of surface-heating effects. The continental influence produces different and out-of-phase seasonal oscillations in the pressure fields over the subtropical continents and middle-latitude oceans throughout the tropo-

sphere. South of the high-pressure cells, pressure drops to a minimum in the circumpolar trough between 60° S and 70° S where four or five minima may occur.

At the beginning of summer, from September to December, pressure drops throughout the southern hemisphere, except over Antarctica (Fig. 11.6). From December to March it rises strongly over the southern continents, while dropping in middle latitudes (particularly over

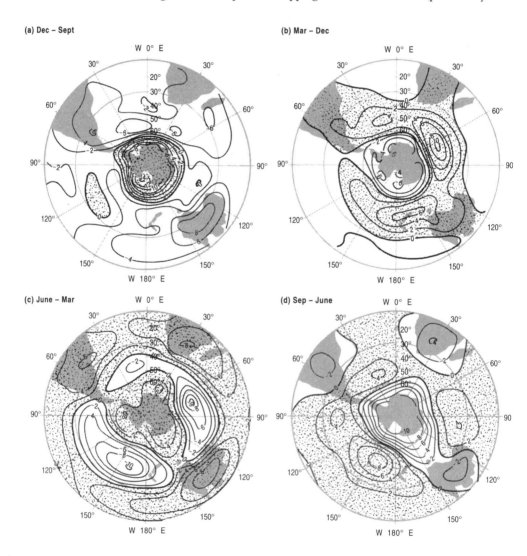

Fig. 11.6 Differences in mean sea-level pressure (hPa) for: (a) December minus September, (b) March minus December, (c) June minus March, and (d) September minus June (after Van Loon and Rogers, 1984).

the Indian Ocean to the east of Marion Island and in the Pacific). Finally, from June to September the pressure falls over the continents and rises over mid-latitude oceanic areas.

These pressure variations occur in response to two seasonal pressure waves. The first is an *annual (yearly) wave* that predominates over subtropical continental and adjacent ocean areas, accounting for over 95 per cent of the annual variation of pressure (at about latitude 30° S) (Fig. 11.7). The wave peaks in winter over the continents when the land is coldest relative to the oceans. To the south and centred

on about 40°–50° S the annual variation in pressure occurs predominantly in a *semi-annual (half-yearly) wave*. Locally the semi-annual cycle may account for over 75 per cent of the annual variance in sea-level pressure (Fig. 11.7). In the vicinity of Gough and Marion Islands the wave manifests itself prominently with equinoctial peaks. This temporal variation is linked to a comparable spatial oscillation in the Antarctic trough as a consequence of the trough moving poleward from December to March and from June to October and equatorward during spring and autumn. Between March and June

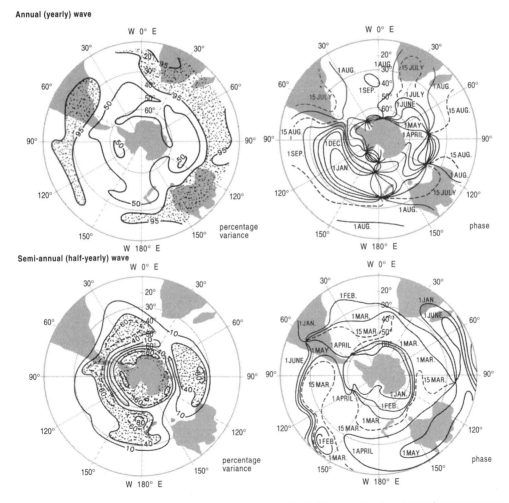

Fig. 11.7 Amplitude (hPa), explained variance (per cent) and phase (date) of the annual and semi-annual pressure waves in the southern hemisphere (after Van Loon and Rogers, 1981, 1984).

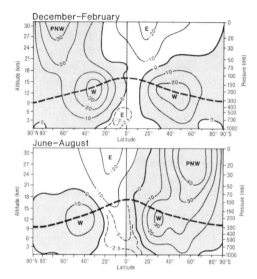

Fig. 11.8 Mean zonal tropospheric and stratospheric winds (m s⁻¹) during December–February and June–August (after Newell et al., 1972). The heavy broken line indicates the tropopause.

the sea-level pressure rises in the region to the south of Gough Island and falls east of Marion Island. In the former area pressure continues rising between June and September, but not as strongly, and in the latter area it begins rising strongly. The fact that the phase of the semi-annual pressure wave reverses from two peaks with equinoctial maxima in middle latitudes to solsticial maxima south of 60° S implies an exchange of mass between polar and middle latitudes that reverses twice a year.

The South Atlantic and South Indian Anti-cyclones vary significantly in position throughout the year. In the Indian Ocean the high-pressure cell undergoes a semi-annual variation in its latitudinal movement (in a response associated with the changes taking place east of Marion Island), but an annual oscillation in its longitudinal displacement. Both latitudinal and longitudinal movements of the high in the Atlantic Ocean show a semi-annual oscillation. On average the South Atlantic Anticyclone is about 3° further north than its Indian Ocean counterpart with both cells moving 5°–6° northward in winter. The annual longitudinal shifts in position of the Atlantic high are of the order of 7° to 13° and do not materially affect the weather of the subcontinent. In contrast, it has long been held that the much greater shift (24° to 30° of longitude) of the South Indian Anticyclone to the west during winter exerts a controlling effect on the weather and climate of the northern and eastern parts of South Africa.

Undoubtedly the strong anticyclonic winter circulation has such an effect. Some research

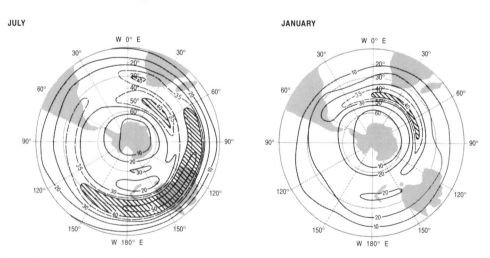

Fig. 11.9 The zonal component of the geostrophic wind (m s⁻¹) at 200 hPa in July and January (after Van Loon et al., 1971).

workers, however, hold the view that the anti-cyclone affecting southern Africa in winter is not actually the South Indian Anticyclone, but one separate from and independent of it, owing its origin to anticyclogenesis in a poleward stream of subsiding air originating in the Indian monsoon system of Asia.

THE WIND FIELD

Mean winds are best considered by resolving them into zonal and meridional components. The zonal component is much larger than its meridional counterpart and in the troposphere reaches a maximum at around the 200 hPa level (Fig. 11.8). Within the stratosphere tropical easterlies are evident throughout the year. They reach a maximum at about the level of the stratopause at a height of about 50 km and are centred over the southern tropical regions from December to February and over the northern tropics from June to August. Owing to the strong cooling of the polar stratosphere in winter, *polar night westerlies* develop at this time of year in direct response to the thermal wind effect (cf. Figs. 11.3 and 11.8). The jet cores are located at heights around 30 km with average general winds exceeding 45 m s^{-1} in the southern hemisphere.

Within the weather-producing troposphere the southern-hemisphere westerly jet stream reaches its maximum in July over the eastern Indian Ocean and Australasia (Fig. 11.9). South of Madagascar it splits into subtropical and polar branches. In July kinematic convergence over South Africa and the eastern Indian Ocean at 200 hPa produces strong subsidence over the region. In January the zonal mean wind field weakens generally, but not everywhere, the core of the westerly jet moves poleward, and the tropical easterlies expand poleward, notably over Africa (see Fig. 10.20). The degree to which the westerlies strengthen and expand northward over the hemisphere as a whole during winter is well illustrated in zonally averaged components (Fig. 11.10). The jet maximum in the upper troposphere occurs at about 30° S in winter. At 200 hPa the westerlies extend equatorward to about 10° S in both seasons. Although the mid-latitude westerlies and tropical easterlies move north and south simultaneously with the sun, this happens with an inverse change in speed. Thus as mid-latitude westerlies weaken, so tropical easterlies tend to strengthen and vice versa.

Just as the pressure field responds to regular annual and semi-annual changes, so too does the wind field (given the geostrophic and gradient wind relationships). The annual cycle in surface mean geostrophic flow is greatest in the tropics and subtropics and exhibits a large inter-annual (year-to-year) variation. The annual wave is associated with maximum westerly flow and minimum easterly flow in winter in the subtropics and maximum westerlies in summer around 50° S and at the solstices north of this latitude.

When compared with zonal components of airflow, meridional components are ageostrophic and small. Nonetheless, they are highly significant in their role of transporting water vapour, mass, heat, and momentum. The meridional structure of the atmosphere is clearly evident in patterns of zonally averaged mass flux (Fig. 11.10). Two major cells are found in each hemisphere. The Hadley cells are thermally direct (warm air rising; cold air sinking). Ascending motion occurs in equatorial regions and subsidence on the poleward extremities of the cells. The Ferrel cells are thermally indirect (cold air rises; warm air sinks). They occur in the middle latitudes and are characterized by subsidence on the equatorward side and ascent on the poleward side of the cell. During the southern summer the equatorial trough and the northern Hadley cell are in the southern hemisphere. The southern Hadley cell is compressed and the region of maximum surface convergence between the two cells occurs at about 10°–15° S. The ascending limbs of the Hadley cells promote in a substantial way the deep convective processes that characterize the tropical atmosphere. The region of maximum

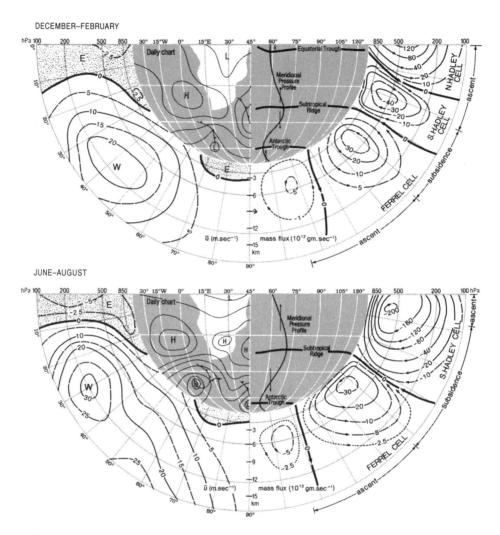

Fig. 11.10 Mean December–February and June–August zonal wind and mass flux in the southern hemisphere (data from Newell et al., 1972).

subsidence and surface divergence between the southern part of the Hadley cell and the northern part of the Ferrel cell occurs south of 30° S. Near the ground, air moves across the equator from the north; in the upper troposphere the flow is reversed. In winter the zone of maximum subsidence is, on average, north of 30° S; the southern Hadley cell expands considerably and the surface convergence zone shifts into the northern hemisphere. Low-level flow takes place from the south across the equator; in the

upper troposphere the movement is reversed and a major mass flux occurs from the northern to the southern hemisphere. In general, the subtropical westerly jet is located above the descending limb of the Hadley cell. The jet owes its existence primarily to the thermal wind effect and to the convergence of the poleward transport of westerly angular momentum in the upper branch of the Hadley cell.

Viewing the atmosphere in terms of a zonally averaged perspective has some advantages.

However, it tends to obscure features of the general circulation possessing strong longitudinal variability. Considerable regional and local variations occur, with the Ferrel cell being strengthened in some localities, the Hadley cell in others. During summer the mean meridional flow is markedly asymmetrical over the southern continents, being poleward over the eastern parts of the land masses and equatorward over the western parts as a consequence of the location of the oceanic subtropical high-pressure cells (Fig. 11.11). Thus near the surface the South Indian Anticyclone produces a northerly component of wind along the east coast of southern Africa which acts in that locality to strengthen the surface manifestation of the Ferrel cell to the south and weaken the Hadley cell to the north. Southerly near-surface winds from the South Atlantic Anticyclone along the Namib coast tend to weaken the Ferrel cell to the south and strengthen the Hadley cell to the north.

INTER-TROPICAL CONVERGENCE

In the region between the semi-permanent subtropical high-pressure cells on their equatorward sides, the mean circulation of the atmosphere is strongly evident in the tropical easterly flows that converge in the region of the thermal equator (Fig. 11.12). Throughout the year the Inter-Tropical Convergence Zone (ITCZ) is in the northern hemisphere in the eastern Pacific and Atlantic regions. Over Africa, the Indian Ocean and western Pacific regions the ITCZ migrates from the southern to the northern hemisphere between January and July. Over the oceans the convergence zone is more clearly delineated than over the continents (Fig. 11.13). Where streamlines with an easterly component cross the equator, the direction of the Coriolis deflection changes from left to right or vice versa. Immediate recurvature of the flow occurs and winds acquire a westerly component, as is

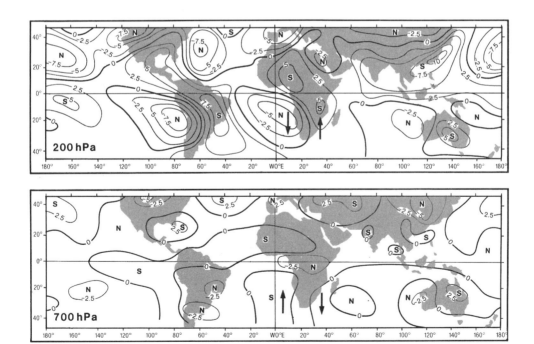

Fig. 11.11 Mean meridional wind components in summer (December–February) at 200 hPa and 700 hPa (after Newell et al., 1972).

shown over Africa and the northern Indian Ocean in July in Figure 11.12. These equatorial westerlies (Fig. 11.14) tend to be shallow, near-surface winds and play an important role in monsoon circulations. (An example of a narrow band of shallow equatorial westerlies is evident in Figure 10.20, *lower right.*)

The ITCZ is a region of pronounced convective activity in which tropical processes are considerably more complicated than was previously thought. Viewed on an annual or seasonal basis the planetary scales appear as structures with well-defined regions of steady convergence and ascent and associated regions of subsidence. On a day-to-day basis the mean convective regions are made up of propagating smaller-scale disturbances with lifetimes measured in days. Conditional instability alone is insufficient as a driving force to maintain the convective nature of tropical disturbances and cumulus convection, although a dominant feature of the tropical atmosphere, is not the only process producing rainfall. The penetration of mid-latitude distur-

bances into the tropics and forcing by sub-synoptic mesoscale disturbances are important additional controls of tropical rainfall. Up to 40 per cent of tropical disturbances appear to be associated with mesoscale circulation features. The structure of the ITCZ is complicated and the surface expression of the feature is seldom associated with rainfall. Instead, maximum convergence, cumulus convection, cloudiness and precipitation often occur equatorward of the major flow discontinuity.

North of about 20° S three major near-surface airstreams are involved in the Inter-Tropical Convergence Zone over Africa south of the equator and affect tropical southern Africa in summer (Fig. 11.15). They are the low-level, recurved south Atlantic air that moves over Africa at about 12° S in the equatorial westerlies; the north-east monsoon air of east Africa that moves south across the equator from the north-east; and the deep tropical easterlies from the Indian Ocean. Vertical transport of air takes place freely in this situation of con-

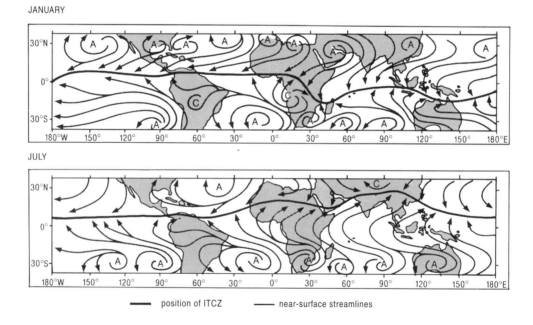

Fig. 11.12 Schematic streamlines of near-surface flow in the global tropics in January and July to show mean inter-tropical convergence. *A* denotes an anticyclonic centre, *C* denotes a cyclonic centre (modified after Hastenrath, 1985).

Fig. 11.13 Meteosat image (visible channel) showing the Inter-Tropical Convergence Zone over the tropical North Atlantic Ocean and West Africa on a July (northern-hemisphere summer) day.

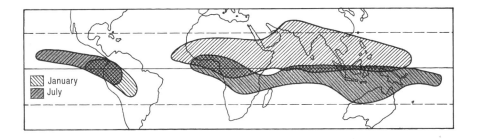

January
July

Fig. 11.14 The distribution of equatorial westerly airflow below 1.5 km (after Flohn, 1960).

SUMMER   WINTER

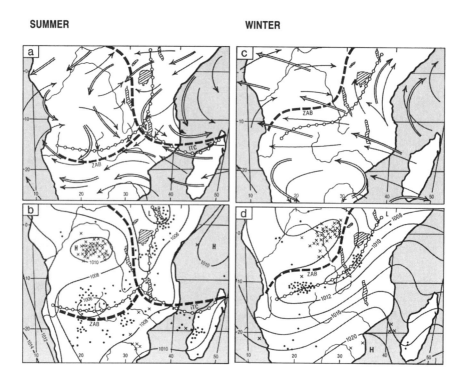

Fig. 11.15 The mean location of the Inter-Tropical Convergence Zone and Congo (formerly Zaire) Air Boundary (ZAB) (heavy broken lines): (a) and (b) during summer (January and February 1958); and (c) and (d) during winter (July and August 1958). Single arrows illustrate low-level flow and double arrows 3 km flow; linked circles mark the location of the major surface trough. Crosses and dots mark centres of individual closed low and high pressures in hPa (after Taljaard, 1972).

vergent winds, notwithstanding local diffluence in places. Closed tropical lows, troughs and ridges form within the convergence zone.

Over southern Botswana and southern Angola, the Congo (formerly Zaire) Air Boundary between Atlantic and Indian Ocean airstreams (Fig. 11.15) fluctuates in position from day to day and marks a region in which low-pressure systems tend to form preferentially. They tend to be much deeper than their heat-low counterparts forming over Kenya to the east of the main north–south axis of inter-tropical convergence. The Botswana lows and easterly waves tend to be quasi-stationary. Occasionally they move westward from Zimbabwe and Zambia towards Botswana and Namibia. When deep and associated with marked troughs extending southward, they may produce copious

general rains in regions east of the trough in the easterlies. Rainfall tends to show a clear diurnal pattern with most rain falling in the late afternoon and early evening.

The movement of the ITCZ over southern Africa shows a clear annual cycle. In winter the zone moves north with the sun and rain belts move accordingly. In summer the ITCZ moves back into the southern hemisphere and constitutes a region of major latent-heat release in the tropical atmosphere of southern Africa. Just as diabatic heating in the tropics is influenced in part by the equatorward penetration of mid-latitude disturbances, so substantial evidence exists, both from observational studies and from theoretical work, to show that tropical disturbances may exert important influences on mid-latitude circulations. This is certainly the case

JAN–MAR JULY–SEPT

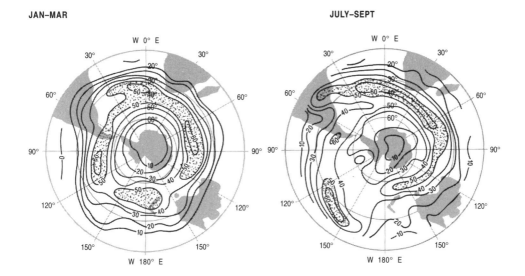

Fig. 11.16 The frequencies of surface fronts (number of times in 100 days when part of a front was situated in an area of 400 000 km^2) in January–March and July–September (after Taljaard, 1972).

over southern Africa, where the interaction between tropical and temperate disturbances has important consequences for the weather of the subcontinent.

FRONTS, CYCLONES AND ANTICYCLONES

The frequent occurrence of fronts is a characteristic of middle latitudes throughout the year (Fig. 11.16). The zone of maximum occurrence changes relatively little with season. To the south and south-east of Africa frontal activity is slightly less in summer than to the south-west and west of Gough Island. In winter the zone of greatest frontal activity stretches more uniformly from west of Gough Island to east of Marion Island.

Both cyclonic and anticyclonic perturbations occur on a daily scale within the mean flow of the southern hemisphere (Fig. 11.17). The zone of maximum anticyclonic frequency occurs within the extreme limits of 23° S to 42° S and average limits of about 27° S to 38° S. The core of highest frequency moves through about 4° to 7° of latitude from summer to winter. The core of

highest anticyclonic frequency is usually farthest equatorward where high-pressure cell concentrations are greatest, for example over the Atlantic Ocean between 0° and 15° W and over the Indian Ocean between 90° E and 105° E. Dissipation of old cells takes place to the west of continents, whereas the preferred areas for the generation of new cells are the subtropical and mid-latitude areas east of the Andes and off the southern and eastern coasts of South Africa. Perhaps the most significant feature of southern-hemisphere anticyclones is the extent to which a tendency is shown for the strongest cells to form well south of the subtropical ridge.

A distinctive feature of the distribution of cyclone centres (Fig. 11.17) is the zone of high frequency surrounding Antarctica in both summer and winter. In winter there is also clear evidence for two spiral bands of high cyclone frequency. The first emanates from South America in the South Atlantic Convergence Zone and the second from the east of Australia in the South Pacific Convergence Zone. Both spiral to the south-east and are coincident with major cloud bands over the two ocean areas. The lows forming over the continents in summer are

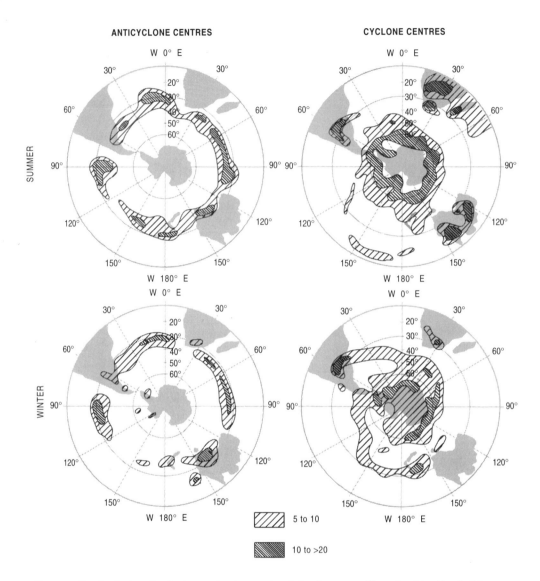

Fig. 11.17 The distribution of anticyclone and cyclone centres during the summer (December–March) and winter (June–September) seasons of the International Geophysical Year (after Taljaard, 1967).

mainly heat lows, though deeper cyclonic circulations may be associated with the easterlies and inter-tropical convergence over Africa.

Blocking anticyclones frequently form at latitudes higher than those where the subtropical highs normally develop and are frequently accompanied by cut-off lows in lower latitudes (see Fig. 10.17). In the southern hemisphere, blocking highs tend to have a shorter lifespan than in the northern hemisphere and they tend to form at somewhat lower latitudes. More often than not, blocking is associated with a split of the westerly jet into two branches. In addition, blocking is usually associated with semi-stationary *wave 3*, first, over New Zealand, secondly, to the south-east of South America, and thirdly, to the south-east of Africa and east of Marion Island.

ZONAL AND MERIDIONAL PATTERNS

From day to day the circulation intensity and degree of perturbation superimposed on the mean flow may vary considerably. More usually, however, the atmosphere settles into states where the level of perturbation may remain quasi-constant for some time. Zonal circulation types occur when the level of perturbation is low and the flow is strongly zonal (Fig. 11.18). Such a state may last for a few days or for longer periods. At the other extreme, meridional circulation types are characterized by a high state of perturbation and weather extremes. Zonal flow patterns produce less variable and more equable conditions. Should the atmosphere adjust to either one of these patterns of flow over an extended period of time, say years or longer, then the consequences for climatic change are obvious.

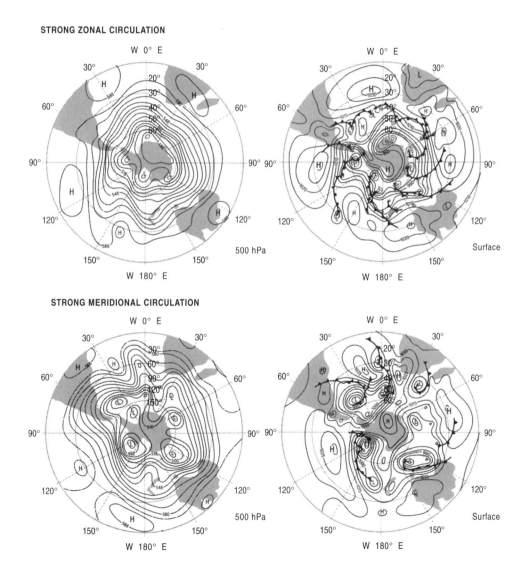

Fig. 11.18 Examples of strong zonal and meridional circulation types.

WAVE STRUCTURE AND SEMI-STATIONARY WAVES

Wave perturbations are important features of the westerlies. It is necessary to distinguish between large-scale, semi-stationary, forced waves and shorter-wavelength, free, travelling waves. The long, semi-stationary, forced waves have an equivalent barotropic structure. They maintain a similar amplitude and location throughout the year and over longer periods of time. The shorter, transient, baroclinic disturbances are steered by the semi-stationary waves and larger-scale flow. Over a short time scale, should the long waves shift in phase (longitudinally) or in amplitude (latitudinally), or should the travelling, short-wave (high frequency) disturbances undergo a change in variance, the effects on weather are immediately apparent. Should this happen over a longer time scale, climates alter accordingly.

If the pressure at equidistant points along a circle of latitude is submitted to harmonic (wave) analysis, it is possible to decompose the zonal pressure distribution into its constituent wave components. The first wave, *wave 1*, has one peak and one trough around the circle of latitude. *Wave 2* has two peaks and two troughs, *wave 3* has three and so on. From all the possible constituent waves it is then possible to isolate those which contribute most to the total or overall pattern. The spatial and temporal characteristics of these waves may then be examined. The results of such a zonal harmonic analysis of 500 hPa geopotential heights allows the partitioning of westerly geostrophic airflow into its constituent waves, some of which are illustrated in Figure 11.19. The lower number waves are quasi-stationary; the higher number waves are transient Rossby waves. The degree to which actual airflow at 200 hPa may exhibit wave motion is illustrated by three circum-hemispheric trajectories of a balloon released from Christchurch, New Zealand (Fig. 11.19). The essential wave-like nature of the airflow is strikingly apparent, as too is the tendency for small waves to be superimposed on larger ones. Attention may now be focused on the long, semi-permanent waves having lower wave numbers.

South of the subtropical highs the pressure drops steeply to its lowest values in the circum-

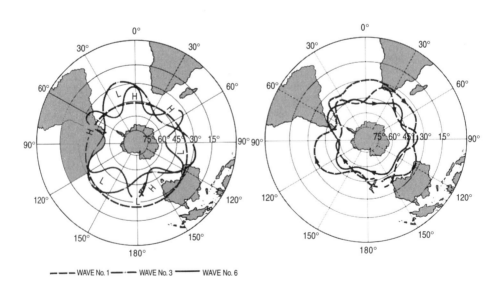

— — — WAVE No. 1 —·—·— WAVE No. 3 ——— WAVE No. 6

Fig. 11.19 *Left:* a schematic illustration of standing waves 1 and 3 and travelling wave 6; *right:* the flight trajectory of a GHOST Project balloon released from Christchurch, New Zealand and flying at 200 hPa (after Dyer, 1975).

polar trough between 60° S and 70° S. At 500 hPa the centre of the southern-hemisphere circumpolar vortex is situated over the Ross Sea in both seasons. The subtropical ridge, which lies 10° north of its surface position in both summer and winter, is somewhat further south in the eastern Atlantic and Indian Ocean sector of the hemisphere. The eccentricity of the circumpolar vortex is described by zonal standing *wave 1*. Together *waves 1* and *3* account for over 85 per cent of the total variance in the annual mean 500 hPa pattern. *Wave 1* and, to a lesser extent, *wave 3* are the dominant features of the mean southern-hemisphere circulation.

Wave 1 exhibits peaks in subtropical and subpolar latitudes (Fig. 11.20). The subtropical feature has a ridge near the Greenwich meridian and a trough in the Pacific Ocean to the north of New Zealand. Between latitudes 20° S and 35° S the amplitude of *wave 1* at 500 hPa over the Atlantic Ocean to the west of the Namib coast reaches 50 m. The wave accounts for more than 50 per cent of the variance in the height of the 500 hPa surface for most of the year and for more than 75 per cent in winter. At 30° S the daily position of the ridge of *wave 1* varies little

in winter; during summer its longitudinal variation is greater, as is its influence on the weather of southern Africa. The subpolar ridge in the Pacific at about 120° W has a maximum amplitude exceeding 120 m and the wave at those latitudes accounts for 95 per cent of the variance. *Wave 1* exhibits its greatest inter-annual variability at 35° S and 65° S; it dominates the African sector and its subtropical expression affects southern Africa significantly.

Standing *wave 3* has an amplitude of about 30 per cent that of *wave 1*; it has an equivalent barotropic vertical structure and it reaches its peak amplitude in middle latitudes at all times of the year. In latitudes where the wave is strongest, the first ridge is to the east and southeast of KwaZulu-Natal. Other ridges affect the Tasman Sea area and southern South America. *Wave 3* responds to the semi-annual oscillation in mid-latitude pressure in the southern hemisphere such that the longitude of the first ridge is at about 65° E in March, 38° E in June and 57° E in September. An important attribute of the wave is the role it plays in governing the location of blocking anticyclones. One of the major blocking areas is to the east of South Africa at 55° E. Inter-annual variability of *wave 3* appears to reach its maximum between 45° S and 55° S.

Southern-hemisphere planetary waves often display a surprising degree of regularity. At times a stationary zonal *wave 3* pattern may dominate the 500 hPa geopotential height field for up to 30 days. Higher-wave-number, higher-frequency, shorter-wavelength travelling disturbances may likewise show persistence.

TRAVELLING WAVE DISTURBANCES

High-frequency travelling *waves in the westerlies* are baroclinic disturbances and, as a result of the conservation of absolute vorticity and changing radius of curvature, are associated with distinctive patterns of surface divergence ahead of troughs and of convergence behind them.

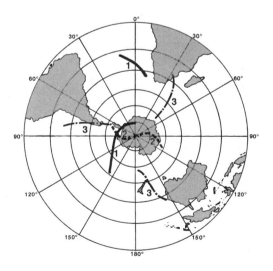

Fig. 11.20 Positions of the ridges of standing waves 1, 2 and 3 in the annual mean height of the 500 hPa surface (after Van Loon and Jenne, 1972).

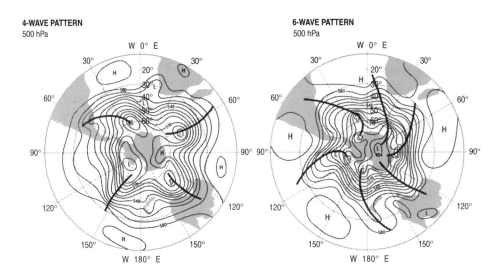

Fig. 11.21 Examples of 4-wave and 6-wave structures in the westerlies at 500 hPa.

Normally they are tilted to the west with height such that upper-level divergence usually overlies surface-level convergence to the rear of the surface trough, thus promoting uplift, cloud and precipitation. The opposite prevails to produce subsidence and fine weather ahead of the surface trough.

On any one day, from four to eight travelling wave perturbations may be present in the westerlies of the southern hemisphere (Fig. 11.21). On a daily basis at 50° S, in any one year *waves 4, 5* and *6* may account for over 25 per cent of the variance of 500 hPa heights in summer and over 35 per cent in winter. High-frequency transient eddies with periods of less than two weeks dominate the total variance and covariance fields of geopotential height, total kinetic energy and mean eddy momentum flux. These transient disturbances are responsible for so much of the low-frequency atmospheric and weather variability in middle latitudes. Baroclinic wave disturbances exhibit a broad spectral range peaking with a period of 2–8 days. Maximum poleward fluxes of momentum in this frequency range are observed at about 40° S. The effect of perturbations with a period of about 6 days is clearly evident in spectra of surface-pressure oscillations

around the coast of southern Africa (Fig. 11.22). Cospectra of fluctuations at Cape Town and east-coast stations reveal that, on average, disturbances take 1.2 days to travel from Cape Town to Port Elizabeth (650 km direct distance), 2.1 days to Durban (1 260 km) and 2.8 days to Maputo (1 580 km). These figures accord well with synoptic experience for the passage of coastal lows and cold fronts. The regions in the vicinity of southern Africa in which such disturbances reached their maximum activity during the International Geophysical Year period are shown in Figure 11.23 and are in broad agreement with the results of other research workers.

Perturbations in the westerlies tend to follow well-defined storm tracks. On the basis of geopotential height variance, mean annual storm tracks appear to occur preferentially at 50° S. Other criteria suggest a less restricted zone of occurrence. The axis of the surface frontal frequency maximum, the axis of minimum frequency of cyclone centres and the location of large-scale cloud bands all show that the pat-tern of storm tracks is more complicated than a single mean measure suggests and that storm tracks show a clear seasonal variation (Fig. 11.24).

Waves in the easterlies have long been thought

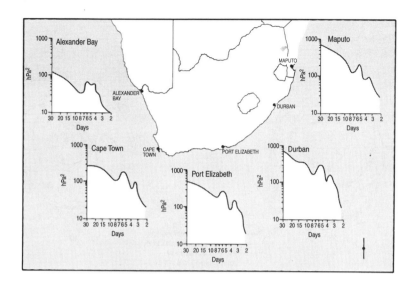

Fig. 11.22 Spectra of surface-pressure fluctuations at stations around southern Africa. The 95 per cent confidence limit is included.

to conform to a model of upward motion, lower-troposphere convergence, upper-level divergence, cloudiness and rainfall to the rear of the surface wave, and fine weather ahead of the wave (see Fig. 10.13, *lower right*). It has been shown, however, that this model is one of several types and that the location of bad weather may even occur ahead of the wave. Experience suggests that over southern Africa the major source of low-level convergence and precipitation does in fact occur to the rear of a wave in a deep easterly current. However, rather than exhibiting all the characteristics of travelling waves, the easterly waves of southern Africa tend to be quasi-stationary features.

MAJOR CONVERGENCE ZONES AND CLOUD BANDS

The factors controlling the formation and variation of major cloud-band convergence zones are not yet completely understood. Undoubted-

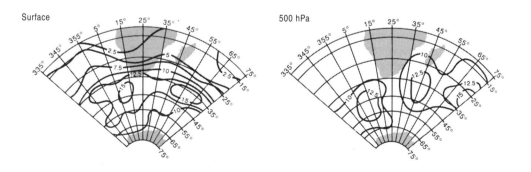

Fig. 11.23 The spatial variation of the percentage variance associated with 6- to 8-day pressure fluctuations during 1958.

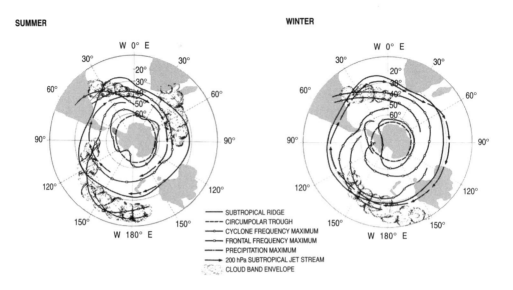

Fig. 11.24 The seasonal variation of storm tracks as defined by the positions of the subtropical ridge, circumpolar trough, cyclone and frontal frequency maxima, the velocity maxima in the subtropical westerly jet and cloud-band envelopes.

ly semi-stationary long waves, induced orographically or by large-scale latent heat release in the tropics, or a combination of both, play a major role in fixing the location of high-tropospheric (500–200 hPa) troughs. It is in association with these troughs that major cloud bands appear to form. When wave perturbations in the westerlies become linked with disturbances in the tropical flow field, extended north-west to south-east convergence zones form, linking the tropics to the middle latitudes. These zones are inferred to be channels for the large-scale transfer of momentum and of energy in the form of heat and moisture from tropical to middle latitudes and are associated with precipitation maxima. The cloud-band convergence zones are usually tilted equatorward of the subtropical jet, may extend to the equator itself, are often linked to southern branches of the ITCZ and always appear to spiral from north-west to south-east towards higher latitudes.

The structure of cloud bands over the southern hemisphere varies from simple to extremely complex. On some occasions it is possible to observe a clear four-band organization; on others up to eight bands may be discerned (Fig.

11.25). Not only do low-numbered organized cloud bands appear on a daily basis, they are also revealed in monthly averages. On some occasions the bands move fairly uniformly along their entire length as they are steered by the larger-scale westerly flow; at other times they become semi-stationary along their subtropical extensions, while continuing to move eastward with the disturbances responsible for their generation on their southern extremities (Fig. 11.26). Any change in the mode of circulation that encourages the slower movement or stagnation of the cloud bands will have an important effect on weather and climate depending on the time scale of the change.

Three sectors of the southern hemisphere are preferred locations for the occurrence of slowly moving or semi-stationary cloud bands: one over the South Atlantic Ocean, one over the Pacific in the South Pacific Convergence Zone and one in the Indian Ocean extending southeast from southern Africa. Illustrations of the South Atlantic and South Pacific convergence zones and Pacific cloud bands are given in Figure 11.27. The Atlantic and Pacific cloud-band convergence zones show little seasonal

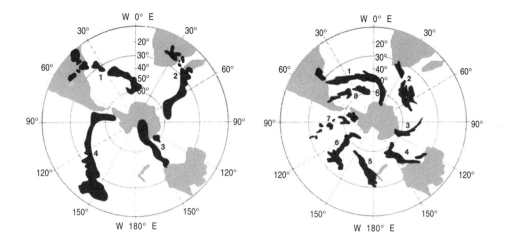

Fig. 11.25 Examples of 4-band (*left*) and 8-band cloud patterns (*right*) in the southern hemisphere. Cloud masses over Antarctica have been omitted.

change in position and frequency of occurrence. The Indian Ocean counterpart shows a marked variation. Over southern Africa it is a summer phenomenon. In winter the divergence field and strong subsidence prevailing over the subcontinent preclude its occurrence. Variation in the location of the preferred occurrence of cloud-band convergence zones over the southern Africa-Indian Ocean sector in summer, both on day-to-day and year-to-year scales, has important implications for rainfall over southern Africa and will be considered in more detail in Chapters 12 and 16.

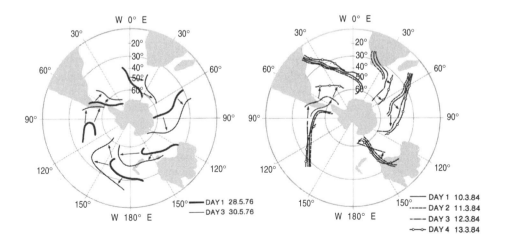

Fig. 11.26 *Left*: uniformly moving cloud bands in a 4-band pattern over a 48-hour period; *right*: semi-stationary cloud bands in a 5-band pattern over a 72-hour period. Only the leading edge of each band is indicated.

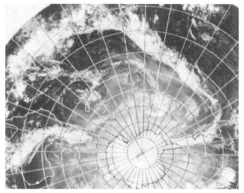

Fig. 11.27 Satellite photographs showing, *left*, the South Atlantic Convergence Zone and, *right*, the South Pacific Convergence Zone and their associated cloud bands.

THE WALKER CIRCULATION

The Walker Circulation is the major zonal circulation of the atmosphere. It originates in the Pacific Ocean and extends across the globe and is associated with El Niño-Southern Oscillation (ENSO) events. Since these are closely related to sea-surface temperature changes, the Walker Circulation will be discussed in full in Chapter 13 in relation to ocean-atmosphere interactions. At this point it is sufficient to note that the Walker Circulation is a major feature of the general circulation of the atmosphere and that no discussion of the general circulation is complete without consideration of the Walker Circulation.

MONSOON SYSTEMS

Large-scale seasonal reversals of wind characterize the monsoon areas of the world (Fig. 11.28). For the most part monsoons are tropical phenomena. They have little direct effect on the weather or climate of southern Africa and will not be discussed in detail here.

The nature of the surface wind reversals associated with the monsoon of Asia is clearly evi-

dent in Figure 11.12. In January, the winter monsoon of the northern hemisphere exhibits anticyclonic curvature. Its winds circulate around centres of high pressure over Asia and it is associated with dry, stable air and little rainfall. In July, the thermal equator, the equatorial trough and the subtropical semi-permanent anticyclones are displaced northward with the sun and the changing pattern of solar heating of the earth. As the southern trades cross the equator they recurve to become north-westerly. The summer north-westerly monsoon exhibits cyclonic curvature into centres of low pressure and is associated with moist, unstable air and rainfall. Earlier explanations of monsoons as being thermally driven in a manner analogous to land and sea-breeze systems with winds blowing out from a surface high in winter and into a surface low in summer are insufficient to explain the workings of the Asian monsoon system. Other means of forcing at both the planetary level, through jet streams, and at the regional level, as a consequence of the Tibetan Plateau, as well as other factors, must be considered. They need not be considered here, though. What is relevant for southern Africa is the involvement of Indian monsoon air in the ITCZ over Africa in the southern-hemisphere

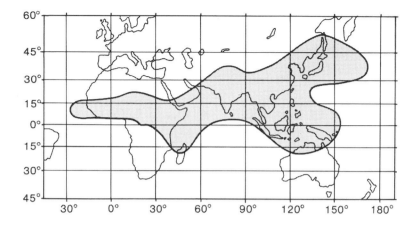

Fig. 11.28 The monsoon regions of the earth (after Critchfield, 1974).

summer (see earlier in this chapter) and the extent to which subsidence from upper-level outflow from the Indian monsoon in July may play a role in enhancing the subtropical anti-cyclones affecting the subcontinent and its dryness in winter.

ADDITIONAL READING

GARSTANG, M. and FITZGARRALD, D. M. 1999. *Observations of Surface to Atmosphere Interactions in the Tropics.* New York: Oxford University Press.

BARRY, R. G. and CHORLEY, R. J. 1998. *Atmosphere, Weather and Climate.* 7th Edition. London: Routledge.

SCHNEIDER, S. H. (ED.). 1996. *Encyclopaedia of Weather and Climate.* Vols. 1 and 2. Oxford: Oxford University Press.

STURMAN, A. S. and TAPPER, N. 1996. *The Weather and Climate of Australia and New Zealand.* Melbourne: Oxford University Press.

TYSON, P. D. 1986. *Climatic Change and Variability in southern Africa.* Cape Town: Oxford University Press.

HASTENRATH, S. 1985. *Climate and Circulation of the Tropics.* Dordrecht: D. Reidel Publishing Company.

RIEHL, H. 1979. *Climate and Weather in the Tropics.* London: Academic Press.

NEWTON, C. W. (ED.). 1972. *Meteorology of the Southern Hemisphere.* Meteorological Monograph 35. Boston: American Meteorological Society.

PALMÉN, E. and NEWTON, C. W. 1969. *Atmospheric Circulation Systems.* New York: Academic Press.

12

ATMOSPHERIC CIRCULATION AND
WEATHER OVER SOUTHERN AFRICA

The climate and weather of southern Africa are influenced by the mean circulation of the atmosphere over the subcontinent and by the perturbations (disturbances or deviations) from that mean. Southern Africa's location in the subtropics ensures that it is affected by circulation systems prevailing in both the tropics to the north and temperate latitudes to the south. At the same time it is dominated by the high-pressure systems that, when averaged, constitute the semi-permanent, subtropical high-pressure cells of the general circulation of the southern hemisphere.

MEAN CIRCULATION PATTERNS

Except near the surface, the mean circulation of the atmosphere over southern Africa is anticyclonic throughout the year. In winter the mean anticyclone intensifies and moves northward. Upper-level circumpolar westerlies expand and displace the upper tropical easterlies equatorward (Fig. 12.1). The near-surface circulation at 850 hPa in January consists of a weak heat low centred over the central interior and linked by a trough across the northern Cape and Botswana to a tropical low north of Botswana. The centre of the latter feature wanders in an east-west direction along the southern branch of the Inter-Tropical Convergence Zone in the region of the Congo Air Boundary and occasionally southward towards Namibia, Botswana, or Zimbabwe. Over the Northern Cape

a weak ridge is to be observed. In contrast to the January situation, the low-level July mean pressure field at 850 hPa is strongly anticyclonic. The change to a single high-pressure cell takes place by March, resulting in a northerly flow of moist air from the tropics over the western parts of southern Africa at a time when the air masses over Zimbabwe, Zambia and northern Botswana and Namibia are still moist. The influx of moist air is largely responsible for the autumn rainfall maximum over the western regions at this time. In the middle and upper troposphere (500 and 200 hPa) during January, easterly flow occurs north of about 23° S. In July the winter increase and equatorward extension of the westerlies is clearly evident. The greatest increase is from April to July, followed by a slow decline until October. The lowest speeds are observed in February or March.

The seasonal changes in the circulation over southern Africa are clearly evident in mean meridional sections through the atmosphere along 20° E. It is apparent that tropical easterlies affect most of southern Africa throughout the year (Fig. 12.2). At about 12° S low-level westerlies from the Atlantic Ocean extend to about 500 hPa in summer; in winter they weaken, become shallower and shift northward. Northward expansion and intensification of the circumpolar westerlies during winter are evident at the 200 hPa level. In winter, meridional components reflect the subsidence occurring over much of southern Africa. In summer,

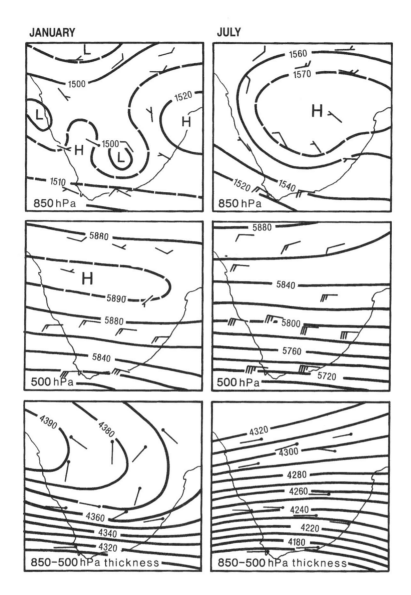

Fig. 12.1 Monthly mean winds, contours of the 850 and 500 hPa surfaces (gpm) and 850–500 hPa thicknesses (gpm) (after Taljaard, 1981).

flow tends to be poleward below about 500 hPa and equatorward above that level; at the 200 hPa level, flow is again poleward.

The degree to which differential heating of land and sea occurs between summer and winter is reflected in 850–500 hPa thickness gradients (Fig. 12.1). Heating of the land during summer clearly affects the circulation in the lower troposphere, not only over the subcontinent, but also over the adjacent oceans where the South Atlantic and Indian Anticyclones undergo seasonal intensification. Anomalies in 500 hPa geopotentials over southern Africa occur out of phase with anomalies over the

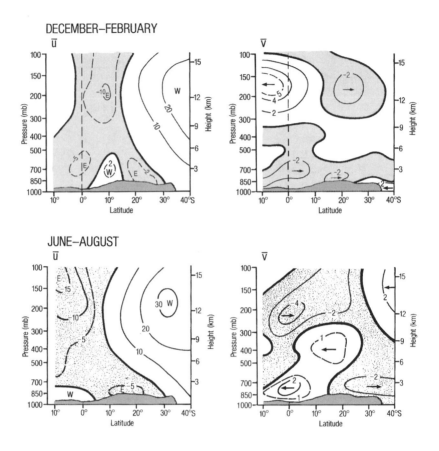

Fig. 12.2　December–February and June–August mean zonal (*u*) and meridional (*v*) wind components (m s⁻¹) at 20° E (after Newell et al., 1972). Easterly components are shaded.

adjacent oceans. The seasonal intensification of the ocean highs in turn affects the ocean currents and upwelling. Thus, when the Atlantic High is strongest in summer, west-coast upwelling is greatest in the Benguela Current and sea- surface temperatures are lowest along the coast.

Whereas the mean circulation of the atmosphere controls to a large extent the climate of the subcontinent, it has little affect on day-to-day weather. This is determined by the synoptic and smaller-scale disturbances that constitute individual weather systems.

SYNOPTIC AND OTHER WEATHER PERTURBATIONS

The main, though not conclusive, elements affecting the day-to-day weather of southern Africa owe their origins to subtropical, tropical and temperate features of the general circulation. The *subtropical control* is effected through the semi-permanent South Indian Anticyclone, the continental high and the South Atlantic Anticyclone (Fig. 12.3), which are elements of the discontinuous high-pressure belt that circles the southern hemisphere at about 30° S.

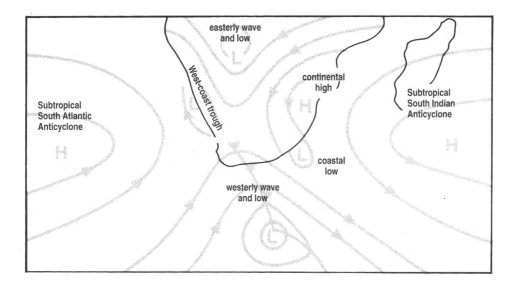

Fig. 12.3 Some important features of the surface atmospheric circulation over southern Africa.

The South Atlantic and Indian Anticyclones vary significantly in position throughout the year (Fig. 12.4). Both cells move about 6° northward in winter. The Atlantic high shows a yearly cycle in its latitudinal position. In contrast, the Indian high undergoes a half-yearly oscillation in its zonal position in response to the semi-annual pressure variations taking place to the east of Marion Island (see Fig. 11.7 for the distribution of the annual and semi-annual pressure waves). The monthly longitudinal shifts in position of the Atlantic high are up to double those of its latitudinal variation and do not materially affect the weather of the subcontinent. On the scale of days, however, the high may ridge eastward and to the south of the continent. Extended ridging leads to breaking off (sometimes referred to as budding) of a separate high, which drifts eastward into the Indian Ocean before being subsumed into the South Indian high. In so doing the freshly formed ridging anticyclone in the westerlies may strongly affect the weather of South Africa. The seasonal east-west shifts of the high in the Indian Ocean (up to 24°) have important effects on the summer and win-

ter weather of the eastern parts of southern Africa.

The *tropical control* of southern African weather is affected through tropical easterly

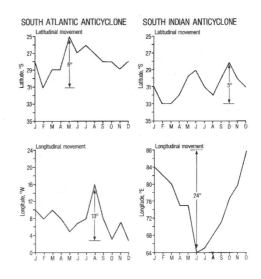

Fig. 12.4 Annual variation in the positions of the South Atlantic and South Indian anticyclones (modified after Vowinckel, 1955; McGee and Hastenrath, 1966).

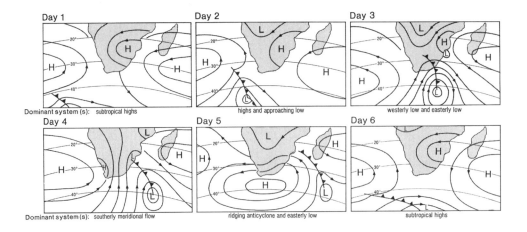

Fig. 12.5   A schematic illustration of a typical surface sequence of weather disturbances over southern Africa.

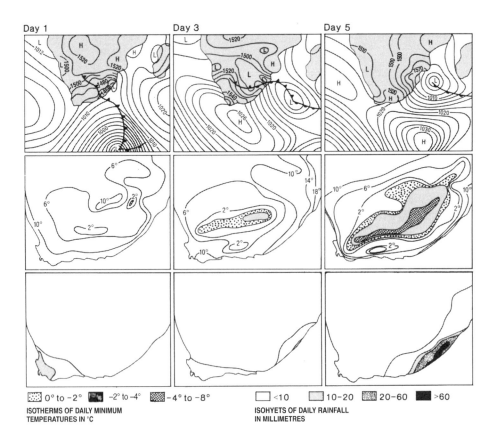

Fig. 12.6   A three-day sequence of circulation, temperature and rainfall changes over southern Africa in winter to illustrate the occurrence of a cold snap and coastal rainfall following the passage of a westerly wave and cold front.

flow and the occurrence of easterly waves and lows. *Temperate control* is exerted through travelling perturbations in the westerlies that take the form of westerly waves and lows. Of a more local nature are west-coast troughs and the smaller-scale and shallow coastal lows.

Underlying a considerable day-to-day variability in weather, it is possible to discern a tendency for the occurrence of typical sequences in the passage of disturbances to the south of the subcontinent. These sequences seldom occur for long enough to impose a predictable regularity on the weather; occasionally they may establish themselves clearly for short periods. Such a 6-day sequence is illustrated schematically in Figure 12.5. On *day 1* the circulation is dominated by subtropical highs. Thereafter a low in the westerlies drifts eastward to trail its cold front across the southern part of South Africa by *day 3*. At the same time an easterly wave develops to the north. By *day 4* the low is moving off the east coast and strong southerly

meridional flow sweeps across the subcontinent. At the same time the Atlantic high begins to ridge eastward towards Cape Town. By *day 5* a large ridging anticyclone has broken free of its parent Atlantic High and is drifting around the tip of Africa and into the Indian Ocean. By *day 6* the ridging high has been amalgamated into the Indian high and the circulation pattern has reverted to one similar to that prevailing originally on *day 1*.

With the passage of a westerly wave and low across South Africa in winter, distinctive changes take place in the surface temperature field as the cold front sweeps north-eastward (Fig. 12.6). Whereas minimum temperatures drop immediately after the passage of the front, lowest minima are usually recorded on the first morning after the cloud associated with the disturbance has cleared. Under these conditions strong surface radiational cooling, combined with cold air advection with a southerly component, produces the lowest temperatures. The

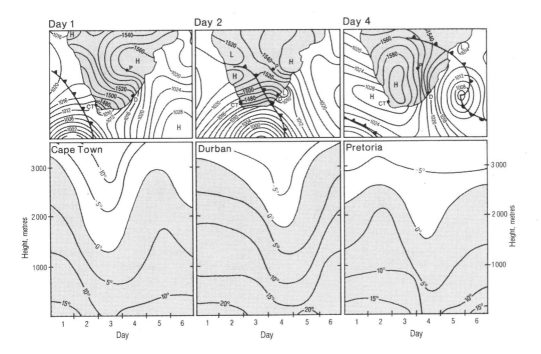

Fig. 12.7 A three-day winter sequence to show the effect of a westerly low and cold front on the vertical distribution of temperature (°C) (after Diab, 1975).

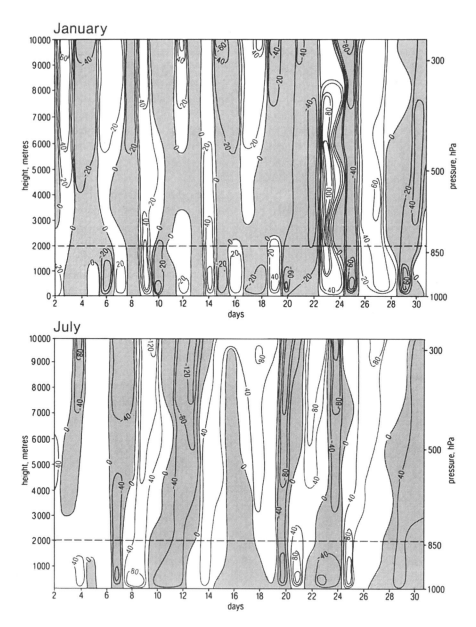

Fig. 12.8 Thirty-day time–height sections to show the effect of passing weather disturbances on pressure departures (hPa) from monthly means at Durban. Shading indicates negative departures.

manner in which coastal rainfall may accompany a westerly disturbance is clearly evident in Figure 12.6.

Surface fields are not the only ones to be disturbed by the passage of cold fronts. Distinctive changes likewise occur in the distribution of temperature in the vertical as isotherms dip towards the ground as cold-cored westerly disturbances sweep across South Africa. A winter example of the way in which the temperature field in the vertical adjusts with the passage of a cold front at Cape Town, Durban and Pretoria is given in Figure 12.7.

Pressure fields likewise vary considerably in the vertical with the passage of westerly disturbances. If daily pressure departures from the mean are plotted from the surface to the 300 hPa level, the quasi-regularity of the pressure perturbations may be clearly discernible, particularly at coastal stations. January and July examples are given for Durban in Figure 12.8.

The winter case shows a greater degree of regularity than that of summer and reflects the equator-ward movement of the circumpolar westerlies and their attendant disturbances in winter.

The extent of the warming and cooling of the atmosphere consequent upon the passage of westerly disturbances, and the changes in atmospheric stability that take place at the same time, are best considered using tephigrams. Again a winter example is used. On *day 1* of the

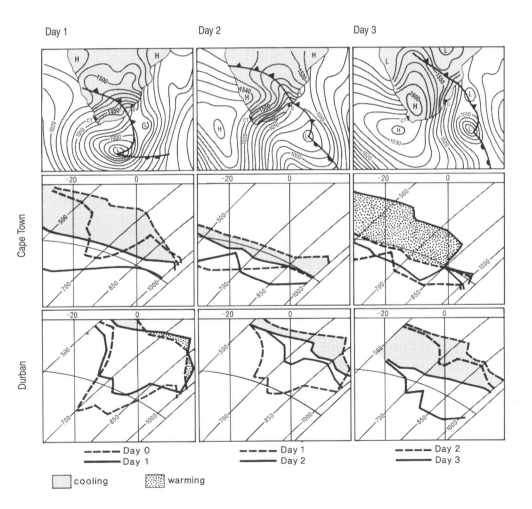

Fig. 12.9 A three-day sequence to show the vertical extent of cooling associated with the passage of a westerly disturbance and cold front. The synoptic charts are for 14:00 and tephigrams for 06:00 of the same days. Lapse rates of dry-bulb and dew-point temperatures are given for each day (after Taljaard et al., 1961).

sequence given in Figure 12.9 a cold front had passed Cape Town. Comparison of the 06:00 lapse rate of dry-bulb temperature on that day with that of the previous day shows that the atmosphere had *cooled* substantially by more than 15 °C above the 700 hPa level and that stability of the atmosphere had diminished to the extent that a deep, conditionally unstable layer would have developed during the day with surface heating. To the north-east Durban was yet to be affected by the front, was still under the influence of an anticyclonic circulation, and there was little difference in the lapse rates of *day 1* and the preceding day. On *day 2* further

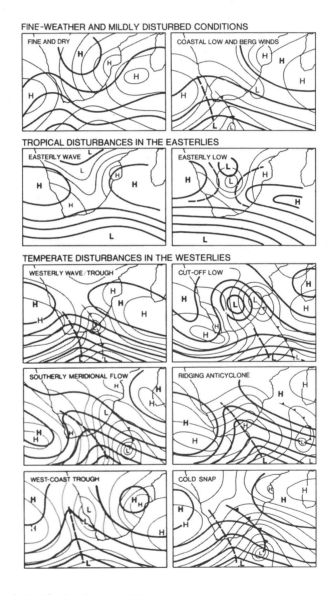

Fig. 12.10   A schematic classification of southern African weather types based on circulation patterns at the surface (light lines) and 500 hPa (heavy lines).

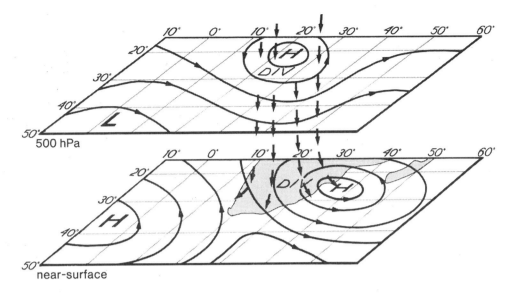

Fig. 12.11 A schematic representation of the near-surface and 500 hPa fine-weather circulation associated with high-pressure systems over southern Africa.

cooling had occurred over Cape Town and the lapse rate was almost wet adiabatic. The front had passed Durban by then and deep cooling had commenced. By *day 3*, as anticyclonic conditions became established over the south-western Cape, *warming* of the atmosphere occurred and a clear subsidence inversion developed at around 850 hPa over Cape Town. Over Durban further cooling took place in the stream of southerly air behind the front and the lapse rate became increasingly unstable.

The use of time–height sections (Figs. 12.7 and 12.8) and tephigrams indicates the need to consider the three-dimensional structure of weather disturbances in explaining the weather of southern Africa. Though useful as an introductory device, the study of surface synoptic sequences alone offers no satisfactory explanation for the changing weather patterns affecting the subcontinent. Instead, it is necessary to examine the relationship between surface and upper-air circulations, the thermal stability structure and the different sources of uplift that promote or inhibit weather processes within various systems. To do this it is useful to develop a framework of circulation types.

CIRCULATION TYPES

In this chapter a classification of synoptic patterns incorporating generalized near-surface circulation types and their attendant upper-level counterparts will be given. Three categories of circulation patterns will be considered: *first*, fine-weather and mildly disturbed conditions, *secondly*, tropical disturbances associated with tropical easterly airflow, and *thirdly*, temperate mid-latitude disturbances associated with westerly airflow (Fig. 12.10).

FINE WEATHER AND MILDLY DISTURBED CONDITIONS

Subtropical anticyclones

Fine-weather and mildly disturbed conditions are invariably associated with the generation of anticyclonic vorticity in large subtropical anticyclones centred over the subcontinent. These systems are deep and are tilted upward from the surface towards the north-west (Fig. 12.11). They are associated with divergence in the near-surface wind field, strong subsidence

Table 12.1 The frequency of occurrence (percentages) of some important synoptic types: (a) on the basis of seven types recognized at Pretoria (after Vowinckel, 1956); (b) on the basis of types producing heavy rainfall of at least 40 mm on one day at more than half the stations within an area of four degrees square or 60 mm on two successive days or 75 mm on three successive days (after Taljaard, 1982). In the analysis of Taljaard only three types met the selection criteria, namely east-wind troughs and lows, west-wind troughs and cut-off lows. Due to the fact that seven types were used by Vowinckel and only three by Taljaard, comparison of the frequencies must be approached with caution.

	J	F	M	A	M	J	J	A	S	O	N	D
Fair-weather anticyclonic conditions												
(a) Continental high (Type 1, Vowinckel, 1956)	18	17	19	25	38	79	79	72	65	38	19	11
Tropical disturbances												
(a) Easterly wave (Type 4, Vowinckel, 1956)	19	16	17	18	14	6	6	10	7	9	12	14
(a) Easterly low (Type 5, Vowinckel, 1956)	27	35	37	26	19	7	3	1	5	15	25	35
(b) Easterly wave and low (East-wind trough and tropical low, Taljaard, 1982)	35	39	23	0	0	0	0	0	7	0	6	19
Westerly disturbances												
(a) Ridging anticyclone (Type 7, Vowinckel, 1956)	7	9	6	8	8	2	2	0	2	11	9	9
(a) Cold snaps (Type 3, Vowinckel, 1956)	0	0	0	3	4	4	5	8	6	3	3	1
(b) Westerly low (West-wind trough, Taljaard, 1982)	19	11	26	10	3	3	3	3	10	16	13	3
(c) Cut-off low (Taljaard, 1982)	3	11	16	27	29	17	13	30	33	26	32	26

FINE-WEATHER CONTINENTAL HIGH

Fig. 12.12 Examples of fine-weather continental high-pressure types prevailing in summer and winter. Light lines show isobars at mean sea level (hPa) over the oceans and contours of the 850 hPa surface (gpm) over the land; heavy lines show contours of the 500 hPa surface (gpm).

throughout a deep layer, the occurrence of inversions, fine clear conditions and little or no rainfall. Surface divergence accompanied by convergence in the uppermost levels of the troposphere, together with descent in the poleward limb of the tropical Hadley cell (see Fig. 11.10), produces subsidence through the 500 hPa surface and into lower layers. The frequency of occurrence of anticyclones reaches a maximum over the interior plateau in June and July (79 per cent) with a minimum during December (11 per cent) (Table 12.1). The dominant effect of winter subsidence is such that, averaged over the year, the mean vertical motion is downward. Clearly there must be times when upward motion dominates, other-

wise no weather would occur. When such motion does occur, it is usually localized within the system producing it and subsidence continues to prevail elsewhere. Subsidence is a dominant characteristic of *subtropical climates*. Examples of summer and winter fine-weather continental high-pressure circulation types associated with strong and deep subsidence are given in Figure 12.12. On occasions such as these the subcontinent may be entirely cloud free (Fig. 12.13).

When prevailing for extended periods in summer, anticyclones produce severe heat waves and desiccation. Most abrupt increases of temperature from one day to the next and exceeding 5 °C are associated with the occur-

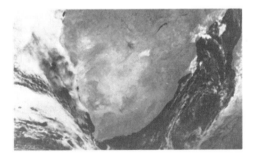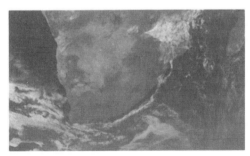

Fig. 12.13 Examples of clear, fine-weather, anticyclonic conditions over southern Africa (note the convection cloud over the Agulhas Current in the right-hand satellite image).

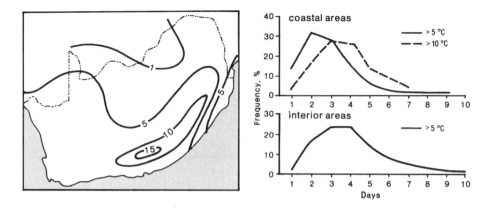

Fig. 12.14 *Left*: the annual number of increases in temperature from one day to the next exceeding 5 °C; and *right*: the duration of warm spells over coastal and interior regions of South Africa (after Schulze, 1984).

rence of strong continental anticyclones and occur in coastal and adjacent areas as a result of the subsidence and warming associated with Berg winds. Over the interior, heat waves with an increase of temperature exceeding 5 °C occur with strong continental anticyclones and often last 4 days (Fig. 12.14). Those lasting more than 7 days are less common, but do occur.

Examples of particular subsidence inversions, as well as the climatological mean distribution and clouds associated with such features, were given in Chapter 10 (Figs. 10.33 and 10.34). However, to consider only the inversions of temperature and not other stable layers associated with anticyclones is to take too restrictive an approach. In Chapters 4 and 5 it was shown that when the environmental lapse rate is less than the wet (saturated) adiabatic lapse rate, the condition of absolute stability is initiated long before the lapse rate reverses sign to increase with height. Thus absolute stability defines a state of the atmosphere that inhibits or prevents upward convection and transport of air and whatever is carried in it. Persistent anticyclonic circulation and the subsidence of air therein increases the stability to a point when absolutely stable layers begin to form. This does not happen uniformly throughout the tropo-

sphere, but in distinctive layers that form preferentially at different levels (Fig. 12.15). Over coastal and adjacent inland areas of South Africa, the first absolutely stable layer develops at ~850 hPa – at or below Escarpment level. Over plateau areas, the mean positions of the stable layers are ~700 hPa, ~500 hPa and ~300 hPa. The layers extend far beyond South Africa, particularly the ~500 hPa layer (Fig. 12.16), and are responsible for trapping aerosols (particulate material) and trace gases and controlling their vertical and horizontal transport in the atmosphere over southern Africa as a whole, a subject considered in detail in Chapter 15. The layers occur throughout the year whenever subsidence occurs and are not confined to anticyclonic circulations per se. They form whenever anticyclonic curvature is present in the windfield and so may occur with the high-pressure cells to the east of easterly waves or the highs ahead of and behind cold fronts. They form most often from autumn to spring, but are by no means absent during summer when they are to be observed on most days when no rain is experienced. In the winter part of the year they may be remarkably persistent. On one occasion, the ~500 hPa layer was observed to oscillate up and down about its mean height of ~500 hPa for 40 days without

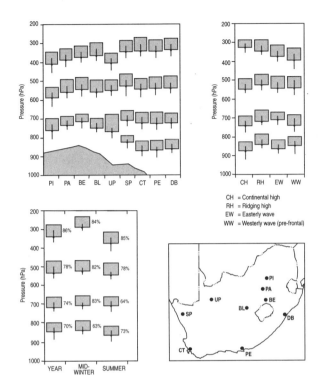

Fig. 12.15   The occurrence of absolutely stable layers (where the observed lapse rate is less than the saturated adiabatic lapse rate) on no-rain days; *upper left*: over South Africa; *upper right*: by circulation type; *lower left*: by time of year (with percentage frequencies of occurrence); and *lower right*: the location of stations for which analyses were completed (PI denotes Pietersburg, PA Pretoria, BE Bethlehem, BL Bloemfontein, UP Upington, SP Springbok, DB Durban, PE Port Elizabeth and CT Cape Town). The results are based on the analysis of a total of 2 925 radiosonde ascents taken over the 7-year period 1986–1992 (after Cosijn and Tyson, 1996). The depth of the layers is indicated by the vertical extent of the shaded boxes; the horizontal extent has no meaning; vertical lines give the standard deviation of base height.

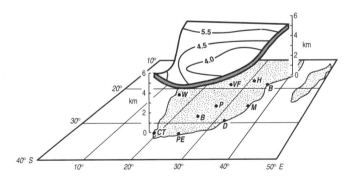

Fig. 12.16   An illustration of the spatial extent and height variation (km) of the ~500 hPa absolutely stable layer during early spring (after Garstang et al., 1996). CT denotes Cape Town, PE Port Elizabeth, B Bloemfontein, D Durban, P Pretoria, W Windhoek, M Maputo, B Beira, H Harare and VF Victoria Falls.

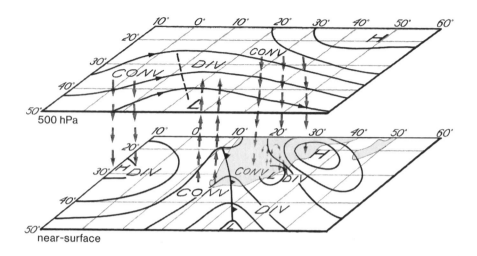

Fig. 12.17   A schematic representation of the near-surface and 500 hPa circulation associated with Berg winds and coastal lows.

being dissipated. Below it, the ~700 hPa layer was broken temporarily about every six or seven days with the passing of cold fronts over South Africa.

Coastal lows and Berg winds

Coastal lows and Berg winds usually occur together and do much to determine the characteristic features of coastal and adjacent inland climates. Coastal lows owe their origin to the generation of cyclonic vorticity with eastward movement of air off the high interior plateau, that is with large-scale flow having a substantial velocity component normal to the coastline (see Chapter 10 and Figure 10.13). Without the topographic configuration of the plateau, escarpment and coastal littoral, they would not occur. Usually coastal lows are initiated on the west coast and propagate southward to Cape Town. Thereafter they move eastward and north-eastward around the coast as internal Kelvin waves trapped vertically beneath a strong low-level subsidence inversion and horizontally on the landward side by

the Escarpment. All coastal lows produce warm offshore airflow ahead of the system and cool onshore airflow behind (Fig. 12.17). The temperature and wind-shift characteristics associated with coastal lows resemble those of cold fronts and often may be mistaken for such. The systems are confined to coastal areas, are shallow (seldom deeper than 850 hPa, i.e., about 1 500 m), are decoupled from the wave usually present above the surface layer and seldom produce localized precipitation in excess of mist or fine drizzle. They certainly may not be recognized as major rain-producing systems. The systems seldom extend inland of the Western Cape mountains and over KwaZulu-Natal often no further inland than the Midlands. On occasions they may reach the Escarpment, but will never move across it. Should the coastal low become coupled to the wave above, it may develop into a coastal depression that will produce rainfall on a larger scale.

Berg winds are important features of coastal climates and are associated with large-scale pre-frontal divergence and dynamic warming of subsiding air moving offshore with anti-

Fig. 12.18 *Left:* winter Berg-wind dust plumes blowing out to sea over the South Atlantic; *right:* a Namib coastal low south of a major stratus fog bank.

cyclonic curvature. The dynamic warming is such that even on the plateau inland of the Escarpment positive temperature departures may be experienced. Additional adiabatic warming of the air as it descends from the interior plateau enhances the warming effect. Berg winds may blow for several days or only for a few hours. The winds are most common in late winter and early spring. They result in the anomaly of highest maximum temperatures being recorded in winter at coastal stations on the east coast. Almost all rapid rises in temperature from day to day in coastal areas are the result of Berg wind warming (Fig. 12.14). On the west coast strong offshore Berg winds may produce clearly visible plumes of dust extending out to sea (Fig. 12.18).

The sequential links between a Berg wind, a coastal low and the westerly wave and cold front that follows the systems are often clearly illustrated in autographic weather records. A spring example is given in Figure 12.19 for Durban. On this occasion the Berg wind was short-lived and weak. The coastal low was heralded by the onset of the southerly *buster*, when the temperature fell rapidly and pressure began to rise. Only at 02:30 on the next day did the cold front pass and the wind veer to south-westerly. The depth of the coastal low did not extend beyond the 850 hPa level. While the temperature trace did not reveal a large further drop in temperature with the passage of the cold front, the comparison of tephigrams for 14:00 on *days 1* and *2* shows that the atmosphere underwent substantial deep cooling as a result of the westerly wave disturbance. The phenomenon that the coastal low produces greater surface cooling than the cold front and that the opposite occurs with deep cooling is of frequent occurrence (see later in Figure 12.38 for a further illustration of this effect).

TROPICAL DISTURBANCES IN THE EASTERLIES

Easterly waves and lows

Disturbances in the tropical easterly flow occurring around the northern sector of subtropical anticyclones take the form of easterly waves and lows. They are usually associated with the Inter-Tropical Convergence Zone and the warm, humid easterly winds between the Zone and the subtropical high-pressure belt. Unlike their counterparts over the Sahel, the tropical North Atlantic Ocean and the Caribbean Sea, *easterly waves* over southern Africa are semi-stationary in character. They form in deep easterly currents in the vicinity of an easterly jet. The waves are barotropic and the perturbations usually take the form of open waves or cool-cored closed lows which are evident at lower levels (850 hPa and 700 hPa),

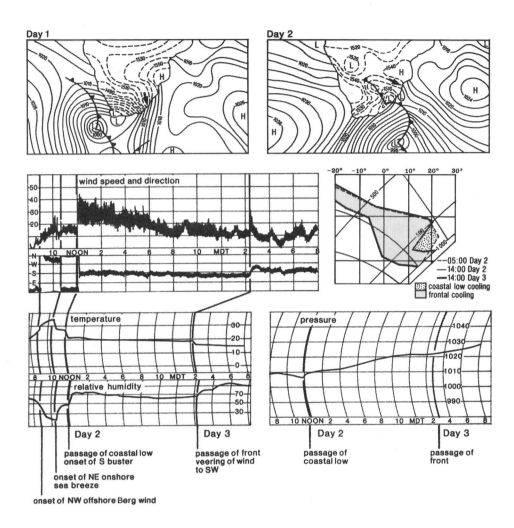

Fig. 12.19 Surface synoptic charts at 14:00, autographic records and tephigrams to show Berg-wind warming and coastal-low cooling effects at Durban on *day 2*. The area affected by the Berg wind is indicated by the heavy arrows on the synoptic charts.

weakly developed at 500 hPa and generally absent in the upper troposphere, where they are replaced by a warm-cored ridge of high pressure (Fig. 12.20). Since the systems are barotropic, their axes are not displaced with height as is the case with their baroclinic counterparts in the westerlies.

Low-level convergence occurs to the east of the surface trough, while at 500 hPa or above the flow is divergent. The consequence is strong uplift, which may sustain rainfall in the absence of pronounced instability. With the presence of unstable air, good rains, often occurring in rainy spells of a few days' duration, may occur over wide areas to the east of the trough in association with winds of a northerly component. Ahead of and to the west of the trough, and consequent upon surface divergence and upper-level convergence, subsidence ensures no rainfall, clear skies and hot conditions.

With an *easterly low*, surface convergence occurs to the east of the low with the diver-

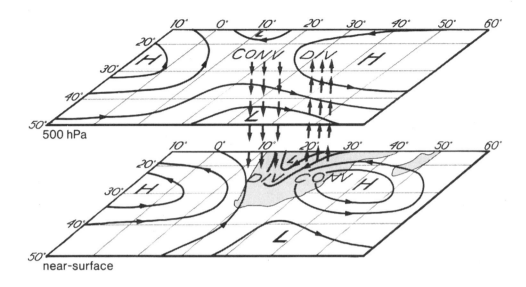

Fig. 12.20 A schematic representation of the near-surface and 500 hPa circulation associated with an easterly wave.

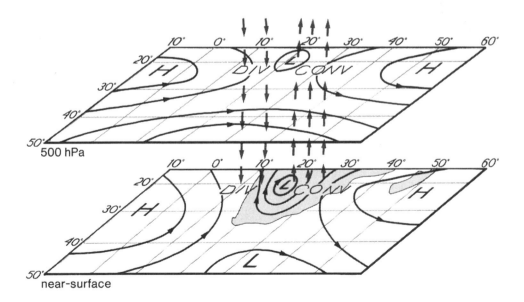

Fig. 12.21 A schematic representation of the near-surface and 500 hPa circulation associated with an easterly low.

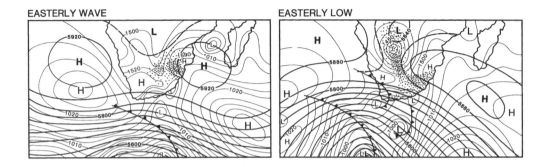

Fig. 12.22 Examples of a summer easterly wave situation (*left*) and a summer easterly low type (*right*). Light lines show isobars at mean sea level (hPa) over the oceans and contours of the 850 hPa surface over the land; heavy lines show contours of the 500 hPa surface. Areas receiving precipitation are stippled.

gence occurring at higher levels in the troposphere than in the case of easterly waves. Strong uplift occurs through the 500 hPa level with commensurate subsidence to the west of the system (Fig. 12.21). Copious rains may occur, again in association with airflow having a northerly component. It has long been realized that general rains over the plateau originate from a northerly direction. It is now appreciated that it is largely the occurrence of these rains which distinguish abnormally wet years from dry years.

Typical 850 hPa and 500 hPa circulation patterns associated with a midsummer tropical-ly induced easterly wave disturbance that produced extensive scattered rainfall over northeastern South Africa are given in Figure 12.22. An example of a deep summer easterly low that produced rainfall of a more general nature is also given. Satellite photographs show typical cloud distributions associated with the uplift to the east of the easterly troughs and lows and the clear air resulting from the subsidence over western areas (Fig. 12.23).

Tropical disturbances are almost exclusively a summer phenomenon and show a clear annual cycle peaking between December and February (Table 12.1). They occur rarely

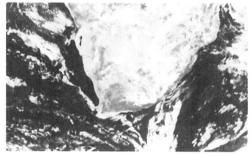

Fig. 12.23 Cloud associated, *left*, with a weak easterly wave and weak upper-level divergence over central South Africa, and *right*, with a deeper wave and stronger upper-level divergence over the same area.

between April and October. The annual cycle of easterly waves and lows does much to control the summer rainfall over the interior.

The surface trough of an easterly wave is usually associated with a well-defined boundary separating moist air to the north-east from drier air to the south-west (see Fig. 9.20). Thunderstorms tend to develop within a convergence zone extending about 200–300 km north-eastward of the moisture discontinuity and often form squall lines that advance from the south-west and are steered by the airflow at about 500 hPa (see later in this chapter).

Tropical disturbances are not only influenced by northward-penetrating mid-latitude disturbances, but also exert a reciprocal influence on the mid-latitude circulation. Both these points will be considered later in relation to the development of tropical-temperate convergence zones and cloud bands.

Subtropical lows

Cyclonic vortices, often limited to the upper troposphere and not evident at 850 hPa, are occasionally observed over southern Africa at times when upper westerly waves are remote and when a strong surface anticyclonic circulation is centred over the eastern subcontinent. Subtropical lows are usually associated with clusters of cumulonimbus clouds extending south from the tropics and are similar to the upper-level cold-cored lows reported elsewhere in the subtropics. They may be associated with heavy rainfall over parts of South Africa and tend to occur mostly between September and March.

TEMPERATE DISTURBANCES IN THE WESTERLIES

Westerly perturbations are more varied than the disturbances that occur in the tropical easterlies. Six classes of baroclinic disturbances have been recognized in the classification given in Figure 12.10. Although they are presented as discretely occurring types, this is often not the case and from day to day one may merge with and change into another.

Westerly waves

Unlike their barotropic easterly counterparts, westerly waves are baroclinic, Rossby waves and normally are tilted westward with height. At the surface closed isobars may be present; in the upper air a closed circulation does not occur. Westerly waves have a peak frequency of occurrence in the range of 2–8 days. The conservation of absolute vorticity and the changing radius of curvature in the wave produce surface convergence to the rear of the trough in south-westerly airflow, whereas at the 500 hPa level and above, divergence occurs to the east and ahead of the trough line (Fig. 12.24). Coincident surface convergence and upper-air divergence provide ideal conditions for sustained and gentle uplift of air. To the rear of the surface trough, cloud and precipitation occur in unstable air, whereas ahead of the feature, subsidence and stable conditions ensure that clear, fine conditions develop. Clouds associated with troughs on the east coast may extend into the interior, but are usually separated from tropical cloud systems to the north-west. The low-pressure system may be deep and may be associated with heavy rain. Although the effects of westerly waves may be experienced as far north as the tropics, heavy rainfall originating from the systems per se seldom occurs inland of the Escarpment. However, inland rainfall may be induced indirectly through organized squall-line convection occurring in association with the fronts that often accompany the waves. An example of a westerly wave is given in Figure 12.25. It is closed at the surface but not at 500 hPa, and is illustrative of systems producing light coastal rainfall during early summer as they traverse the southern coast of South Africa.

Westerly troughs producing heavy rains occur most frequently between October and

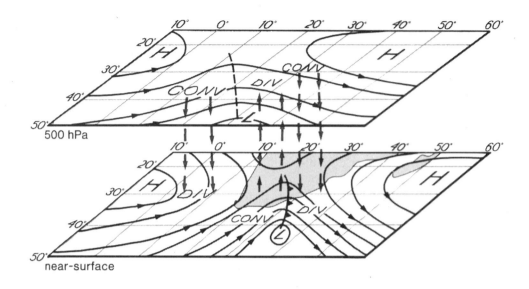

Fig. 12.24 A schematic representation of the near-surface and 500 hPa circulation associated with a westerly wave.

April and seldom between May and September. They appear to be modulated on a semi-annual cycle with the early summer and late autumn periods showing maximum frequencies of occurrence (Table 12.1).

Cut-off lows

A more intense form of westerly trough is the cut-off low, a feature that is a cold-cored depression, which starts as a trough in the upper

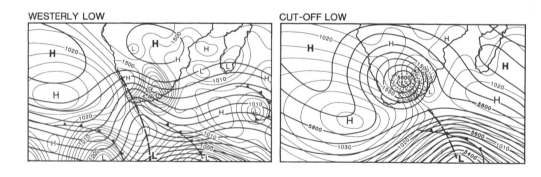

Fig. 12.25 Examples of a westerly wave and/or low type in early summer (*left*) and a late winter-early spring cut-off low type (*right*) (the latter taken from Hayward and van den Berg, 1970). Light lines show isobars at mean sea level (hPa) over the oceans and contours of the 850 hPa surface (gpm) over the land; heavy lines show contours of the 500 hPa surface (gpm). Areas receiving precipitation are stippled.

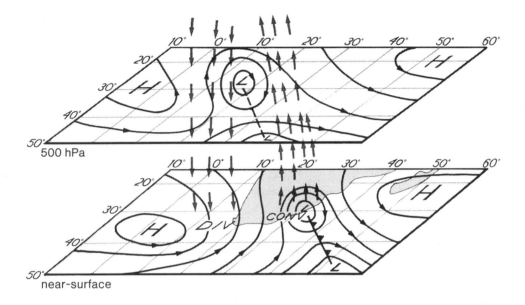

Fig. 12.26 A schematic representation of the near-surface and 500 hPa circulation associated with a cut-off low.

westerlies and deepens into a closed circulation extending downward to the surface and which becomes displaced equatorward out of the basic westerly current. Cut-off lows are unstable, baroclinic systems that slope to the west with increasing height and are associated with strong convergence and vertical motion, particularly while they are deepening. The characteristic vertical motion associated with these features is illustrated schematically in Figure 12.26. The source of major divergence necessary to act together with the surface convergence to produce the deep uplift that is observed in cut-off lows occurs at a level much higher than the 500 hPa surface. Cut-off lows account for many of the flood-producing rains observed over South Africa. Such was the case with the Laingsburg floods of 1981 and the KwaZulu-Natal flood disaster of 1987.

It is clear from Table 12.1 that the frequency of cut-off lows producing heavy rains shows a semi-annual variation with peaks in March to May and September to November. The lowest frequencies of cut-off lows occur between December and February. In Figure 12.25 an exam-ple is given of a strong, deep spring cut-off low that produced flooding over many parts of eastern South Africa. The satellite photographs given in Figure 12.27 illustrate the cloud patterns associated with westerly lows and cut-off lows of the kinds illustrated in Figure 12.25.

Disturbances in the westerlies are often associated with clearly defined comma-cloud patterns originating in the interaction between warm and cold conveyors in the manner discussed previously in Chapter 10. A spectacular example of a comma cloud to the south of Cape Town is given in Figure 12.28. A sequence of examples illustrating the stages in the evolution and decay of a depression in the westerlies in the southern African sector of the southern hemisphere is shown in Figure 12.29.

Southerly meridional flow

The southerly meridional flow type takes its name from the surface circulation pattern over the ocean to the south of the subcontinent. This pattern exhibits a strong zonal pressure

 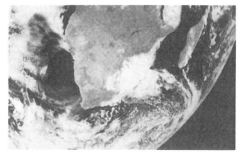

Fig. 12.27 Cloud distributions over eastern South Africa resulting from a westerly wave (*left*) and a deep cut-off low (*right*).

Fig. 12.28 A mature depression associated with a pronounced comma cloud south of southern Africa in autumn. The distinct cold front and more diffuse warm front to the east of KwaZulu-Natal are clearly discernible.

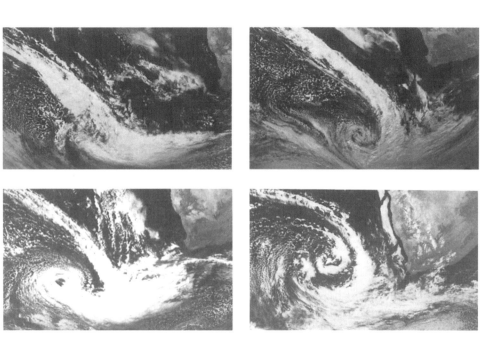

Fig. 12.29 Examples of stages of the development of westerly lows affecting southern Africa.

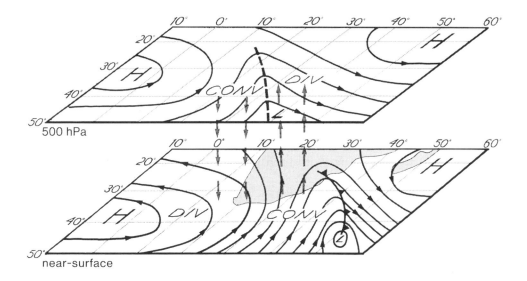

Fig. 12.30 A schematic representation of the near-surface and 500 hPa circulation associated with southerly meridional flow.

gradient between a high to the west and a low to the east. The 500 hPa circulation shows a clear westerly wave in which the region of upper-level divergence ahead of the trough overlies the convergence zone west of the surface low and cold front (Fig. 12.30). The resulting vertical motion sustains the rainfall, which tends to be confined to coastal regions and to the Lowveld of eastern South Africa where the

cyclonic curvature of the contours of the 850 hPa surface enhances convective activity.

A southerly meridional circulation type producing light spring rainfall over southern regions is illustrated in Figure 12.31. The Lowveld also experienced rain on this occasion. Under the influence of this type of synoptic situation temperatures drop sharply over most of South Africa, particularly in southern regions.

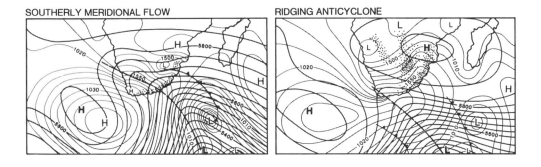

Fig. 12.31 An example of a spring southerly meridional flow type (*left*) and a summer ridging anticyclone type (*right*). Light lines show isobars at mean sea level (hPa) over the oceans and contours of the 850 hPa surface (gpm) over the land; heavy lines show contours of the 500 hPa surface (gpm). Areas receiving precipitation are stippled.

Ridging anticyclones

A ridging anticyclone associated with a westerly wave at 500 hPa often produces widespread general rainfall over the eastern parts of southern Africa. At the surface over the Indian Ocean and adjacent inland areas, steep pressure gradients promote strong advection of moist, unstable air over the land. Weakening inland pressure gradients, the changing curvature of the flow, mesoscale orographic forcing, and upper-level divergence in the westerly wave combine to produce widespread uplift over eastern regions (Fig. 12.32). General rains or thunderstorm activity may result. Simultaneously over the Western Cape the anticyclonic conditions and attendant subsidence bring clear, fine and hot weather, often accompanied by strong south-easters. The contrast between the fine, south-easter weather over the Western Cape and the inland plateau rains gives rise to the popular (and correct) perception of a link between the two conditions. Ridging anticyclones produce extensive cloud along the southern and eastern coastal and adjacent inland areas of South Africa (Fig. 12.33, *left*). Sometimes the cloud is not deep enough to rise above the Drakensberg mountains (Fig. 12.33, *right*).

Analysis of the surface expression of ridging anticyclones reveals that this circulation type brings rainfall to eastern South Africa throughout the summer rainfall season from October to May, but with a slight tendency for maximum frequencies of occurrence in October and February (Table 12.1).

A summer example of a ridging anticyclone at the surface (with an attendant upper westerly wave and ridge over north-eastern areas at 500 hPa) is given in Figure 12.31. The system brought good rains to KwaZulu-Natal and other eastern and central areas.

West-coast troughs

It has long been known that the occurrence of a surface trough of low pressure over the west coast and an upper-tropospheric westerly wave to the west of the continent result in a situation conducive to widespread rains over the western parts of South Africa. Surface convergence and

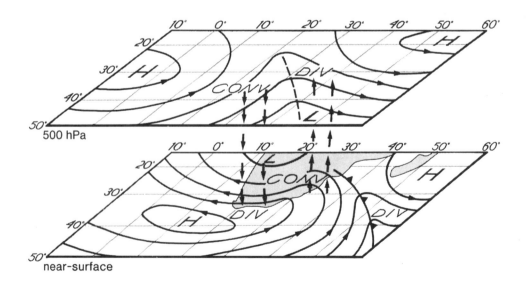

Fig. 12.32 A schematic representation of the near-surface and 500 hPa circulation associated with a ridging anticyclone.

Fig. 12.33 *Left:* cloud associated with a deep ridging anticyclone; *right:* stratocumulus cloud below the Drakensberg having formed in a shallow airstream with a ridging anticyclone (photo: D. Dodds).

upper-level divergence ahead of the trough combine to produce the conditions necessary for general upward vertical motion (Fig. 12.34). Cyclonic disturbances originating as west-coast troughs tend to follow a south-easterly trajectory as they move into the Indian Ocean. They also appear to develop only at times of active tropical convection, suggesting a dynamic link with the easterlies in their formation. Examples of early-summer and autumn west-coast troughs producing rainfall over central and western parts of South Africa are given in Figure 12.35.

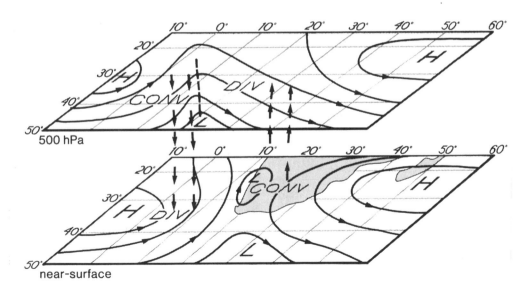

Fig. 12.34 A schematic representation of the near-surface and 500 hPa circulation associated with a west-coast trough.

WEST-COAST TROUGH

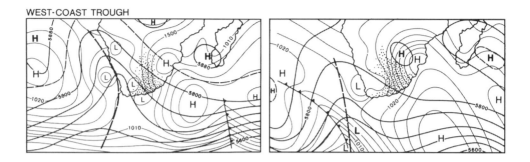

Fig. 12.35 Examples of the west-coast trough type; *left:* strongly developed on one summer occasion; and *right:* less well developed on another occasion in late summer. Light lines show isobars at mean sea level (hPa) over the oceans and contours of the 850 hPa surface (gpm) over the land; heavy lines show contours of the 500 hPa surface (gpm). Areas receiving precipitation are stippled.

Cold fronts

The final synoptic type in the classification is the cold front. These have been referred to in many connections already in this chapter and they are important determinants of the weather and climate of southern Africa, particularly southern South Africa. Cold fronts occur together with westerly waves, depressions, or cut-off lows and should not be considered in isolation from these systems. However, as fronts are associated with invasions of cold air from the south and south-west that produce characteristic cold snaps of a few days' duration,

it is useful to discuss them separately. They occur most frequently in winter when the amplitude of westerly disturbances is greatest. They also produce conditions favourable for the promotion of convection to the rear of the front where low-level convergence in airflow with a marked southerly component is at a maximum. Ahead of the front, in airflow with a pronounced northerly component, divergence and subsidence are responsible for stable and generally cloud-free conditions. Cold fronts are associated with distinctive bands of clouds which may extend far inland (Fig. 12.36). The wind shift across the front is usually associated with

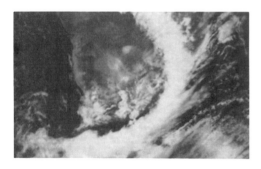
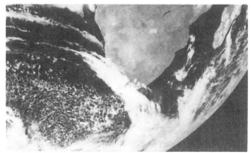

Fig. 12.36 Cold-front cloud formation associated with spring depressions. In both cases the clear conditions in the subsiding air ahead of the fronts and the suppressed cumulus-cloud formation in similarly sinking air to the rear of the features are apparent.

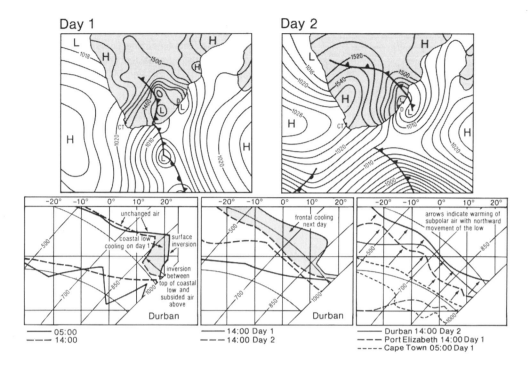

Fig. 12.37 The vertical extent of cold-front cooling associated with the passage of a late-winter westerly low (*lower left* and *centre*), together with the warming of the air (*lower right*) as the system moves to the north-east.

a pronounced reversal of direction and changes in wind speed and gustiness, although, as was pointed out earlier, the passage of the preceding coastal low may produce even more pronounced shifts at the surface.

Tephigrams to show the passage of a typical cold front at Durban in late winter are given in Figure 12.37. On *day 1* the movement of the coastal low had produced pronounced cooling to only the 850 hPa level by 14:00. By the same time the next day the cold front had produced deep cooling to above the 500 hPa level. The extent of the modification and warming of the post-frontal air as the front moved towards the north-east is also evident in this example. Part of this warming would have been due to radiational heating and part to mixing with warmer air being advected in from the Indian Ocean. (A further example of deep cold-front cooling was included in Figure 12.9.)

At and near the coast, the surface contrasts between pre-frontal Berg wind conditions on the one hand and post-coastal-low and post-frontal conditions on the other may be most pronounced, as is shown by the example of Pietermaritzburg autographic records given in Figure 12.38. Similar contrasts may also be experienced further inland. However, in such cases, the coastal low effect is rapidly diminished with increasing distance inland, so that away from coastal areas the contrast is usually between pre-frontal Berg wind conditions and the post-frontal conditions behind the cold front. In both cases the change from warm (or hot) to cool (or cold) conditions may be spectacular in the regions affected (Table 12.2).

The eastern parts of South Africa are most affected by rapid drops in temperature from one day to the next (Fig. 12.39). Cold snaps produced by cold fronts last about two days on

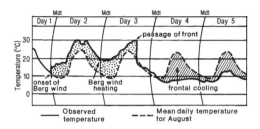

Fig. 12.38 Continuous temperature records to show the onset and occurrence of Berg-wind warming and the initiation and occurrence of coastal-low/cold-front cooling over Pietermaritzburg in late winter.

Table 12.2 Departures from mean daily maximum and minimum temperatures (°C) in pre-frontal and post-frontal conditions over eastern South Africa with the passage of a spring cold front over the region.

	Pre-frontal		Post-frontal	
	From max.	From min.	From max.	From min.
Durban	5	2	−2	−3
Pietermaritzburg	9	4	−11	−10
Newcastle	7	5	−12	−7
Bethlehem	6	1	1	0

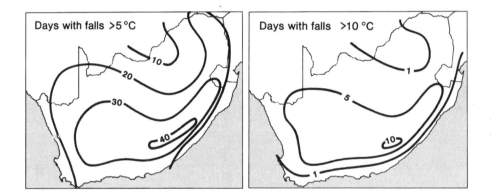

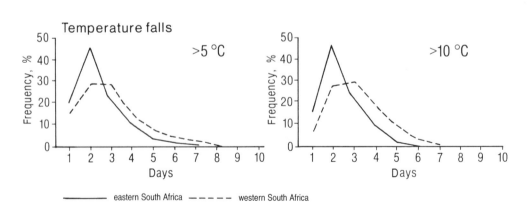

Fig. 12.39 *Upper:* days per annum recording falls of temperature exceeding 5 °C and 10 °C; *lower:* frequencies of duration of cold snaps with temperature falls greater than 5 °C and 10° C in eastern and western South Africa (after Schulze, 1984).

between 40 and 50 per cent of occasions; on about 10 per cent of occasions they last four days. In most cases the cold snaps are of somewhat longer duration in western parts of South Africa. In general, cold snaps are of shorter duration than heat waves (cf. Figs. 12.14 and 12.39).

The reality of summer cold fronts and their passage across South Africa has long been a matter for debate. However, it is increasingly being accepted that such fronts may be associated with severe weather conditions over the interior. They are associated with marked mesoscale wind shears and low-level convergence. Although changes in pressure and frontal temperature discontinuities may be small, the effect of frontal penetration into the tropics may be considerable.

OTHER IMPORTANT RAIN-PRODUCING SYSTEMS

Thunderstorms

The mechanism and nature of cumulus convection were presented in Chapter 9, where a good deal of the material related directly to southern Africa. A similar discussion does not need to be presented here. Instead, this section should be read in conjunction with the earlier chapter.

Much of the rainfall received in the summer-rainfall region is of convective origin. The degree of convectivity is modulated not only by the diurnal heating cycle of the near-surface air, but also by the prevailing dynamics of particular synoptic conditions and by mesoscale and local effects. Owing to the strong influence of diurnal heating and atmospheric instability on the development of the convective process, rainfall over much of South Africa shows a clear diurnal variability. In general, over inland areas rain falls most frequently during the afternoon and early evening. Storms with high rainfall rates tend to show this pattern of variation most clearly. Disturbances producing low rainfall rates show far less diurnal variation. This fact is well illustrated over the south-western and southern parts of South Africa where precipitation tends to be most frequent at night or in the early morning. In the coastal regions of KwaZulu-Natal local offshore mountain–plain winds blowing from the Drakensberg to the sea may produce late-evening rainfall peaks.

A measure of the spatial distribution of cumulus convective activity may be inferred from the distribution of thunder and hail days (Fig. 12.40). The eastern high plateau areas are

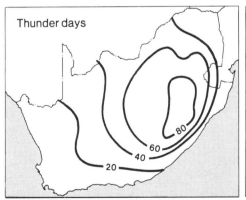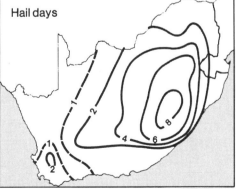

Fig. 12.40   Generalized distributions of thunder (*left*) and hail days (*right*) per year (after Schulze, 1972).

affected most. In the mountainous areas of south-western South Africa hail tends to be soft and of the graupel type. In contrast, over the Highveld hard hail is more usual. Hail falls mostly during the late afternoon and early evening. Over the Highveld the average number of hailstorms per year at any one point is about five. These tend to occur most frequently in late spring, usually November, when lapse rates are steep and temperatures high. Over Lesotho the point frequency of hail days rises to over eight occurrences per year. Areal frequencies of hail days may be considerably higher with values of about 100 days of hail per year likely over the high plateau of KwaZulu-Natal and Lesotho, in comparison with 20 to 30 days over the coastal area of KwaZulu-Natal. In the vicinity of Johannesburg and Pretoria, hail occurs on 69 days of the year on average. On half of these days hailstones are no larger than 1 cm in diameter. Usually hailstones are not larger than 3 cm and are highly variable in occurrence. Severe storms occur most frequently early in the season and those that produce the largest hailstones have the greatest areal extent. On any one hail day within the Johannesburg-Pretoria network the number of storm cells may exceed seven, and over thirty cells may be observed within a three-hour period. The incidence of hail varies with topography, not only with the general relief of southern Africa, but also on a local scale, as for example over Johannesburg and lower areas to the north. In addition, more hail is observed over urban areas than in rural surroundings owing to urban heat-island effects.

Hailfall at a point may show extreme variability owing to the movement of irregularly shaped cells, repeated expansion and contraction of cells and the passage of more than one cell past a given point. Over the Witwatersrand area the average duration of point hailfall is 7.3 minutes as storms move downwind along their individual tracks. Individual storm cells last 18–30 minutes and move at speeds of about 25 km h^{-1}. Hail tracks may be irregular, but more often produce clearly defined quasi-linear

swathes. Patchiness within these swathes is common and illustrates the unsteadiness of the hail production process.

The greatest intensities of rainfall generally occur in the storms of the Highveld and Escarpment where rainfall exceeding 10 mm per hour occurs in more than 7 per cent of all storms. Intensities in excess of 100 mm in one hour have been recorded in some instances, for example a maximum rate of 118 mm per hour was recorded in the Port Elizabeth storm of 1 September 1968 when 552 mm fell in 24 hours. During the Laingsburg floods of 25 January 1981, more than 180 mm of rain fell in one day. The dynamics of such rare severe storms are always much more complicated than those of normal cumulus convection. Cumulus convection is a dominant rainfall-producing process over most parts of southern Africa. However, it is a process controlled not only by the cloud microphysical processes, diurnal heating and mesoscale forcing, but also by synoptic conditions that determine the dynamic control (vertical motion) and thermal environment (stability) in which convective systems can grow. Many different types of synoptic situation are conducive to storm formation. All are such that appropriate divergence fields provide the necessary uplift to realize thermal instability and to sustain cloud growth. A common synoptic situation promoting storm formation is a trough in the tropical easterlies. Frontal passages also frequently trigger storms. Often no single mechanism is responsible for a major storm and usually synoptic and mesoscale forcing as well as the local heating cycle combine to produce severe storm generation (see Fig. 9.10).

Storms over South Africa are frequently organized in a north-west to south-east alignment that may often be observed extending from the Namib to the east coast. Part of a band of cumulus and cumulonimbus clouds associated with such a storm belt is illustrated in Figure 12.41. One of the earliest thunderstorm models for southern Africa was that proposed by Taljaard for line storms developing along a sur-

Fig. 12.41 Part of a midsummer north-west to south-east belt of mid-afternoon cumulus and cumulonimbus clouds extending from the Namib to the east coast. (Notice the stratus cloud over the Benguela current.)

face moisture discontinuity in the presence of strong vertical wind shear (see Fig. 9.20). The model anticipated many of the features of the Ludlam-Browning and subsequent models for supercell storms and has been used to explain some South African storms. While not common, supercell storms do occur from time to time over the Highveld. Taljaard envisaged storms being fed by moist, unstable, surface air with a northerly component and steered by mid-tropospheric winds, most often from the south-west. Subsequent research has confirmed these ideas. Between 1971 and 1975, two-thirds of all new cells formed on the equatorward side of existing cells (54 per cent to the left; 23 per cent ahead) (see Fig. 9.11). Average storm speed was 30 km h⁻¹ and the deviation from the 500 hPa winds was 57 per cent to the left and 27 per cent to the right. In the northern hemisphere deviation is generally to the right of mid-level winds. Over the Highveld of South Africa scattered thunderstorms occur on about 54 per cent and isolated storms on about 39 per cent of storm days. Both types show little organized spatial development. Squall-line storms are least frequent, occur on about 7 per cent of storm days and are highly organized spatially. Multi-cellular structures occur most frequently with scattered storms (72 per cent). Multi-cell isolated storms only occur on 29 per cent of storm days. Multi-cell storms of one kind or another account for the majority of storms on more than half of all storm days over the Witwatersrand-Pretoria area.

Despite the fact that they are least frequent, line storms are important contributors to the rainfall of plateau regions of southern Africa, particularly from November to January. Warm, moist, northerly surface air fuels the convection; dry upper air with a southerly component steers the storms and allows cooling through evaporation of precipitation in the main convective towers and anvils of the cumulonimbus clouds; the vertical shear in the horizontal wind is instrumental in maintaining the necessary patterns of convergence and divergence and propagation of the squall line, and invigoration of cells occurs with the onset of precipitation and the initiation of downdraughts. Whereas individual cells may have lifetimes of an hour or less, the system as a whole may last much longer. Most storms over central South Africa move in westerly to south-westerly tracks (see Figs. 9.11 and 9.14). Squall-line systems generally behave differently from the sum of the component thunderstorm cells, owing to their greater degree of synoptic forcing. Scattered and isolated thunderstorms are less the result of synoptic forcing and more the consequence of other factors. In all cases a non-suppressive synoptic situation is a necessary condition for storm generation. The limited degree to which entirely local convective processes may sustain rainfall is illustrated by an example from the Mpumalanga Escarpment area near Nelspruit where, during the 1984–85 season, stratiform synoptic conditions produced rain on 32 per cent of days, convective synoptic situations were responsible for 23 per cent of rain days and only 22 per cent of rain days could be classified as resulting entirely from local convective conditions. The remaining 23 per cent of rain days were days when non-significant rainfall was recorded.

Tropical cyclones

Tropical cyclones regularly form over the Indian Ocean in summer. They usually recurve to the south before reaching land, often moving south along the Mozambique Channel (see Fig. 10.36). The annual frequency of cyclones occurring in the south-west Indian Ocean

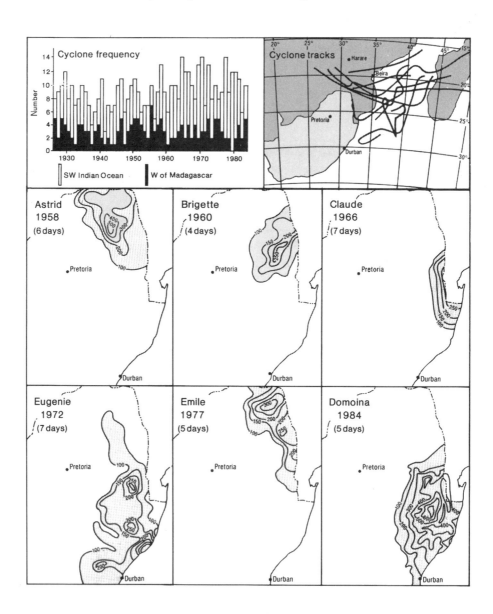

Fig. 12.42 *Upper left:* annual frequencies of tropical cyclones in the south-west Indian Ocean and those passing to the west of Madagascar; *upper right:* tracks of cyclones causing significant rains over eastern South Africa since 1950; *centre* and *lower:* rainfall amounts recorded during rainy spells produced by individual cyclones between 1955 and 1984 (after Dunn, 1985).

between 1927 and 1984 is given in Figure 12.42 (*upper*). Usually between six and twelve storms are observed in any one year, with one to five passing south between Madagascar and Africa. The lowest cyclone activity occurred in 1938, 1942 and 1957, with the highest frequencies in 1959, 1965, 1971 and 1978. Most cyclones occur in January or February. Those recurving south in the Mozambique Channel seldom penetrate to 30° S. Occasionally the storms cross the coast, but seldom move inland of the Escarpment. The storms are always associated with copious rainfall and flooding and are maintained by the release of latent heat in the manner described in Chapter 9. Once they move inland, the latent energy source diminishes as evaporation decreases and the storms decay rapidly. Nonetheless, the inland penetration of decaying tropical storms may have devastating effects, as was the case with the flooding produced by cyclones Domoina and Imboa in 1983/84 (see Fig. 10.36 for a satellite image of Domoina). Rainfall distributions associated with the six major cyclones between the mid-fifties and mid-eighties are given in Figure 12.42 (*lower*). On all occasions maximum rainfall in excess of 250 mm was precipitated in rainy spells of four to seven days. With Astrid in 1958 and Domoina in 1984 maximum precipitation exceeded 500 mm over six- and five-day spells respectively. Heavy rainfall occurs over east coast and adjacent inland areas only. To the west, over the central and western plateau, fine and dry conditions prevail.

NECESSITY AND SUFFICIENCY OF SYNOPTIC AND THERMAL FORCING

Not all situations associated with the various circulation types discussed in this chapter will produce rainfall. Much depends on the moisture fluxes, the degree of instability and the divergence field prevailing at any one time and place. On some occasions thermal instability in moist air may be adequate to produce rainfall, but uplift may be insufficient to realize the

instability. At other times surface convergence and upper-level divergence may not be strong enough to produce sufficient uplift in stable air to result in precipitation, and only cloud will form. On yet other occasions all of the many factors necessary to produce precipitation may act together to result in heavy rains. If Figure 12.35 is considered, then from a reading of the surface and 500 hPa circulation fields alone it is reasonable to assume that the deeper west coast trough of 19 November should have produced heavier rainfalls than did the less-developed trough of 28 March; in fact the opposite occurred. The circulation fields were necessary to cause the uplift, but they were not sufficient to ensure that the expected amounts of rainfall occurred.

Frequently an easterly wave at the surface appears conducive to rain, whereas the upper circulation and divergence fields are not and no rain falls. Two summer examples illustrate the point in Figure 12.43. On both days the atmosphere showed conditional instability to above the 300 hPa level. On the rain day the surface convergence to the east of the 850 hPa trough combined with the upper-level divergence in the westerly wave at 500 hPa to produce conditions ideal for uplift. This uplift realized the conditional instability and widespread rain occurred. (On some occasions the uplift may be sufficient to sustain stratiform-cloud development and precipitation, even though instability may not be present.) On the no-rain day conditional instability was present, the surface circulation pattern was almost identical to that of the rain day (with convergence to the east of the wave) and yet no rain fell because the upper-air flow field was inimical to such an occurrence. At 500 hPa the ridge of high pressure produced convergence to the east of that ridge and this inhibited the development of the required uplift. In this instance, the surface convergence and instability were necessary but not sufficient conditions for rainfall.

That the upper-level flow field is a vital determinant of rainfall is further illustrated for the eastern Free State region. (Fig. 12.44).

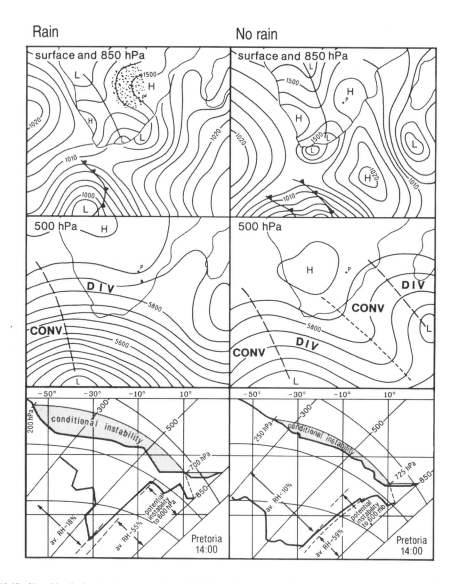

Fig. 12.43 Near-identical easterly wave surface circulation patterns on a rain day and on a no-rain day (area of rain shaded on 850 hPa chart) when the amounts of conditional and potential instability present were approximately similar. On the rain day 500 hPa divergence in a wave promoted uplift to realize the instability; on the no-rain day 500 hPa convergence in a ridge suppressed vertical motion and the occurrence of rainfall. Dry-bulb and dew-point temperatures are given on the tephigrams for Pretoria.

Above-normal rainfall (defined by the specific precipitation density (SPD) ratio of mean daily precipitation for a specific circulation type to the mean daily precipitation for all days) occurs with wave disturbances in the 300 hPa wester-

lies that are positioned such that divergence occurs over the region at that level. The more pronounced the cyclonic curvature sustaining the upper-level *divergence*, the greater is the positive rainfall anomaly (cf. 118 per cent of

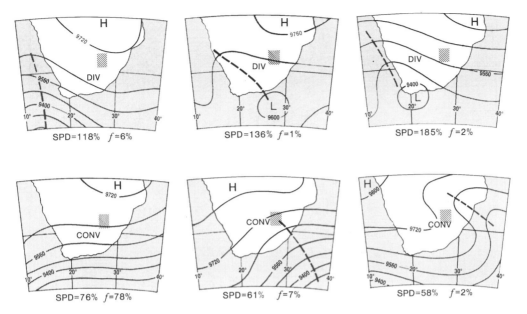

Fig. 12.44 Upper level, 300 hPa, summer circulation patterns associated with specific precipitation densities (SPD as given by the ratio of mean daily precipitation for days of specific pattern to mean daily precipitation for all days) occurring with above- and below-normal rainfall over the eastern Free State (after Steyn, 1984). Frequency of occurrence, *f*, is given for each circulation type.

normal rainfall with the *top left* pattern in Figure 12.44 and 185 per cent with the *top right*). In direct contrast, with below-normal rainfall, in each case convergence prevailed at 300 hPa over the area. Lowest rainfall was recorded with the most pronounced convergent conditions (Fig. 12.44, *bottom right*). It bears repeating that heavy rainfall will only occur when low-level convergence and upper-level divergence occur together.

COMPOSITE SYNOPTIC TYPES

The circulation types that have been included in the synoptic classification given earlier in this chapter have all been illustrated with easily classified, single-type examples. In practice various circulation types tend to occur together in complex, composite situations, often making classification difficult, if not impossible. When systems act in conjunction, heavy and widespread precipitation may occur. Many different combinations of basic circulation types are possible. Examples of four common ones will be presented here. They are a cut-off low and ridging anticyclone, a westerly wave and easterly wave, an easterly low and cut-off low, and finally, an easterly wave and low together with a ridging anticyclone (Fig. 12.45). In the first case, snow and rain fell over the South African Highveld in response to a deep cut-off low occurring in association with a ridging anticyclone to the south of the continent. Good rains were also recorded in KwaZulu-Natal (Fig. 12.45, *upper left*). In the second, the alignment of an easterly wave with its westerly counterpart produced general rains over southeastern parts of southern Africa (Fig. 12.45, *upper right*). Thirdly, to the north of South Africa a westerly trough may rapidly transform a stable tropical air mass into strong disturbances. An easterly low occurring together with a cut-off low at 500 hPa may act to produce widespread heavy rainfall to the east of the surface trough (Fig. 12.45, *lower left*).

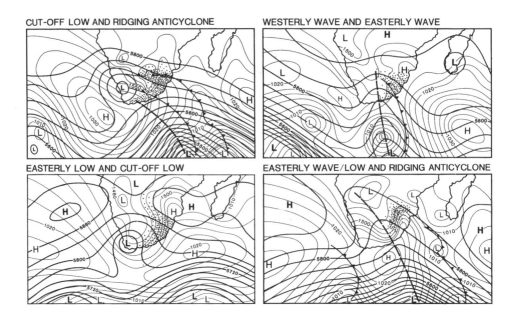

Fig. 12.45 Composite circulation types. Light lines show isobars at mean sea level (hPa) over the oceans and contours of the 850 hPa surface (gpm) over the land; heavy lines show contours of the 500 hPa surface (gpm). Areas receiving precipitation are stippled.

Finally, easterly waves and lows occurring in association with ridging anticyclones at the surface often produce good rainfall over eastern areas of South Africa (Fig. 12.45, *lower right*).

CLOUD BANDS

The interaction of westerly disturbances with those in the tropical easterlies needs to be considered further. The effect of westerly troughs on the tropical circulation may be considerable. They offer a mechanism for the extraction of heat and momentum from the tropical zone to maintain the meridional temperature gradients that drive the Hadley circulations. These strengthen during the times of active tropical convection when latent heat release increases. At such times the interaction between tropical easterly waves and mid-latitude westerly waves assumes an even more

important role. On occasions the effects of surface westerly troughs may be discernible as far north as Zambia and Zimbabwe.

The appropriate conjunction of a tropical disturbance in the form of an easterly wave or low and a westerly wave perturbation or cut-off low, and the establishment of an elongated tropical-temperate trough and convergence zone at the surface and divergence zone aloft, create ideal conditions for the development of upward vertical motion and for the formation of extended cloud bands linking tropical and middle latitudes over southern Africa. Similar sorts of cloud bands are to be found in the South Atlantic and elsewhere in the South Pacific Oceans. They mark zones of strong confluence and convergence of the low-level flow in baroclinically unstable conditions and often occur in association with surface frontal discontinuities. An example of a massive cloud band connecting tropical and temperate circulations over southern

Africa is given in Figure 12.46.

Easterly waves show an annual cycle of variation with a maximum occurrence in mid-summer; westerly troughs vary on a semi-annual cycle with maxima in March to May and September to November. Notwithstanding the latter, the combined tropical-temperate troughs and attendant cloud bands vary on the annual cycle. Two examples of cloud-band formation over southern Africa will be considered. In the first, a March easterly wave and tropical low over northern Namibia, Botswana and Angola at 850 hPa and a ridge of high pressure at 500 hPa provided the appropriate divergence fields to force tropical convection and latent-heat release over those areas (Fig. 12.47, *left*). Uplift to the east of the wave over Botswana and central South Africa sustained widespread cloud formation in those areas. To the south, convergence behind the cold front at the surface and upper-level

Fig. 12.46 A major early-summer continental-scale cloud band over southern Africa and the adjacent Indian Ocean.

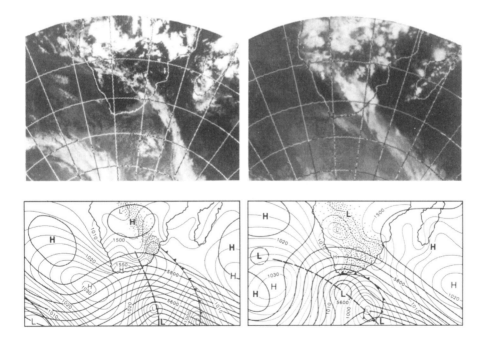

Fig. 12.47 Major summer cloud bands over southern Africa and the synoptic circulations with which they were associated. Light lines show isobars at mean sea level (hPa) over the oceans and contours of the 850 hPa surface (gpm) over the land; heavy lines show contours of the 500 hPa surface (gpm). Areas receiving precipitation are stippled.

divergence ahead of the upper wave forced extensive cloud development. Once linked, the two systems allowed the formation of a distinctive cloud band along which momentum and energy were transferred southward. Considerable precipitation occurred from this cloud band, which extended through at least 35° of latitude and 65° of longitude and constituted a major feature of the southern hemisphere general circulation on that day. The rainfall distribution over southern Africa showed a clear correlation with the cloud band.

The second example shows a distinctive cloud band formed in association with a January tropical easterly low and a westerly wave and cut-off low (Fig. 12.47, *right*). Again the cloud band was a major feature of the general circulation on that day and was associated with a similar precipitation distribution.

CIRCULATION TYPES, CLOUD SYSTEMS AND RAINFALL

Using satellite imagery and principal component analysis of rainfall, a classification of circulation has been developed to link circulation types with typical attendant cloud distributions and rainfall over South Africa. The most important type in the classification defines the major cloud bands that form in association with the tropical-temperate troughs that link tropical lows and westerly waves (Fig. 12.48). Truncated cloud bands are variants of such bands and occur when the tropical-temperate troughs are confined to land and terminate at the coast. Subtropical cyclones, comprising upper-air cold-cored lows imbedded in the easterly flow and developed independently of tropical lows over central Africa, may produce extensive convective cloud cover, often in the form of subtropical cloud clusters. West-coast troughs occurring together with active tropical convergence may be associated with west-coast bands. All these circulation types are linked to barotropic tropical circulations and produce widespread cloud

cover. Circulation types linked to the baroclinic westerly circulation exert a more local effect. Thus surface depressions on the coast, occurring together with upper-air westerly waves, produce coastal cloud-masses. Finally, east-coast troughs with their attendant cold fronts produce frontal bands of cloud. Examples of the patterns of cloud cover resulting from these circulation types are given in Figure 12.49.

Clearly the contributions to annual rainfall from the circulation types as defined above will vary from region to region over southern Africa. Coastal regions will experience very different characteristics from those of plateau regions. Over the central plateau it is the

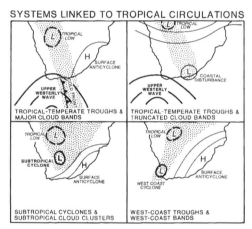

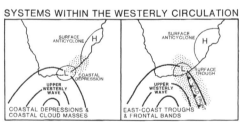

Fig. 12.48 A generalized classification of circulation types and their attendant cloud systems (after Harrison, 1986). Lower-level circulation is shown by light lines; upper-level by heavy lines. Cloud cover is shown by stippling.

tropical-temperate troughs and major cloud bands which contribute most to annual rainfall (Fig. 12.50). Thus 39 per cent of annual eastern Free State rainfall is produced by the weather systems supporting these bands. The circulation systems maintaining truncated, west-coast and frontal bands and subtropical cloud clusters and coastal cloud masses do not account individually for more than 14 per cent each; collectively they produce 46 per cent. The remaining 15 per cent of the rainfall is more local and isolated in origin.

Over the central plateau the daily amount of rainfall from each system is much the same; it is the number of rain days that determines total annual rainfall, and it is the major cloud bands that dominate the distribution of rain days (Fig. 12.51). Over the Free State these bands may contribute between 50 and 90 per cent of the observed January rainfall.

The contribution to total annual rainfall by the tropical-temperate troughs supporting major cloud bands over the central plateau of South Africa shows a clear annual cycle with

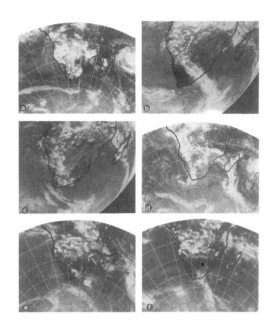

Fig. 12.49 Examples of cloud systems showing (a) a *subtropical cloud cluster*, (b) a *major cloud band*, (c) a *truncated band*, (d) a *west-coast band*, (e) a *coastal cloud mass*, and (f) a *frontal band*.

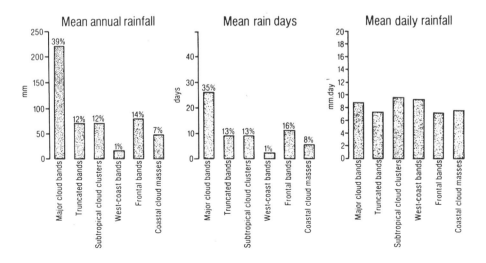

Fig. 12.50 Contributions to annual rainfall over the eastern Free State from circulation types supporting particular cloud systems (modified after Harrison, 1984). The percentages given on the tops of columns refer to the contribution to annual rainfall.

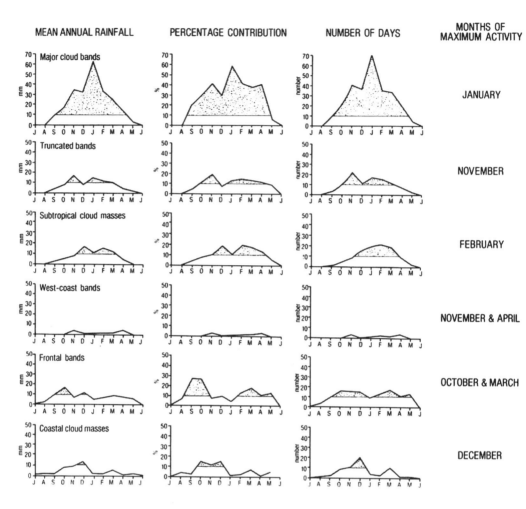

Fig. 12.51 The annual march of significant rainfall resulting from particular cloud systems over the eastern Free State (modified after Harrison, 1984). Stippling is for effect only and indicates values exceeding 10 mm, 10 per cent or a frequency of 10.

one peak per year (Fig. 12.51), in contrast with the contributions to rainfall by most other systems. These, in various degree, show tendencies to exhibit a semi-annual variation with two peaks per year, for example the east-coast troughs and coastal depressions supporting frontal bands and coastal cloud masses. It is clear that *tropically induced, rain-producing*

disturbances show a *yearly cycle* peaking in summer, whereas the *temperate disturbances* in the westerlies tend to exhibit a *half-yearly cycle* peaking in autumn and spring. The peak occurrences of westerly perturbations are in phase with the half-yearly wave in surface, 500 hPa and 200 hPa pressure in southern-hemisphere mid-latitudes.

ADDITIONAL READING

TALJAARD, J. J. 1987. 'The anomalous climate and weather systems over South Africa during summer 1975–1976'. *South African Weather Bureau Technical Paper*, No. 16.

TYSON, P. D. 1986. *Climatic Change and Variability in southern Africa*. Cape Town: Oxford University Press.

TALJAARD, J. J. 1985. Cut-off lows in the South African region. *South African Weather Bureau Technical Paper*, No. 14.

SCHULZE, B. R. 1984. *Climate of South Africa: Part 8: General Survey, WB28*. 5th Edition. Pretoria: South African Weather Bureau.

TALJAARD, J. J. 1981. The anomalous climate and weather systems of January to March 1974. *South African Weather Bureau Technical Paper*, No. 9.

SCHULZE, B. R. 1972. 'South Africa' in Griffiths, J. F. (ed.). *Climates of Africa*. World Survey of Climatology, Vol. 10. Amsterdam: Elsevier.

13

OCEAN–ATMOSPHERE INTERACTIONS

The coupling between the oceans and the atmosphere exerts a profound effect on weather and climate. The exchange of energy, matter and momentum at the ocean–atmosphere interface ensures that the atmosphere and ocean systems rarely behave independently of each other. The feedbacks (see Chapter 17) between the two systems act in both directions, are complicated and operate on a variety of time scales ranging from seconds to millennia. In general, the oceans circulate slowly, whereas the atmosphere does so rapidly. Heat exchange in the oceans is slow; in the atmosphere it is rapid. Nonetheless, the interaction between the two systems produces responses that not only influence climatic change and variability, but also day-to-day weather events. This coupled variability will be examined in this chapter. In so doing it must be remembered that both air and water obey the same laws of fluid dynamics. Thus, for instance, geostrophic balance governs movement in the same way in both the atmosphere and the oceans.

WIND AND OCEAN MOVEMENT

The effect of the wind stress on the surface of the ocean is to set water moving in the uppermost layer. Movement is to the left of the wind in the southern hemisphere owing to Coriolis deflection (Fig. 13.1). The movement of the top layer of water sets the layer beneath in motion, again to the left of the driving force, and so on.

The net result is an *Ekman spiral* of water movement in the current initiated by the wind stress. The current weakens with depth in the Ekman layer to the point where it will be moving slow-

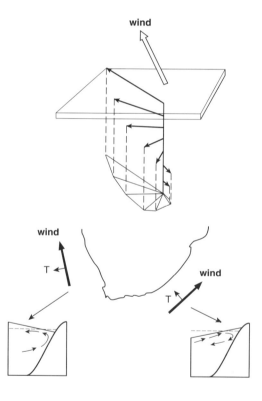

Fig. 13.1 *Upper:* the Ekman spiral in ocean water movement; *lower:* deformation of the surface of the ocean by Ekman transport and the consequent upwelling and sinking of water off the coasts of South Africa. T denotes the transport vector.

ly in the opposite direction to the initiating wind at the surface. The ocean-current spiral is analogous to that in the variation of wind in the atmosphere near the surface, but is directed downwards, whereas that in the atmosphere is directed upwards (see Chapter 8). The Ekman mass transport vector acts at right angles to the wind, to the left in the southern hemisphere and to the right in the northern.

Given that the transport of water in a current is to the left of the wind in the southern hemisphere, it follows that water will either be removed from a coastline or piled up against it depending on the orientation of the wind to the coast. Along the west coast of southern Africa water is transported away from the coast with winds of a southerly component, the sea surface is depressed near the coast and an upwelling of cool water occurs in compensation (Fig. 13.1, *lower left*). By the same token, south-westerly winds along the south-east coast of South Africa will produce an accumulation of water along the coast, an elevation of sea level along the shore and a resulting sinking of surface water in compensation (Fig. 13.1, *lower right*).

Once the surface of the ocean is raised or lowered, slopes are produced and geostrophic motion is initiated parallel to the slope, such that warmer, lighter water is to the left of the wind and cooler, denser water is to the right of the motion in the southern hemisphere (and the opposite in the northern hemisphere) (Fig. 13.2). The slope of the sea surface contributes to the barotropic component of the pressure gradient that drives the geostrophic current; the difference in densities between the adjacent water masses produce the baroclinic component. Thus the current is driven by both the slope and the density gradient. With the occurrence of large circular oceanic *gyres* or small-scale *eddies*, gradient flow under balanced forces is initiated, as in the atmosphere with cyclones and anticyclones. In such circumstances, large anticyclonic gyres show a tendency to become warm cored owing to inward Ekman transport of warm surface water to the left (Fig. 13.2, *lower*). Cyclonic ocean circulation systems tend to become cold

cored since the transport of warmer surface water is away from the centre. In the case of ocean eddies occurring on the scale of kilometres to hundreds of kilometres this does not hold. They will be cyclonic or anticyclonic depending on the density of the water they contain and hence the pressure gradient between them and the ambient water in which they occur.

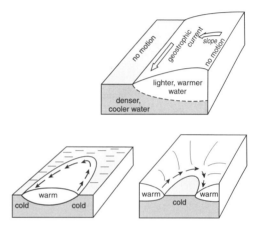

Fig. 13.2 *Upper:* geostrophic flow parallel to a sloping sea surface in the southern hemisphere; *lower:* warm and cold pools and anticyclonic gyres and cyclonic vortices.

The major planetary wind systems over the earth produce equally major deformations of the ocean surface over the globe as a whole (Fig. 13.3, *left*). Thus the oceans in the regions of the subtropical semi-permanent highs are a few tens of centimetres higher than their counterparts in the equatorial and mid-latitude regions. Where confluence occurs in atmospheric air streams or ocean movements, convergence of surface water takes place and downward convection results (Fig. 13.3, *top right*). With diffluence in the motion field at the surface, the divergence of surface water produces upwelling to satisfy continuity. Convergence and divergence of surface waters produce ocean fronts across which, as in the atmosphere, horizontal temperature gradients may be pronounced. A number of fronts are to be observed in the

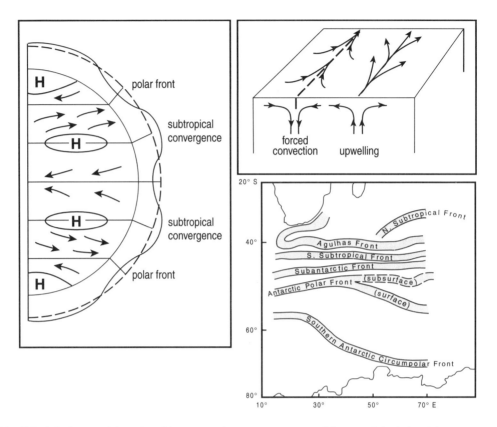

Fig. 13.3 *Left:* planetary deformation of the ocean surface as a consequence of the general circulation of the atmosphere (after Defant, 1961); *right upper:* confluence and diffluence and downward forced convection and upwelling; *right lower:* ocean fronts in the vicinity of South Africa (modified after Sparrow et al., 1996).

southern African sector of the oceans (Fig. 13.3, *lower right*), the most important of which for South African weather and climate are the subtropical and polar fronts.

The South Equatorial Current forms the northern limb of the oceanic gyre in the South Indian Ocean. It is driven by the South Indian Anticyclone in the atmosphere. The current bifurcates when it reaches the east coast of Madagascar (Fig. 13.4). Water moves north in the northern limb of the East Madagascar Current; the rest proceeds south in the southern limb of the current. The waters of the northern limb round the northern tip of Madagascar and proceed west to the African coast where most turn north. Those that move south form the weak and highly variable Mozambique Current

on the western side of the Mozambique Channel. This current contributes a small amount of water to the Agulhas current to the south (south of the Mozambique–South Africa border, the southward-moving current is known as the Agulhas Current). The southern limb of the East Madagascar Current likewise contributes little to the Agulhas system. Instead it retroflects back towards the central South Indian Ocean. The Agulhas Current is supplied with most of its water by a large recirculation gyre in the South-West Indian Ocean that extends to about 60° E. The current flows south along the edge of the continental shelf before converging with the Antarctic Circumpolar Current and bending back on itself in the Agulhas Retroflection to become the Agulhas

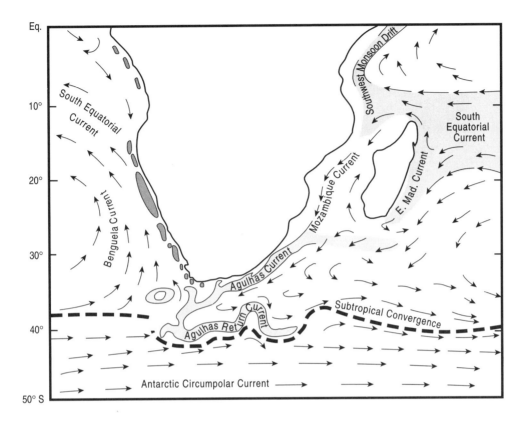

Fig. 13.4 Ocean currents around South Africa (modified after Niiler, 1992). The shaded areas off the west coast represent cells of upwelling of cold water in the Benguela upwelling system.

Return Current along the subtropical convergence. The area to the south of South Africa is characterized by turbulence, by high current variability, the generation of eddies and shedding of rings as warm and cooler water mix. It is a region of considerable ocean variability.

On the west coast, the cold Benguela Current is driven by the South Atlantic Anticyclone. Offshore Ekman transport of water ensures upwelling in discrete cells along the coast of sub-surface cold South Atlantic Central Water from depths of 200–300 m. Colder Antarctic Intermediate Water (see Fig. 13.10 to follow) is usually found at around 1 000 m and is rarely involved in the upwelling. The upwelling of nutrient-rich water supports the rich fishing grounds associated with the Benguela upwelling system. Satellite images show the warm

Agulhas Current and cold Benguela system upwelling clearly (Fig. 13.5). A clear distinction needs to be made between the Benguela Current, which is an offshore, wide, slow, northward drift of water driven by the South Atlantic Anticyclone, and the Benguela upwelling system, which, in contrast, is a localized, cellular, wind-induced, coastal upwelling on a very different scale. Much of the cold water produced by the upwelling system eventually enters the Benguela Current, but the two phenomena are essentially separate.

A vertical section through the Agulhas Current off the east coast of South Africa shows that the current is clearly defined to a distance exceeding 90 km offshore (Fig. 13.6, *upper*). Maximum velocities exceeding 150 cm s^{-1} occur at the surface at around 30 km from the coast,

Fig. 13.5 *Left:* cold water (light coloured) in the Benguela upwelling system off the west coast of southern Africa; *right:* the warm Agulhas Current (dark coloured) and mixing in the retroflection area to the south of the Western Cape (NOAA satellite image provided by J. R. E. Lutjeharms).

closer inshore where the continental shelf is narrow and further out where it widens to the south. Directly beneath the current, a weak undercurrent flows in the opposite direction. Around 150 km offshore, at depths of 1 500–3 500 m, Antarctic Intermediate Water may be observed moving slowly northward.

The structure of the Benguela upwelling system in the vertical on the west coast is more complicated. Upwelling tends to occur in discrete cells. The upwelling zone extends to around 50 km offshore as South Atlantic Central Water rises to the surface (Fig. 13.6, *lower*). Further out to sea, some downward convection occurs within the upwelling cells in a mixed zone extending out to 125 km. Beyond that limit the Benguela Current is generally undisturbed in its northward passage.

The retroflection region where the Agulhas, Benguela and Antarctic Circumpolar Current systems interact south of the Western Cape is a region characterized by considerable variation in sea-surface temperature, fluxes of heat and moisture from the sea surface to the atmosphere and the generation of ocean eddies. The region is one of the most variable in the world oceans. Changing ocean-surface topography illustrates the variability (Fig. 13.7).

The ocean currents of the world transport vast amounts of heat that are crucial in balancing the ocean-atmosphere heat budget (see Fig. 11.2). Estimates of the heat transport are given in Figure 13.8 and show how much of the heat maintaining the north Atlantic Ocean system comes from the Pacific Ocean through the Drake Passage to the south of South America and from the Indian Ocean via South Africa.

THERMOHALINE CIRCULATION

Not all circulation in the oceans is wind driven. Some circulation results from convection initiated by changes in density, which arise from salinity and/or temperature variations at the surface. The degree of downward convection is governed by the strength of the *thermocline*. The absorption of solar radiation at or near the surface produces heating of surface water. Below cooler conditions prevail. Between the warmer surface water and the cooler water mass below, the strong temperature gradient with depth between the two is known as the thermocline. In low latitudes, the thermocline is a pronounced feature owing to strong surface heating by solar radiation. Any process acting to

Agulhas Current

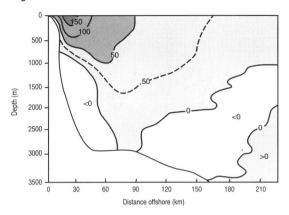

Benguela upwelling system

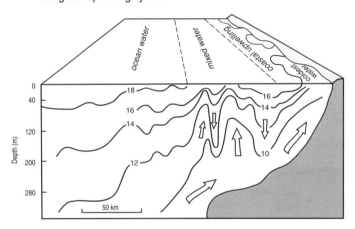

Fig. 13.6 *Upper:* a vertical section through the Agulhas Current off the east coast of South Africa to show the speed of the current in cm s^{-1} (after Pollard and Smythe-Wright, 1996); *lower:* a vertical section through the Benguela Current off the east coast of South Africa to show the temperature structure, isotherms in °C, and coastal upwelling (after Bang, 1971).

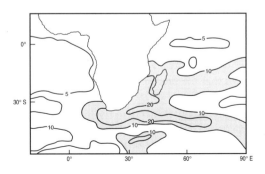

Fig. 13.7 Sea-level variability (cm) in the oceans adjacent to southern Africa from October 1992 to October 1994 as determined from the TOPEX/POSEIDON altimeter (after Wunsch, 1994).

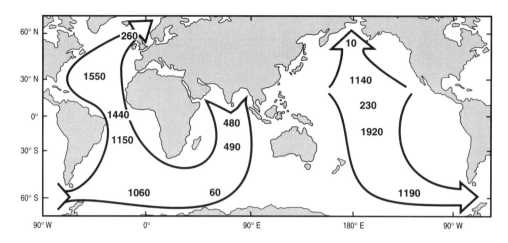

Fig. 13.8 Estimates of ocean heat transport (10^{12} W) to show the linkages between oceans (after Woods, 1984).

increase the density of the surface water in these latitudes, such as cooling by evaporation, initiates downward convection to a depth at which the sinking water finds an equal density level. This will usually be above the thermocline and the convection will be shallow (Fig. 13.9, *upper left*). In contrast, at high latitudes, the net radiation balance is negative and cooling processes increase the density of surface water. In addition, and in many ways more importantly, as pack ice is formed by freezing in winter, the salinity of the sea water from which the ice derives increases. The increase in salinity is accompanied by an increase in density. Since the thermocline is weak or absent without strong radiational heating of surface waters, deep convection may be initiated, with sinking taking place to considerable depths, or even to the ocean floor. In general, radiational heating in the tropics and cooling and salinity-induced density increases of water on the ice margins in polar regions will initiate a slow thermohaline circulation component in the general circulation of the oceans in the manner envisaged in Figure 13.9, *lower*).

A second way in which thermohaline circulation may be initiated is by the mixing of water masses (analogous to air masses) of different temperatures and salinities. It is best to examine the mixing process on a temperature–salinity (T–S) diagram (in some ways to oceanographers what a tephigram is to meteorologists). In the oceans, the density of water is a function of both temperature and salinity, unlike the situation in the atmosphere where air density is a function of temperature alone. On a T–S diagram lines of constant density, *isopycnals*, are plotted as a function of temperature and density and are curved lines. They increase in value with depth; that is, towards the bottom right of the diagram (Fig. 13.9, *upper right*). If two water masses of the same density (A and B in Figure 13.9, *upper right*) mix, then the resulting mixture has a density C along the straight line joining the two. Given the curvature of the isopycnals, the density of C is higher than the density of either of the parent water masses A and B and the resulting mixture will sink to the level where it finds its own density level. Thereafter, it will be transported as a sub-surface current or water mass at that level. This process only occurs with low temperatures and high salinities. This means that the process is only observed at high latitudes.

One of the best examples of the initiation of a mixing-induced thermohaline circulation of this type is to be found in the North Atlantic Ocean when cool, less saline water moving

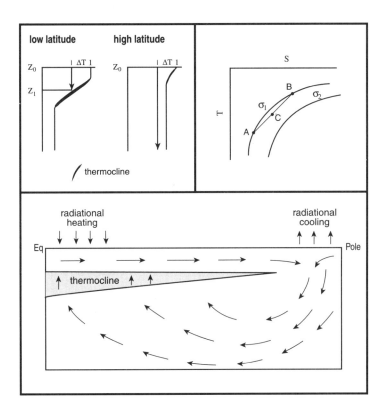

Fig. 13.9 *Upper*: temperature–depth profiles in low and high latitudes to show the thermocline in each case (*left*) and isopycnals (marked σ_1 and σ_2, where $\sigma_1 < \sigma_2$) on a temperature–salinity (T–S) diagram; *lower*: hemispheric thermohaline circulation resulting from radiational heating in tropical-equatorial regions and cooling near the poles.

equatorward south of Greenland from polar regions meets and mixes with warmer, more saline water moving northward in the Gulf Stream off North America. The mixing takes place in the region off Newfoundland and the resultant North Atlantic Deep Water moves south at around 2 500 m eventually to exit the Atlantic Ocean south of South Africa.

Thermohaline circulations may be initiated in subtropical latitudes if ocean-surface, evap-oration-induced salinity increases are strong enough. Strong summer evaporation from the eastern Mediterranean Sea causes the salinity of surface water to increase in the Levantine Basin. This is followed by cooling in the subsequent winter. The resulting increase in density and negative buoyancy causes the water to sink and start moving westward until it moves out to the

Atlantic Ocean over the sill across the Straits of Gibraltar. The relatively warm saline water begins to drain out of the Mediterranean and down the continental shelf to the west until it finds its neutral buoyancy level. Thereafter, the thermohaline current of Mediterranean Water spreads far south in the Atlantic Ocean (Fig. 13.10). Above the saline outflow, fresh water flows into the basin to maintain continuity. Adjacent seas in high-evaporation regions often exhibit such currents. Another example is the subsurface flow out of the Red Sea into the Indian Ocean. In low-evaporation areas, the reverse patterns of inflow and out-flow may take place out of adjacent seas. For example, fresh surface water flows out of the Baltic Sea into the North Sea, whereas at depth more saline water flows into the basin.

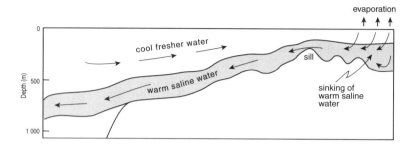

Fig. 13.10 The thermohaline outflow of warm saline water from the Mediterranean Basin (modified after Defant, 1961).

The generalized subsurface circulation in the Atlantic Ocean off western southern Africa is dominated by thermohaline circulations (Fig. 13.11). North Atlantic Deep Water drifts as far south as the Southern Ocean. Antarctic Bottom Water drifts far into the northern parts of the Atlantic. Water converging along the Antarctic Polar Front sinks to intermediate depths to form cool Antarctic Intermediate Water, which then spreads northward into the Atlantic, Indian and Pacific Oceans at depths centred around 600–1 000 m. A more explicit hypothesis of thermohaline circulation forced by convection in the north Atlantic Ocean is

given in the Broecker conveyor-belt model of subsurface flow (Fig. 13.12). In this proposition, deep thermohaline flow takes place out of the Atlantic Ocean through the Indian to the Pacific Ocean before the belt rises to return whence it originated. Since the major forcing of the conveyor is in the high-latitude region of the Atlantic Ocean where the thermocline is weak, only small changes in climate in that region may be needed to inhibit the necessary sinking of water to produce the subsurface flow of intermediate water to the south. Under such circumstances the conveyor would stop or reverse. In this way a small climatic change in

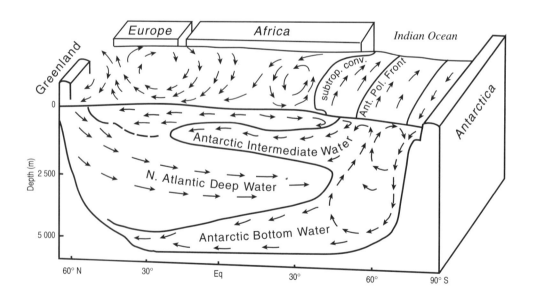

Fig. 13.11 Idealized thermohaline circulations in the Atlantic Ocean system (modified after Defant, 1961).

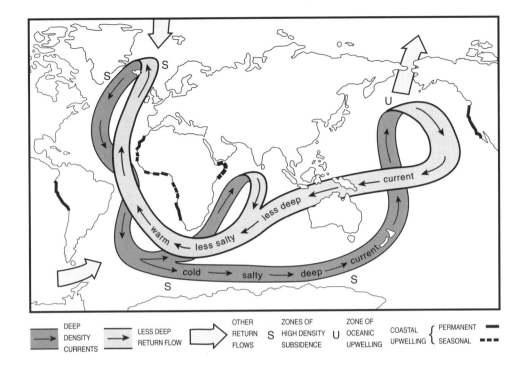

Fig. 13.12 Broecker's concept of the deep-ocean thermohaline conveyor belt (after Broecker and Denton, 1990).

one region could produce a major global climatic event by altering the equilibrium transport patterns of oceanic heat transport referred to earlier in Figure 13.8. This topic is explored further in Chapter 16.

THE OCEAN HEAT ENGINE

The oceans contain a vast amount of heat energy which is released into the atmosphere by boundary-layer exchange processes operating at the surface (see Chapters 7 and 14). The heat flux to the atmosphere is by sensible and latent heat exchange. Both are dependent on the sea-surface temperature. An example of the latent-heat flux in the oceans around southern Africa is given in Figure 13.13.

Air flowing over a relatively warm ocean will acquire additional heat energy and moisture from the underlying surface, thus enhancing the

capacity of that air to precipitate moisture. In this way ocean and atmosphere variability become linked.

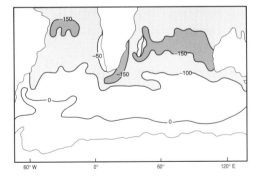

Fig. 13.13 January latent-heat flux (W m^{-2}) from the oceans adjacent to southern Africa (Taylor and Kent, 1996).

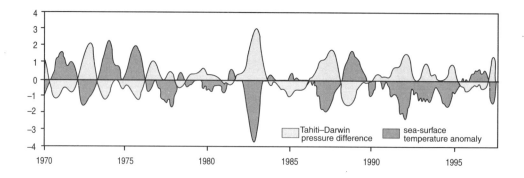

Fig. 13.14 Coupled ocean–atmosphere variability as shown by generally inverse sea-surface temperature variations in the Pacific Ocean in the area 5° N–5° S, 90°–150° W and variations in the sea-level pressure difference between Tahiti and Darwin (after Shukla, 1998).

COUPLED OCEAN–ATMOSPHERE VARIABILITY

In many areas of the oceans, notably in the Pacific Ocean, changes in the sea-surface temperature field are linked to changes in lower-tropospheric pressure fields and the atmospheric circulations associated with them. This is particularly evident in the region of the tropics where changes in the surface pressure difference between Darwin and Tahiti (i.e., changes in the surface windfield) cause changes in sea-surface temperature from 150° W to 90° W (Fig. 13.14). The degree of inverse correlation between the ocean and atmospheric changes in the region is striking and is responsible for the phenomenon known as the Walker Circulation.

THE WALKER CIRCULATION AND ENSO

Most standing circulation cells in the atmosphere tend to be meridional (e.g., the Hadley and Ferrel cells). A linked series of zonal cells, and a significant feature of the southern hemisphere circulation that has long been known to have important consequences throughout the atmosphere, is the Walker Circulation (Fig. 13.15). The Walker Circulation is directly responsive to sea-surface temperatures over the eastern and western Pacific Ocean. As these change from time to time a major pressure oscillation is set up in the atmosphere. This swaying of pressure backward and forward between the Indian and Pacific Oceans (as it was described by the discoverer of the phenomenon, Sir Gilbert Walker, in 1923) has been termed the Southern Oscillation and is measured by observing the pressure differences between Tahiti and Darwin. When the difference is negative, the Southern Oscillation is in *low phase*, sea-surface temperatures are anomalously low over the Indonesian region and pressure rises. Over the eastern Pacific Ocean sea-surface temperatures are anomalously high and the pressure falls. The Walker Circulation then develops an ascending limb over the eastern Pacific and a descending limb over Indonesia. The circulation intensifies and near-surface westerlies and upper easterlies complete the cell. Similar cells are set up over the Indian and Atlantic Ocean areas with descending limbs over eastern southern Africa and South America. This is the situation prevailing during an El Niño period and is often referred to as an El Niño-Southern Oscillation (ENSO) event.

During the *high phase* of the Oscillation, a La Niña or non-ENSO event, the surface ocean water over the eastern Pacific Ocean undergoes cooling to the extent that the pressure gradient between that region and Indonesia reverses. Surface pressure over Indonesia is now relatively lower than over the eastern Pacific. The Pacific Walker Circulation weakens or reverses, surface easterly winds decrease in intensity, on occasions even to the point of reversing, while at the 200 hPa level westerly wind anomalies develop. The air begins to converge over Indonesia and to diverge over the eastern Pacific. As the Walker Circulation weakens and begins to reverse, so the South Pacific cloud-band convergence zone moves westward causing a decrease in rainfall in the western Pacific dry zone and increased rainfall over Indonesia. Contemporaneous changes occur in the Indian and Atlantic cells. A zone of convergence and ascent develops over eastern southern Africa. Similarly, the nature of the convergence and divergence regions over South America undergo changes

which, as everywhere else, materially alter the patterns of rainfall, floods and droughts.

In a typical situation with a falling Southern Oscillation index and a developing El Niño, pressure rises over Indonesia and the pressure gradient along the equator decreases or reverses. The wind stress on the ocean weakens or reverses and sets in motion a complex dynamical response in the ocean (Fig. 13.16). A fast Kelvin wave propagates eastward through the surface water to the South American coast and then is reflected as a slower westward-moving and laterally spreading Rossby wave. The combined effect is to produce a deeper mixed layer and warmer ocean surface off Ecuador and northern Peru. The warm-water El Niño anomaly is associated with a depression of the thermocline. These changes have immediate feedback effects on atmospheric convective and circulation

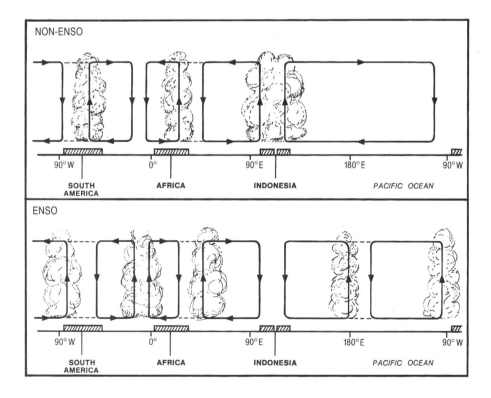

Fig. 13.15 The Walker Circulation during, *upper:* non-ENSO (normal or La Niña) events; *lower:* ENSO (El Niño) events (modified after Lindesay, 1987).

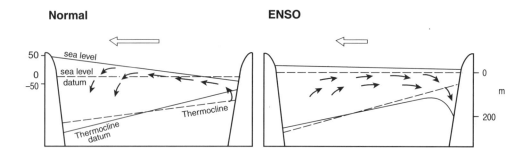

Fig. 13.16 The relationship between sea level, the thermocline and the strength of the tropical easterlies during ENSO and non-ENSO events (after Wyrtki, 1982).

processes. During a non-ENSO or La Niña event the sea level in the Indonesian region is anomalously high, the thermocline is raised over the eastern Pacific, and upwelling is enhanced off the South American coast.

It is instructive to examine the development of the temperature anomaly fields over the eastern Pacific associated with one of the strongest El Niños of the twentieth century, the 1997 event. A typical normal pattern of sea-surface temperature anomalies prevailed between December 1996 and February of 1997 over the eastern Pacific (Fig. 13.17). The El Niño region experienced below-normal conditions. Later that year, by September–November of 1997, the anomaly pattern had reversed to show anomalously cool conditions in the Indonesian region and severely above-normal El Niño conditions in the eastern Pacific. The El Niño warming of the ocean was the most severe recorded in the twentieth century. Over the same period, the sub-surface water was initially warmer at a

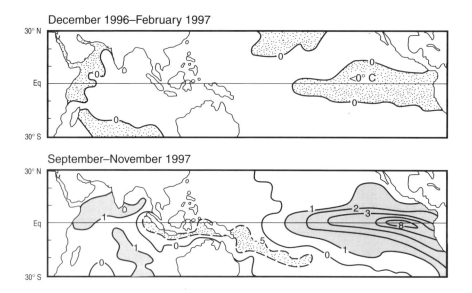

Fig. 13.17 Sea-surface temperature anomalies (°C) for normal conditions obtaining between December 1996 and February 1997 and the strong El Niño conditions between September and November of 1997 (after WMO, 1997).

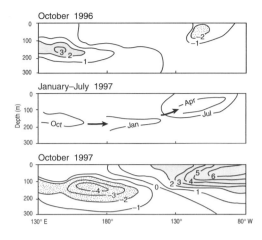

Fig. 13.18 Pacific Ocean sub-surface temperature anomalies (°C) associated with the development of the 1997 El Niño (after Oberhuber et al., 1998).

depth of around 150 m in the western region (Fig. 13.18, *upper*). Thereafter, the zone of subsurface anomalous heating rose toward the surface as it moved eastward. By April of 1997 it had risen to near the surface in the eastern Pacific and by October 1997 had reached its maximum surface expression (Fig. 13.18, *lower*). At that time the configuration of sub-surface heating and cooling had reversed completely in comparison to the conditions a year before.

El Niño events represent the maximum reversals of the Walker Circulation. They set up

teleconnections (spatial correlations) and bring about a chain of extreme events all over the world (Fig. 13.19). Given the global nature of the Walker Circulation within the tropics, it is not surprising that distinctive teleconnections are reported between far-distant places for different phases of the Southern Oscillation and that not all places experience like changes. Thermal forcing in the tropics produces upper-air tropospheric perturbations, taking the form of Rossby wave-trains of alternating high and low pressure, that, after radiating outward from the tropics, follow approximately great circle distributions and affect both tropical and extra-tropical areas. Low-wave-number stationary waves are transmitted to high latitudes: high-wave-number disturbances are confined to the tropics. There is little doubt that variations in the Walker Circulation have mid-latitude implications. Of specific importance for South African climate is the fact that the Southern Oscillation exerts a modulating effect on standing waves in the atmosphere, particularly on the position of the 500 hPa semi-stationary waves, notably *waves 1* and *3*, at 35° S.

The extent to which tropically induced barotropic waves may propagate out of the tropics appears to be closely related to the nature of the wind shear in the vertical. When the Pacific upper-level tropical heat source is embedded in weak upper-level westerlies, teleconnections

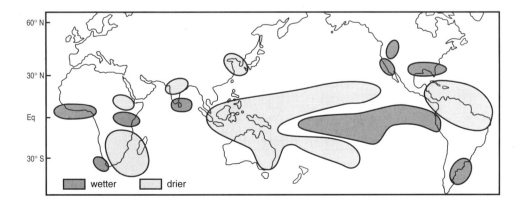

Fig. 13.19 Areas showing teleconnections associated with El Niño (modified after Allan et al., 1996).

develop with extra-tropical regions, poleward and downstream of the forcing. With the heat source in the easterlies, teleconnections tend to be confined to the tropics. The situation is complicated by the fact that the tropical heat release itself may be triggered by Rossby waves emanating from middle latitudes.

During the high phase of the Southern Oscillation (non-ENSO) there is a tendency for the summer rainfall regions of South Africa, Australia and South America to be wetter syn-

chronously (however, the correlation is not strong). Similarly, as pressures rise over the central Pacific and fall over Indonesia, so they tend to fall over South Africa. Negative-pressure anomalies tend to occur over South Africa and positive anomalies develop over the Gough Island region. Zonal westerlies strengthen over the Atlantic and Indian oceans and tropical easterly wind anomalies develop at 200 hPa over much of the subtropics, including the southern Africa region.

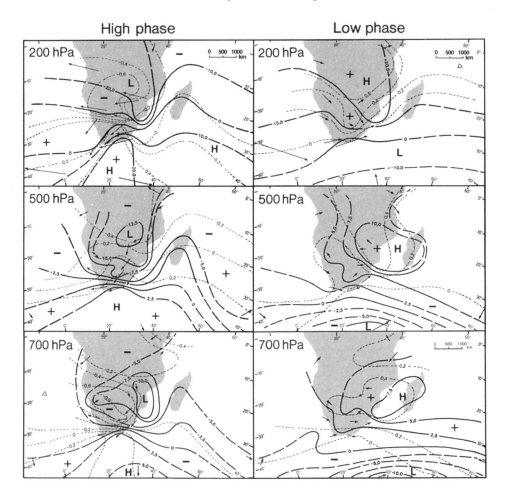

Fig. 13.20 Late summer (January–March) geopotential (solid lines) and temperature (broken lines) deviations (in gpm and °C) from the 1957–1983 means, together with wind anomalies at 200 hPa, 500 hPa and 700 hPa for the high (non-ENSO) and low (ENSO) phases of the Southern Oscillation. Wind anomaly vectors have lengths proportional to magnitude (after Lindesay, 1987).

ENSO AND SOUTH AFRICAN RAINFALL

Events initiated in the South Pacific Ocean alter the temperature, pressure and wind fields over southern Africa (Fig. 13.20). This happens throughout the atmosphere and produces very different conditions during high and low phases of the Southern Oscillation. In general, by affecting the location of major cloud bands over southern Africa, the Southern Oscillation exerts a modulating effect (but by no means a complete one) on the rainfall of the subcontinent (Fig. 13.21). During the El Niño low phase of

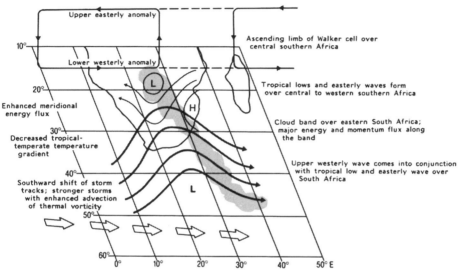

HIGH PHASE southern African rainfall above normal

Upper easterly anomaly

Ascending limb of Walker cell over central southern Africa

Lower westerly anomaly

Tropical lows and easterly waves form over central to western southern Africa

Enhanced meridional energy flux

Cloud band over eastern South Africa; major energy and momentum flux along the band

Decreased tropical-temperate temperature gradient

Upper westerly wave comes into conjunction with tropical low and easterly wave over South Africa

Southward shift of storm tracks; stronger storms with enhanced advection of thermal vorticity

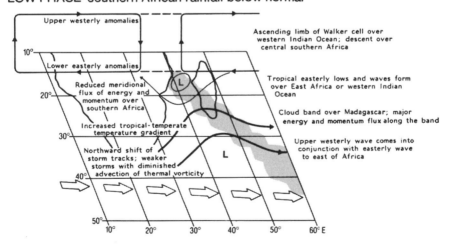

LOW PHASE southern African rainfall below normal

Upper westerly amomalies

Ascending limb of Walker cell over western Indian Ocean; descent over central southern Africa

Lower easterly anomalies

Reduced meridional flux of energy and momentum over southern Africa

Tropical easterly lows and waves form over East Africa or western Indian Ocean

Increased tropical-temperate temperature gradient

Cloud band over Madagascar; major energy and momentum flux along the band

Northward shift of storm tracks; weaker storms with diminished advection of thermal vorticity

Upper westerly wave comes into conjunction with easterly wave to east of Africa

Fig. 13.21 Schematic representation of the anomalous Walker Circulation over southern Africa; *upper:* during the high, non-ENSO, phase of the Southern Oscillation when above-normal rainfall may occur over southern Africa; *lower:* during the low, ENSO, phase when droughts may be expected (after Harrison, 1986). Light lines indicate surface flow; heavy lines denote upper tropospheric conditions.

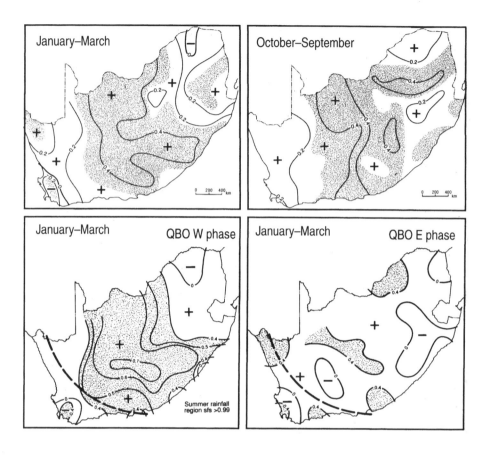

Fig. 13.22 *Upper:* the correlation between rainfall and the Southern Oscillation Index over the period 1935–1983, *left,* in late summer, *right,* for the whole summer rainfall year (after Lindesay, 1987); *lower:* changes in the late-summer correlation with the phase of the QBO, *left,* when the QBO is westerly, *right,* when the QBO is easterly (after Mason, 1992).

the Oscillation the cloud-band convergence zone moves offshore and with it the highest rainfalls. During the high phase cloud bands locate preferentially over southern Africa and rainfall is higher. Two of the most extreme El Niño events recorded, those in 1877 and 1982/83, were associated with catastrophic droughts in South Africa. The 1997 El Niño, which produced extreme climate anomalies all over the globe, including some in northern parts of southern Africa, had relatively little impact on South Africa. While much is known about the effect of ENSO on South Africa, much remains to be investigated, as the 1997 El Niño has shown.

The link between South Africa's rainfall and the Southern Oscillation is clear, but not always strong. The correlation is higher in the late summer season than earlier and reaches 0.5 in places (i.e., up to 25 per cent of the rainfall variability may be explained by Pacific Ocean ENSO forcing) (Fig 13.22, *upper*). However, this is not the whole story, since the initial condition of the atmosphere, as determined by whether the stratospheric Quasi-Biennial Oscillation (QBO) is in its easterly or westerly phase plays an important, but not well understood, modulating role on the effects of ENSO in southern Africa. The QBO is the quasi-periodic reversal of equatorial stratospheric

Modulation of the ENSO effect by the QBO

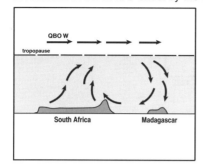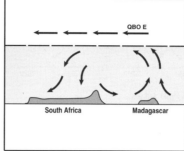

Fig. 13.23   Modulation of ENSO in the region of southern Africa by the QBO (after Jury et al., 1994).

winds in the region of the 50 hPa level. When the QBO is westerly, over large areas of the summer rainfall region more than 36 per cent of the inter-annual variability in late summer (January–March) may be ascribed to the effect of ENSO (Fig. 13.22, *lower*). When the QBO is easterly, the ENSO-rainfall association weakens to account for, at maximum, 16 per cent of the variability and ceases to be statistically significant. The modulation of the ENSO response over southern Africa by the QBO is thought to result through the enhancing of convergence and divergence fields in the manner illustrated in Figure 13.23.

The primary drive for the tropical zonal Walker Circulation lies in zonal variations of latent-heat release. The existence of the high release areas in the ascending limbs of the Walker cells is essential to the maintenance of the upper troughs (and cloud bands) over the South Atlantic, Indian and Pacific Oceans. The Australian/Indonesian centre of heat release and convergence region is the strongest and least sedentary. Any change in its position has consequential effects around the globe. The area of maximum heat release in the vicinity of Africa is not only a function of land-induced convective processes, but is also modulated by the position of the ascending limb of the African Walker cell, which in turn is modulated by sea-surface

temperatures and large-scale ocean–atmospheric interaction in the eastern Pacific in association with El Niño.

ENSO exhibits considerable inter-annual and lower frequency variability (Fig. 13.24). Fluctuations with quasi-periodicities less than 6 years are confined to the El Niño area within the Pacific region. Those with quasi-periodicities greater than 6 years cover a wider area. Organized inter-annual variability of southern African rainfall occurs on a variety of scales. A marked 2–3 year oscillation associated with the QBO occurs over central South Africa. More extensive spatially is a band of oscillations in the range of 3–5 years, and primarily at about 3.5 years, that is associated with ENSO variability (Fig. 13.25). The most pronounced oscillation in South African rainfall occurs at the inter-decadal scale at around 18 years. The reasons for this oscillation are complicated. Part of the explanation lies in fluctuations in adjacent ocean conditions, part in a link with Pacific Ocean sea-surface temperature changes (Fig. 13.26), part in a combination of atmospheric circulation changes in and beyond the region. A more detailed discussion is given in Chapter 16.

The manner in which an ENSO event may alter sea-surface temperatures and horizontal and vertical circulation fields over the Indian Ocean to the east of southern Africa is illustrat-

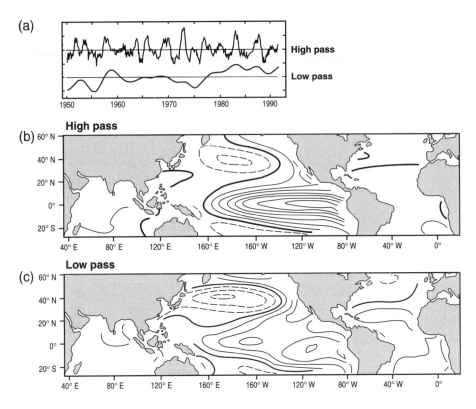

Fig. 13.24 *Upper*: the amplitude of the leading principle component of sea-surface temperature in the Pacific Ocean after filtering with a high-pass (HP) filter to show inter-annual variability with periods less than 6 years and after filtering with a low-pass (LP) filter to show variability with periods greater than 6 years (the ordinate is in units of 1 standard deviation); *middle*: the spatial pattern of high-frequency ENSO variability (contour interval is 0.1 °C per standard deviation with negative contours being given in broken lines); *lower*: the spatial pattern of low-frequency ENSO variability (after Zhang et al., 1997).

ed in Figure 13.27. The warming of the surface of the ocean changes the stream function which defines the cyclonicity and anticyclonicity of the flow. Changes in velocity potential in turn define the changes taking place in the divergence field and the regions within which convergence is likely to be enhanced and those in which subsidence will increase. Thus with an ENSO event, enhanced convergence occurs offshore of South Africa and not over the subcontinent. Of particular relevance for circulation and rainfall over southern Africa during an El Niño is the warming of the ocean north of Madagascar. This is linked to the development of an associated circulation anomaly that

inhibits airflow to the subcontinent via the tropical easterlies and their attendant disturbances.

COUPLED OCEAN–ATMOSPHERE VARIABILITY IN THE VICINITY OF SOUTHERN AFRICA

A number of ocean areas around southern Africa have been identified, using principal components analysis, as regions having a particular influence on the rainfall of the subcontinent (Fig. 13.28). Should air be transported over these areas, then the heat and moisture content of the air is sufficiently modified to affect rain-

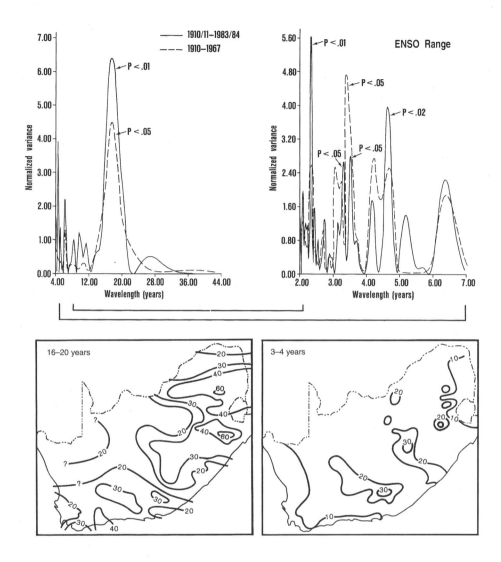

Fig. 13.25   *Upper:* the spectrum of rainfall fluctuations for the summer rainfall region of South Africa over the period 1910–1984. Those associated with high-frequency ENSO variability occur in the range of 3–5 years, *lower:* the spatial pattern of variability associated with the ~18-year and 3–4-year fluctuations in South African rainfall.

fall significantly, either positively or negatively, providing the atmospheric circulation fields are conducive to rainfall production. An illustration of the localized advective control of the moisture field in the lower atmosphere is given in Figure 13.29. With offshore flow, low dew-point temperatures indicate the dryness of the air. With reversal to onshore flow across the warm Agulhas Current, dew-point tempera-

tures rise rapidly as moist near-surface air from over the ocean is transported inland. The same distinction is obvious on a larger scale, where it has been shown that maximum moisture transport takes place at around the 700 hPa level. In midsummer (January) on rain days, transport to central southern Africa is from the warm tropical western Indian Ocean north of Madagascar (Fig. 13.30). On no-rain days in midsummer,

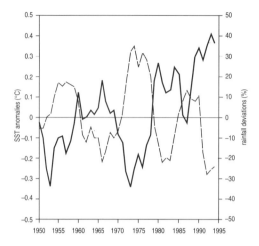

Fig. 13.26 Seven-year running mean sea-surface temperature anomalies averaged over the eastern equatorial Pacific Ocean (180°–90° W, 10° S–5° N) (solid line) and the percentage of normal rainfall over South Africa (dashed line) (after Mason and Jury, 1997).

the transport is in a subsiding stream from the south Atlantic Ocean. Originally the moisture in these air streams would have been derived from surface ocean–atmosphere interactions before vertical transport into the middle lower troposphere and horizontal transport to central southern Africa.

The warming and cooling of the Agulhas system may play a role in modulating tropical-temperate troughs (Fig. 13.31). With positive sea-surface temperature anomalies the formation of deep troughs may be facilitated. Likewise it is possible that during the high (low) phase of the Southern Oscillation, when rainfall is high (low), higher (lower) sea-surface temperatures in the Agulhas region would tend to diminish (enhance) rainfall over the subcontinent.

Correlations between the specific areas of sea-surface temperature influence determined

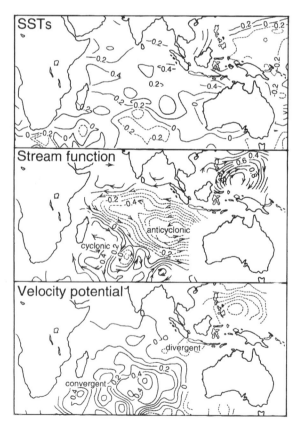

Fig. 13.27 Indian Ocean sea-surface temperature anomalies, stream function (non-divergent component of flow to show cyclonicity and anticyclonicity) and velocity potential (divergent component of flow to show areas of convergence and divergence) for ENSO events (after Lindesay and Allan, 1992).

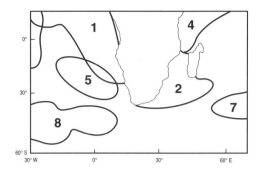

Fig. 13.28 Core areas of sea-surface temperature coherence in the oceans adjacent to southern Africa, as determined from principal components analysis (after Mason, 1992).

by principle components (PC) analysis, and designated in Figure 13.28, and rainfall over South Africa illustrate the important links between rainfall variability and regional ocean–atmosphere interactions. Four such associations are illustrated in Figure 13.32. Sea-surface temperature changes in the Agulhas Current region (PC 2 area) correlate positively, but weakly, with late-summer rainfall (i.e., as sea-surface temperature increases over the current so rainfall increases over most of South Africa and significantly so over eastern and northern areas). In contrast, changes in sea-surface temperature in the western tropical Indian Ocean to the north of Madagascar (PC 4

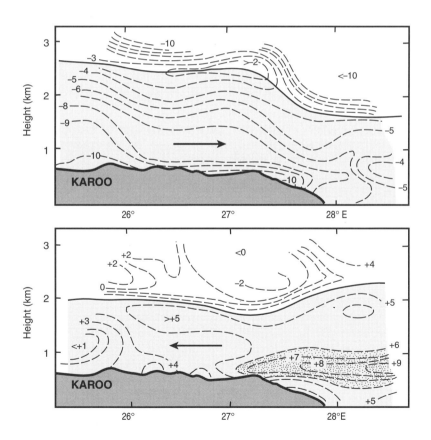

Fig. 13.29 Examples of boundary-layer moisture field associated with: *upper,* the offshore transport of dry continental air over eastern South Africa (dew-point temperatures as low as –10 °C); *lower,* the onshore transport of moist air over the warm Agulhas Current (dew-point temperatures as high as 8 °C) (after Jury et al., 1997).

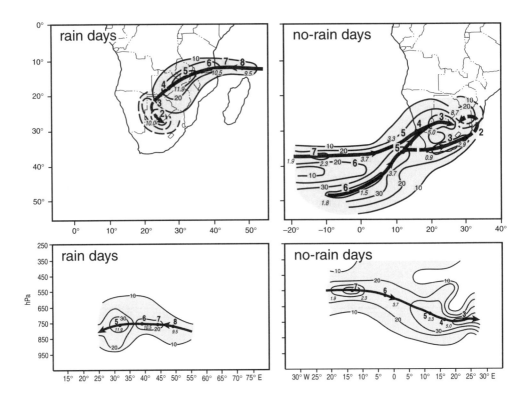

Fig. 13.30 Moisture-transport fields to show air arriving over central southern Africa at 700 hPa on rain and no-rain days in January. The bold numbers along the axis of maximum transport indicate days from the reception point; specific humidities (g kg^{-1}) are indicated in italics along the maximum transport pathway.

region) correlate negatively (i.e., as the sea warms in that area so rainfall diminishes over much of South Africa). The same pattern of negative correlations is evident as a consequence of sea-surface temperature changes in the eastern South Atlantic Ocean off Namibia (PC 5). As sea-surface temperatures increase in the subtropical convergence area of the ocean to the south-west of Cape Town (PC 8), so rainfall shows a tendency to increase over the interior. However, the situation is more complicated than suggested, since the QBO modulates these effects, as it does in the case of the effect of ENSO on South African rainfall.

In the case of the effect of the Agulhas system on rainfall, when the QBO is easterly, sea-surface temperatures correlate poorly with rainfall over South Africa and both negatively and

positively depending on locality (Figure 13.33). When the QBO is westerly, increasing ocean temperatures over the ocean off the east coast have a uniformly positive, weak effect on late summer rainfall. The QBO exerts little effect on the influence of sea-surface temperatures north of Madagascar on South African rainfall. However, changes in sea-surface temperature to the east of KwaZulu-Natal (in the PC 7 area of the Indian Ocean) correlate most strongly with South African rainfall when the QBO is easterly (Fig. 13.34). At such times, warming of the surface waters is associated with a diminution of rainfall. In contrast, changes in sea-surface temperatures in the subtropical convergence zone of the South Atlantic Ocean (PC 8 region) produce the strongest influence on rainfall when the QBO is westerly.

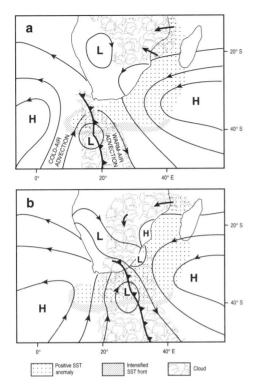

Fig. 13.31 A schematic representation of enhanced tropical and extratropical circulations associated with Agulhas warm events; *upper*: the enhancement of individual tropical and mid-latitude temperate lows; *lower*: subsequent coupling to form a tropical temperate trough (after Walker, 1990).

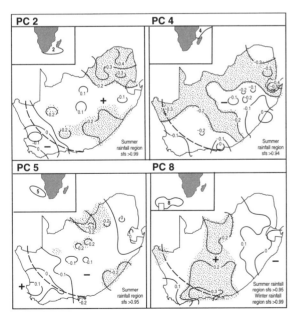

Fig. 13.32 Correlations between January–March sea-surface temperature principal component scores for designated core regions given in Figure 13.28 and rainfall over South Africa. Areas significant at 90 per cent are shaded; where the summer rainfall region as a whole has a field significance exceeding 90 per cent this has been indicated (after Mason, 1992).

Examples of the way in which sea-surface temperatures may modify the circulation patterns to affect rainfall are given in Figure 13.35. The warming of the ocean may cause modifications to the atmospheric circulation and, for example, alter the westerly baroclinic wave configuration of the circulation to produce drier conditions over South Africa (Fig. 13.35 *upper* and *lower left*). However, instead of being the cause of atmospheric circulation changes, sea-surface temperature variability may be the effect of atmospheric circulation changes. Thus enhanced anticyclonicity may result in clearer skies, a greater incidence of solar radiation and a warming of surface water (Fig. 13.35 *upper* and *lower right*). Both processes may operate sequentially to produce perplexing ocean–atmosphere interactions. In general, ocean–atmosphere interactions are complicated and not completely understood. Part of the difficulty experienced is that relationships between the ocean and the atmosphere may change with time with slowly changing ocean conditions. For instance, evidence is accumulating to suggest that responses in Indian Ocean sea-surface temperature patterns to Pacific Ocean ENSO changes have changed since the late 1970s. This in part explains why the 1997 El Niño did not produce the expected severe droughts over South Africa.

If the combined effects of the forcing of rainfall by sea-surface temperature changes in all the areas of influence in the oceans adjacent to southern Africa and elsewhere are considered, then the changes in the ocean temperatures around southern Africa appear to exert a greater influence on year-to-year rainfall variability than does ENSO, except when pronounced El Niños and pronounced La Niñas occur. In the early part of the summer (October–December), more variability over a wider area of South Africa can be explained by the combined sea-surface temperature effects when the QBO is easterly than when it is westerly. In contrast, in the second half of the season (January–March) a greater amount of variability over a greater area can be accounted for when the QBO is westerly.

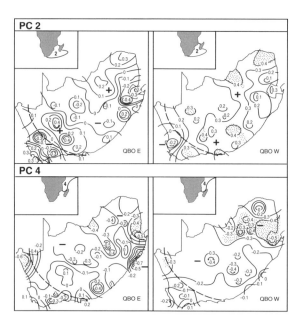

Fig. 13.33 QBO modulation of the rainfall variability associated with the sea-surface temperature variability over the Agulhas Current and in the region to the north of Madagascar (Mason, 1992).

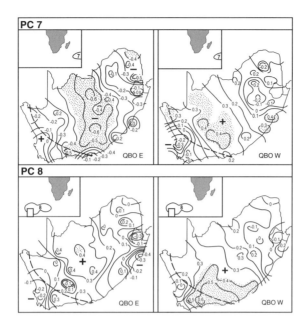

Fig. 13.34 QBO modulation of the rainfall variability associated with the sea-surface temperature variability over the western Indian Ocean off KwaZulu-Natal and in the region of the subtropical convergence to the south west of South Africa (Mason, 1992).

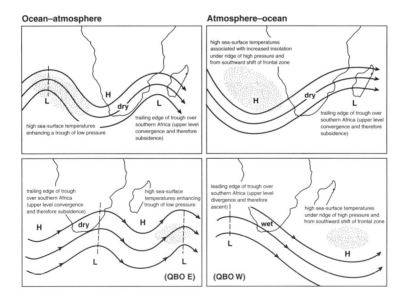

Fig. 13.35 Two-way cause and effect feedbacks involved in ocean–atmosphere interactions. With sea-surface-temperature warm events (high-temperature anomalies) in the shaded areas, the two *left* illustrations give a positive causal feedback from ocean to atmosphere; the two *right* illustrations show how a change in the atmospheric wave motion may induce a positive feedback from atmosphere to ocean (modified after Mason, 1992).

Care must be taken to distinguish between the variability of sea-surface temperature fields per se and the effect this variability has on rainfall over southern Africa. The greatest sea-surface temperature variability is observed over the Benguela system, followed by that over the Agulhas system. It does not follow that the sea-surface temperature variability over these regions most affects South African rainfall as a consequence. It does not. Changes over the Benguela system have a negligible effect on present-day rainfall over South Africa and Namibia. Ranking the rainfall response to sea-surface temperature variations around southern Africa reveals that, if the effect of the QBO is ignored, then changes in temperature in the central South Atlantic Ocean (the PC 5 region in Figure 13.28), those in the South Atlantic Ocean subtropical convergence zone (PC 8) and those in the western tropical Indian Ocean (PC 4) have the greatest effect on South African rainfall. However, if the modulating effect of the QBO is taken into account, then the ranking of areas of significant influence on rainfall changes. Under these conditions, changes over the Indian Ocean to the south-east of South Africa (PC 7–QBO easterly) most affect South African rainfall, followed by changes over the central south Atlantic (PC 5–unstratified), the South Atlantic subtropical convergence zone (PC 8–unstratified), the western tropical Indian Ocean (PC 4–unstratified) and the Agulhas system (PC 2–QBO westerly).

The effects of ocean–atmosphere interactions in the vicinity of southern Africa and afar exert a profound effect on the circulation of the atmosphere over the region. The weather and climate respond accordingly.

ADDITIONAL READING

ALLAN, R. A., LINDESAY, J. A. and PARKER, D. 1996. *El Niño, Southern Oscillation and Climatic Variability.* CSIRO, Australia.

SCHNEIDER, S. H. (eds.). 1996. *Encyclopaedia of Weather and Climate,* Vols. 1 and 2. Oxford: Oxford University Press.

BIGG, G. R. 1994. *The Oceans and Climate.* Cambridge: Cambridge University Press.

HARTMANN, D. L. 1994. *Global Physical Climatology.* San Diego: Academic Press

DIAZ, H. F. and MARKGRAF, V. (eds.). 1992. *El Niño, Historical and Palaeoclimatic Aspects of the Southern Oscillation.* Cambridge: Cambridge University Press.

PHILANDER, S. G. 1990. *El Niño, La Niña and the Southern Oscillation.* San Diego: Academic Press.

GILL, A. E. 1982. *Atmosphere–Ocean Dynamics.* London: Academic Press.

PERRY, A. H. and WALKER, J. M. 1977. *The Ocean–Atmosphere System.* London: Longman.

KING, C. A. M. 1962. *Oceanography for Geographers.* London: Edward Arnold.

KING, C. A. M. 1975. *Introduction to Physical and Biological Oceanography.* London: Edward Arnold.

VON ARX, W. S. 1962. *Introduction to Physical Oceanography.* Reading, Massachusetts: Addison Wesley.

DEFANT, A. 1961. *Physical Oceanography,* Vol. 1. London: Pergamon.

14

BOUNDARY-LAYER PHENOMENA

In the preceding chapters consideration has been given to macroscale atmospheric phenomena, which extend through the depth of the atmosphere and occur over time scales ranging from days to decades. The discussion of the atmosphere, weather and climate over southern Africa, however, would be incomplete without examining some of the phenomena occurring in the atmosphere immediately above the earth's surface. This *atmospheric boundary layer* accommodates phenomena that are tied to the daily heat cycle and that occur at time and space scales ranging from one second and one millimetre to one day and a hundred kilometres.

The air in contact with the surface to a depth of a few millimetres is characterized by the transfer of heat and momentum by molecular action and is called the *laminar boundary sublayer* (Fig. 14.1). The mixing of air by eddies takes place immediately above this layer in the *turbulent boundary layer*. The nature of turbulent flow is of central importance in the boundary layer as turbulence is responsible for the vertical fluxes of heat, matter and momentum. Turbulent eddies occur at varying time and space scales, with large eddies continuously breaking down into ever smaller ones. By day, vigorous small-scale eddies are generated by heating and may extend the boundary layer to depths of several hundred metres or more; by night, turbulence and mixing are suppressed by the development of highly stable air near the ground. Consequently the turbulent mixing layer may shrink to a few metres or disappear

altogether. At a larger scale within the boundary layer, diurnal circulations such as land-sea breezes and mountain–plain winds may extend their fetch for over 100 km and the overall boundary layer may deepen by day due to convection and frictional drag to above 1 km. By night it may shrink to zero.

In this chapter the effect of turbulence in the boundary layer will be considered. Thereafter examples of boundary-layer phenomena that have a direct impact on society will be examined.

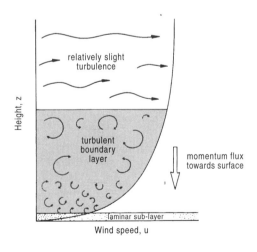

Fig. 14.1 Boundary-layer flow over a smooth surface.

TURBULENCE IN THE BOUNDARY LAYER

The eddies that characterize turbulent motion in the atmosphere are vortices that are generally transverse to the mean wind. The eddies result from accelerations in flow and produce deviations in speed and direction from the mean steady wind that results from motion under balanced forces (see Chapter 8). In some ways eddy motion may be thought of as analogous to molecular motion; in fact the mixing-length hypothesis is based on the notion that the mixing length of eddies is the atmospheric equivalent of the mean free path of molecules. Needless to say, eddies occur at a vastly larger scale. They cause mixing throughout the turbulent boundary layer. Turbulent motion differs from smooth laminar flow in that there is no net exchange of mass between adjacent layers in the latter; in turbulent motion the exchange of mass causes the mixing that characterizes such motion. The convection responsible for the generation of turbulence is usually a mixture of both *forced* and *free convection*. Forced convection depends on the roughness of the surface, wind speed and wind shear, and is most important within about 2 metres of the surface. Free convection, involving the mixing of air in a buoyant environment and under unstable conditions, is the dominant convective mechanism for the vertical transport of momentum, heat and moisture in the lower atmosphere (see Chapter 7).

Momentum flux

Due to the viscosity of air, the surface exerts a frictional drag on the mean wind speed with a downward transmitted force produced by the shearing stress, τ, of the surface on the overlying air. The shearing force is transmitted as a flux of momentum.

In the *laminar boundary sub-layer* momentum flux takes place through molecular agitation with upward-moving molecules increasing the momentum of the upper layers and vice versa. The transfer of horizontal momentum is given by

$$\tau = \rho v \, \frac{du}{dz} \qquad (14.1)$$

where ρ is air density, v the kinetic viscosity and du/dz the variation of wind with height within the layer.

In the *turbulent boundary layer*, eddies moving vertically conserve their horizontal momentum. Assuming an increase in wind velocity with height near the surface, a turbulent eddy moving from a low to a higher level will carry its horizontal momentum to that level and thereby import a net decrease in velocity to that level. This is usually sensed as a lull in the horizontal wind speed. A downward-moving eddy would increase the velocity at a lower level in the form of a gust. The transfer of horizontal momentum is given by

$$\tau = \rho K_m \frac{du}{dz} \qquad (14.2)$$

where K_m is the eddy viscosity. The net result of both upward- and downward-moving turbulent eddies produces a wind profile the shape of which is related to surface roughness and the stability of the atmosphere.

Variations in the turbulence regime

Convective mixing is maintained or enhanced in an unstable environment and dampened in a stable environment (see earlier, equation 7.31). It therefore undergoes a diurnal variation under fine conditions as the stability regime near the ground undergoes a similar variation. The wind traces in Figure 14.2 illustrate three different types of turbulent regime, each imposing a characteristic imprint on wind-direction traces. With strong instability wind direction varies constantly with short-period fluctuations being superimposed on larger ones and those in turn being superimposed on yet longer ones and so on. The shortest fluctuations tend to be small, roughness-generated eddies associated with forced convection. The fluctuations with periods of about 1.5 minutes and longer (up to a few hours) are associated with large buoyancy-generated eddies and free convection. With a reasonable wind speed, eddies generated by forced convection tend to dominate the momentum flux below 2 km. Above this level, however, free convection becomes increasingly important by day as buoyant parcels of surface-heated air break away from the surface. A layer of well-mixed air characterized by gustiness develops as cooler air, carrying the horizontal momentum of faster-moving upper-level air, descends to take the place of rising air parcels.

As conditions change from moderately unstable to stable, so the degree of turbulence is damped, first by suppression of free and then of forced mixing. During the day, wind speed and turbulence increase near the ground as instability increases. The turbulent eddies generated by

day exert a frictional drag on the airflow in the upper boundary layer such that the diurnal variation of wind speed is the inverse of that near the ground (Fig. 14.2, *lower right*). By night the air near the ground cools by radiative flux divergence, air density increases, lapse rates become stable and surface inversions develop. The work needed to lift the air against the gravitational field is not available, turbulence dies down, wind speed decreases and flow tends to become more laminar. Should a strong wind develop during stable nocturnal conditions, turbulent transfer of heat will be toward the ground and the inversion will be destroyed by turbulent warming (equation 7.31). In contrast, during the day turbulent heat transfer in unstable conditions will be predominantly upward and will cause cooling. The vertical transfer of sensible heat by eddies is clearly illustrated in Figure 14.3 for an unstable, daytime, two-minute period. Simultaneous measurements of the vertical velocity component, temperature and sensible heat flux show that updraughts (positive w') are associated with

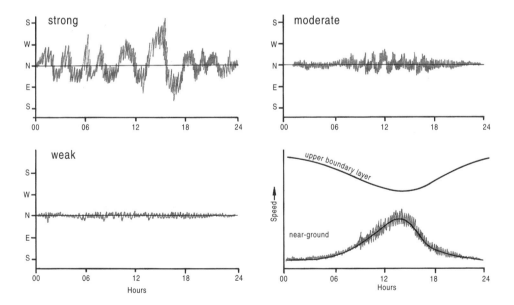

Fig. 14.2 The effect of strong, moderate and weak atmospheric turbulence shown on wind-direction traces, together with the diurnal variation of wind speed near the ground and in the upper boundary layer.

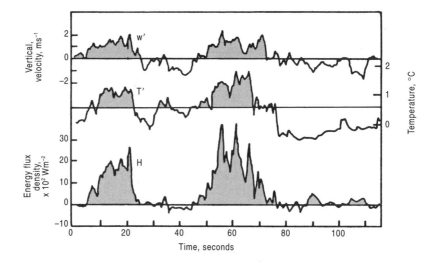

Fig. 14.3 The effect of turbulence on instantaneous vertical velocity, w', temperature, T' and sensible heat flux, H, over a two-minute period (after Priestley, 1959).

an increase in air temperature and an instantaneous sensible heat flux. Downdraughts, in contrast, are associated with a suppression of the heat flux and with negative temperature departures. Most of the heat transfer occurs in bursts coinciding with upward movement of buoyant thermals.

Wind-speed profiles

The measurement of mean wind speed at several heights above a uniform, extensive surface produces a wind profile the form of which provides information on the state of turbulence in the atmosphere (Fig. 14.4).

Under unstable conditions the gradient wind speed is weakened by the vertical mixing of momentum that takes place between faster-moving upper air and slower-moving air near the surface. The wind speed gradient is weakened accordingly (Fig. 14.4, *upper right*). Under stable conditions diminished mixing steepens the wind gradient (i.e., increases the wind shear). A condition of neutral stability exists when the variation of mean wind speed, u, with height, z, accurately describes a linear logarith-

mic decay curve (Fig. 14.4, *lower left*). The *logarithmic wind profile* is given by

$$u_z = \frac{u_*}{k} \ln \frac{z}{z_0} \tag{14.3}$$

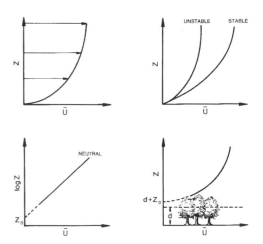

Fig. 14.4 Wind profiles near the ground over open, flat terrain (*upper left*), showing the effects of stability (*upper right*), plotted with a logarithmic height scale (*lower left*) and illustrating the notion of a zero-plane displacement of height d (*lower right*) (adapted after Oke, 1978).

where u_* is the *friction velocity* (m s^{-1}), k is von Karmen's constant (0.40) and z_0 is the roughness length (m). Most surfaces are aerodynamically rough and the length z_0, which defines the height at which the wind profile, in the absence of buoyancy, extrapolates to a zero wind speed, is therefore a function of the height of roughness elements of the surface. The quantity z_0 varies from 0.1 mm for smooth ice to 1–6 m for a forest. Where the surface is very rough, as is the case for a forest, the downward extrapolation of the wind profile describes the behaviour of airflow as if the surface were located at some height near the top of the vegetated surface rather than at the ground (Fig. 14.4, *lower right*). Consequently an additional parameter is required to describe surface conditions. It is known as the *zero-plane displacement*, d, and is accommodated in the logarithmic wind-profile equation in the form

$$u_z = \frac{u_*}{k} \ln \frac{z - d}{z_0} \qquad (14.4)$$

The friction velocity relates to the drag effect of surface friction on the mean wind speed and may be evaluated from measurements of the mean wind gradient (du/dz) by

$$u_* = kz \frac{du}{dz} \qquad (14.5)$$

Alternative, wind profiles may be used to describe the variation of wind speed with height in the boundary layer. One such is the *power profile* in which

$$u_1 = u_2 \left(\frac{z_1}{z_2} \right)^p \qquad (14.6)$$

where u_1 is the mean wind at height z_1, and u_2 that at height z_2. The power p varies with stability between 0 and 1 and is about 0.18 in neutral conditions. More sophisticated profiles incorporate the Richardson number and the ideas of Monin and Obukov; they need not be considered here.

The Richardson number

The form of the wind profile gives a measure of turbulence present in the atmosphere. More specifically, the degree of turbulence is given by the Richardson number

$$R_i = \frac{g}{T} \frac{dT/dz}{(du/dz)^2} \qquad (14.7)$$

where g is the force of gravity and T is the mean temperature in the layer in which the temperature and wind gradients dT/dz and du/dz are observed. The number is dimensional and compares the relative roles of free convection by buoyancy forces (the numerator) to forced convection by frictional forces (the denominator). Put another way, the Richardson number distinguishes between thermal and mechanical turbulence. Thus in a lapse condition free convection dominates and R_i is a negative number (because temperature is decreasing with height), which increases with a steepening of the lapse rate, but is reduced by an increase in the wind speed gradient. In an inversion R_i is positive, whereas in neutral conditions it approaches zero.

Heat flux

The flux of sensible heat was considered in detail in Chapter 7 in connection with the heating of the atmosphere and its energy balance. The reader is referred to the earlier section for a full discussion of the determination of sensible heat fluxes. In the turbulent boundary layer the sensible heat flux may be given by equations (7.30) to (7.34). Of these, equation (7.32) is identical in form to equation (14.2) for the transfer of momentum.

By night, radiative flux divergence causes the air immediately above the surface to cool below the temperature of the overlying air. The temperature gradient becomes positive (inversion) and the sensible heat flux, H, reverses from upward to downward. If the nocturnal winds and the turbulence they produce are strong enough, the downward flux of heat may succeed in destroying the inversion altogether.

After sunrise H changes sign as convective mixing directs the sensible heat flux upward from the surface. The temperature gradient reverses to a negative (lapse) gradient as warm buoyant parcels of air rise to be replaced by cooler air aloft. This process begins in a shallow air layer directly above the surface and steadily erodes the inversion from below. The process of erosion continues until the inversion is eliminated. In the late afternoon the near-surface air begins to cool once again and a new cycle begins.

The mixing depth may be estimated by the height at which the dry adiabat through the surface temperature intersects the environmental lapse rate aloft. The maximum mixing depth is reached at the time of greatest surface heating (maximum temperature) in the afternoon. Since mixing is dependent upon surface heating and the flux of sensible heat, the mixing depth undergoes a diurnal variation along with the temperature profile in the boundary layer as shown in Figure 14.5. Within the mixing layer the lapse rate approaches the dry adiabatic in accordance with equation (7.31). Midday mixing depths in winter vary from under 100 m to over 2 500 m and are somewhat higher in summer over southern Africa. Over plateau areas the annual variation in midday mixing depth is pronounced as a result of summer convective processes. Over coastal areas mixing depths tend to be much more uniform throughout the year. In winter, particularly, atmospheric pollution becomes trapped within the mixing layer to produce a pall of pollution with a clear upper boundary (Fig. 14.6).

Moisture flux

Atmospheric humidity is determined by the exchange of moisture between the surface and the atmosphere. The transport of vapour by eddy diffusion is similar to that for sensible heat and always involves latent-heat changes. These

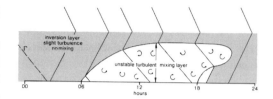

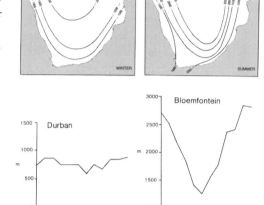

Fig. 14.5 *Upper*: the daytime development of a turbulent mixing layer; *centre*: early-afternoon mixing depths (*m*) over southern Africa in winter and summer; *lower*: examples of the annual variation of mixing depths at a coastal station (Durban) and inland plateau site (Bloemfontein) (adapted after Diab, 1975).

changes and the role of latent-heat exchanges were discussed in Chapter 7.

Evaporation from free water or ground surfaces may be measured directly using evaporation pans of standard sizes, from lysimeters in which a tank containing soil, plants and water gives evaporation by the weight loss of water, or using soil evaporimeters in which a similar tank is kept at field capacity (i.e., maximum water content) and evaporation is determined from the water balance of the tank. All these methods suffer from the disadvantages of the need for buffer zones, the effect of vapour caps developing above the tanks or pans and enhanced turbulence being created by the tank rims. The greatest problem is the difficulty of extrapolat-

Fig. 14.6 The tops of the winter turbulent boundary layer and midday mixing layers over Cape Town, *left*, and Johannesburg, *right* (photos: P. D. Tyson).

ing from a pan or tank to a large free-water surface such as a dam.

One of the simplest ways of determining evaporation is Dalton's method in which the evaporation, E, is given by

$$E = a(1 + bu)\,(e_s - e_a) \tag{14.8}$$

where a and b are regression coefficients, u is mean wind speed, e_s is the saturated water vapour pressure and e_a is the actual vapour pressure. Other methods of determining evaporation are those given by the flux gradient, eddy correlation and aerodynamic models specified in equations (7.35) to (7.38). The best method is probably that of Penman. However, its discussion falls beyond the scope of this book.

The diurnal variation of water-vapour concentration takes place in a similar manner to the diurnal temperature change. After sunrise, water vapour is added to the lower atmosphere by evapotranspiration. This causes a sharp increase in the humidity of the air. The resulting lapse conditions (Fig. 14.7) become most pronounced at the time of maximum surface heating due to convective mixing and subsequent dilution of the vapour concentration. By late afternoon, however, convection wanes as the air near the ground becomes stable. Evaporation continues to supply water vapour to the air above the surface, but the rate of dilution due to mixing slows down and the lapse rate tends toward isothermal. At night, radiative cooling of the air below the dew-point temperature causes dew to form on the ground. The extraction of

water vapour from the overlying air causes an inversion to form in the water-vapour profile. The depth and strength of this inversion is determined by the downward flux of water vapour in a suitably turbulent environment. The level of turbulence is critical. If it is too low (i.e., calm conditions), dew ceases to form, since the ground cannot be replenished by water vapour from above; if it is too high, mixing inhibits surface radiative cooling to below dew-point temperature. Near the surface in the early afternoon, even with strong evapotranspiration, turbulence transfers moisture away from the surface so rapidly that specific humidity usually falls to an early-afternoon minimum. The vertically transported water vapour then produces a humidity maximum at the same time in the upper boundary layer (i.e., the converse of what happens in the case of wind speed).

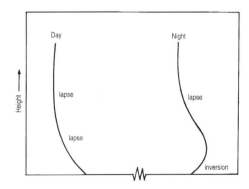

Fig. 14.7 Near-ground lapse rates of water-vapour content by day and by night.

BOUNDARY-LAYER MODIFICATION BY URBAN AREAS

The buildings, roads, motor vehicles and other expressions of an urban environment greatly modify the climate of the urban landscape by changing the radiative, thermal, moisture and aerodynamic characteristics of the environment.

Urban–rural contrasts

Urban–rural climatic contrasts may be attributed to the following factors:

Surface materials
The wide expanse of concrete and tar that characterize central urban areas have higher thermal capacities and thermal conductivities than the soils of the surrounding countryside. The daytime heat storage over these surfaces is, therefore, greater than over adjoining rural surfaces. In addition, the many glass-windowed buildings increase the interior warmth and heat storage of the city.

Surface form and composition
The many surfaces presented by walls and roofs of buildings of varying design and function cause a complex system of energy exchanges. Some surfaces are heated by the sun, while others remain cool in the shade. Infrared radiation emitted from surfaces in the deep canyons of some urban streets is absorbed and re-radiated by adjoining walls. The energy that finally escapes is therefore related to the sky-view factor in the canyon. The many different building materials found in urban areas have different albedos, heat capacities, thermal diffusivities and molecular heat conductivities. The different heights of buildings cause surface roughness to vary appreciably. All these things cause the energy balance to be modified.

Artificial heat
Cities produce prodigious amounts of artificial heat. Major heat sources derive from industrial and domestic heating and heat generated by motor vehicles. In winter, the warming of the air by artificial heat storage over large mid-latitude cities rivals energy received from the sun.

Moisture distribution
Unless moist air is advected into the city, the moisture content of urban air is usually lower than over the surrounding countryside. This is largely due to the relative lack of vegetation and, as a consequence, the reduced input of moisture into the air by transpiration. Moisture input into the air by evaporation is also lower owing to the wide expanses of paved surfaces and the rapid removal of precipitation by stormwater drainage systems.

Atmospheric pollution
The air over cities is frequently contaminated by a wide variety of pollutants derived from industrial wastes, domestic heating and transport systems. Pollution raises the attenuation of short-wave radiation, but it is also responsible for increased re-emission of long-wave counter radiation. In addition, pollution contributes to the atmospheric supply of condensation nuclei necessary for cloud droplet formation and may thus contribute toward increased precipitation over cities.

Airflow around buildings

Buildings constitute an obstacle to airflow and the resulting airflow patterns contribute to the complexity of the microclimate in the urban canopy layer. In the case of a high-standing, flat-roofed building, maximum wind pressure occurs on the upper middle portion of the windward wall. Pressure decreases outward from this *stagnation point* with the lowest pressures occurring along the outside edges of the building. The sides, roof and leeward wall then become areas characterized by a reverse flow as air moves toward lower pressure in the *suction zone* created by the separation of airflow along the sharp building edges (Fig. 14.8). At ground

Oblique view

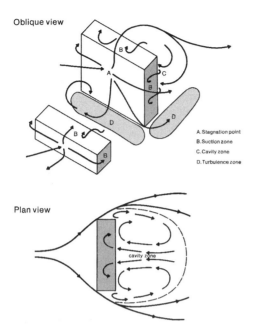

A. Stagnation point
B. Suction zone
C. Cavity zone
D. Turbulence zone

Plan view

cavity zone

Fig. 14.8 *Upper:* flow patterns around sharp-edged buildings to show side and front views; *lower:* a plan view of flow patterns (adapted after Oke, 1978).

level the incorporation of the side- and lee-wall suction zones leads to a double eddy circulation in the *cavity zone* leeward of the building. Increased turbulence is a feature of this area.

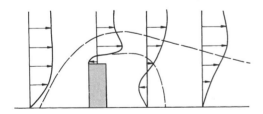

Fig. 14.9 Wind profiles on and adjacent to a sharp-edged building to show separation of flow and the development of boundary layers (after Oke, 1978).

The disturbance of the wind profile is shown in Figure 14.9. A jet of high-velocity air character-

istically occurs in the convergence region above the building with a low-velocity reverse flow immediately above the roof and in the cavity zone. Downwind of this zone the profile slowly adjusts to its upwind shape.

The great variety of building shapes and circulations in a city lead to complex wind patterns with the development of vortex flows between buildings and turbulent winds at street-level corners and below the stagnation point. For city dwellers the increased turbulence in these areas can lead to considerable discomfort. Severe problems may also arise if pollution is released in the suction zone of a flat-roofed building. The downwash in the cavity zone may bring pollution to the surface. An understanding of the nature of airflow around buildings allows the alleviation of the problems caused by turbulence and smoke plumes. For example, a taller stack ejects smoke above the suction zone and thereby eliminates the downwash problem.

The urban heat island

The complex exchange of fluxes in the building-air volume can be expressed in energy-budget terms as

$$R_n + F = H + LE + G + \Delta A \qquad (14.9)$$

where R_n is the net all-wave radiation, F is the total artificial heat generation in the city, H is the sensible heat flux, LE is the latent heat flux, G is the net heat storage in the urban fabric and ΔA is the advection of sensible and latent energy from the rural to the urban surface. Net all-wave radiation is defined by

$$Rn = (Q + q)(1 - \alpha) - (I\uparrow - I\downarrow) \qquad (14.10)$$

where Q is direct solar radiation, q is diffuse radiation, α is the surface albedo and $I\uparrow$ and $I\downarrow$ are the upward and downward components of infrared radiation respectively. Day and night changes in the direction of these fluxes are shown in Figure 14.10. By day, incoming radiation exceeds outgoing radiation to provide a downward flux of net radiation: by night, the reverse occurs and the air in the urban canopy layer and urban boundary layer cools accordingly. The directions of G, H and LE are also shown to be subject to diurnal variation.

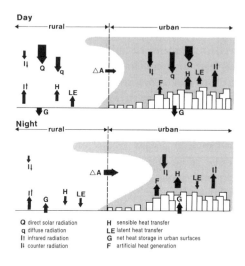

Day

Night

Q	direct solar radiation	H	sensible heat transfer
q	diffuse radiation	LE	latent heat transfer
I↑	infrared radiation	G	net heat storage in urban surfaces
I↓	counter radiation	F	artificial heat generation

Fig. 14.10 Schematic two-dimensional energy balances over urban and rural surfaces (modified after Fuggle and Oke, 1970).

The relative difference in energy flux and storage terms between the urban canopy layer and surrounding rural areas is responsible for the formation of the urban heat island. Likely causes for the increased warmth in the urban canopy layer relative to surrounding rural areas are attributed to changes in energy-budget terms at the surface. These are summarized in Table 14.1.

The form and intensity of heat islands undergoes day-to-day variations. The various causes identified in Table 14.1 may also operate more effectively in specific seasons (e.g., point 5 in mid-latitude winters), or in the urban boundary layer rather than in the urban canopy layer (e.g., point 1). By day, points 3, 4, 5 and 6 serve to create and store heat in the canopy layer, whereas at night points 2 and 7 operate to slow the

Table 14.1 Likely causes of the urban-canopy-layer heat island (after Oke, 1978).

1. Increased counter-radiation due to absorption of outgoing long-wave radiation and re-emission by polluted urban atmosphere.
2. Decreased net long-wave radiation loss from urban canyons due to a reduction in their sky-view factor by buildings.
3. Greater short-wave radiation absorption due to the effect of the canyon geometry on the albedo.
4. Greater daytime heat storage due to the thermal properties of urban materials, and to nocturnal release.
5. Anthropogenic heat from building sides.
6. Decreased evaporation due to the removal of vegetation and the surface waterproofing of the city.
7. Decreased loss of sensible heat due to the reduction of wind speed in the urban canopy.

loss of heat, thereby keeping urban areas warmer than the surrounding countryside.

The net result of all the above complex processes is to produce heat islands that may have a surprisingly simple surface-temperature structure (Fig. 14.11). Within the overall urban heat island, parks, rivers, and other building-free areas manifest themselves as cool spots, whereas car parks, shopping centres, and other areas in which buildings are concentrated produce locally strengthened heat islands.

Urban heat islands are easily identified in cities located in valleys where the natural tendency is for enhanced nocturnal cooling to occur by the katabatic drainage of cold air. The city warming effect is immediately discernible in contrast to the situation of cities located on ridges which tend anyway to be warmer by night than adjacent valleys. Islands in oceans behave like cities in creating heat islands, except that the features are clearest by day, unlike cities which exhibit the clearest heat-island characteristics by night. Given their similarity in producing heat islands, ocean-island studies often provide information of great use in understanding urban heat islands.

London

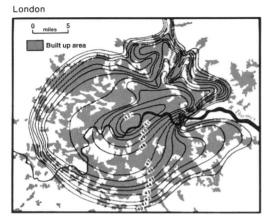

Pietermaritzburg

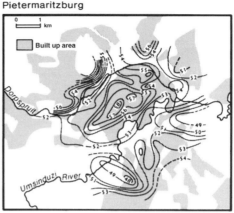

Fig. 14.11 Surface urban heat islands, *left*, over London on a summer night as evidenced by mean minimum temperatures (after Chandler, 1965), and *right*, over Pietermaritzburg at 22:00 on a winter night (after Wilton, 1971). Temperatures in °F.

In the absence of strong geostrophic or gradient winds, an urban heat island induces an inward flow of cool air toward the city centre. Low-level mesoscale convergence occurs and uplifting of air takes place over central areas. At a height of a few hundred metres above the surface the air diverges to produce a compensating outflow aloft (Fig. 14.12).

In the presence of light geostrophic or gradient flow, as air passes from a rural to an urban environment it must adjust to the boundary conditions of the new environment. This adjustment begins at the upwind urban–rural boundary (known as the leading edge) and extends downwind as a deepening internal boundary layer (Fig. 14.13, *upper*). The layer that extends

from street level to rooftop level is known as the *urban canopy layer*, with the *urban boundary layer* extending upward from rooftop level. The two layers are distinguished by the scale of the climatic processes operating in them. The sum of a complex set of microclimates accounts for the modification and adjustment of rural air in the urban canopy layer, whereas mesoscale phenomena modify the rural air in the urban boundary layer. The effect of a synoptically induced wind (or a local wind of a non-urban origin) is to cause the nocturnal urban boundary layer to deform downwind in an *urban heat plume*, examples of which may be quite spectacular (Fig. 14.13, *middle* and *lower*). The urban heat plume is sustained as follows.

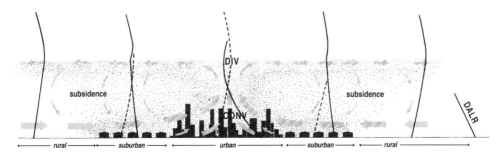

Fig. 14.12 The low-level city breeze and overturning of air in an urban heat island.

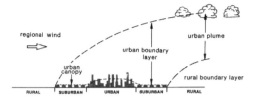

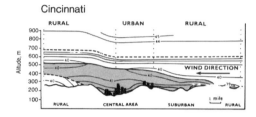

Cincinnati

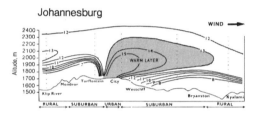

Johannesburg

Fig. 14.13 *Upper:* the development of an urban boundary layer and urban heat plume; and observations to show such a plume, *centre:* over Cincinnati, Ohio, in early summer, temperatures in °F (after Clarke, 1969); and *lower:* over Johannesburg at 05:30 on a late-winter morning, temperatures in °C.

The turbulent mixing of air will cause a downwind transfer of sensible heat in the rural environment, H_r. In the urban area the upward heat transfer will be a function of turbulent mixing, radiative processes, the difference between the temperatures T_u and T_r of urban and rural areas and the artificial heat input into the urban atmosphere. In Figure 14.14, as inversion air moves into a suburban region, the transfer of heat down the vertical temperature gradient is checked by artificial heat input, F. At distance x_1 from the edge of the built-up area $H_r = F$. At that point the heat island begins and continues to grow until at distance x, as F diminishes downwind from the city centre, the downward transfer of heat from the atmosphere balances F. At this point the heat island ceases.

The diurnal variation of heat islands is best illustrated taking an example of an ocean heat island over Barbados (Fig. 14.15). The bent-over nature of the plume is clearly evident as a consequence of easterly airflow over the island. As daytime heating of the surface and the air above proceed, so the plume develops from before 10:00 to after 15:00. By sunset only a remnant of the plume is apparent at a height of about 500 m and during the night no heat island effect is discernible. The effect of advection across a heat island is to displace the zone of surface convergence shown in Figure 14.12

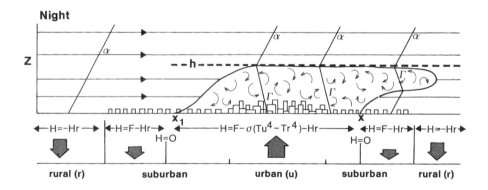

Fig. 14.14 A schematic diagram to show the development of an adiabatic mixing layer with an urban heat plume (modified after Leahey, 1969).

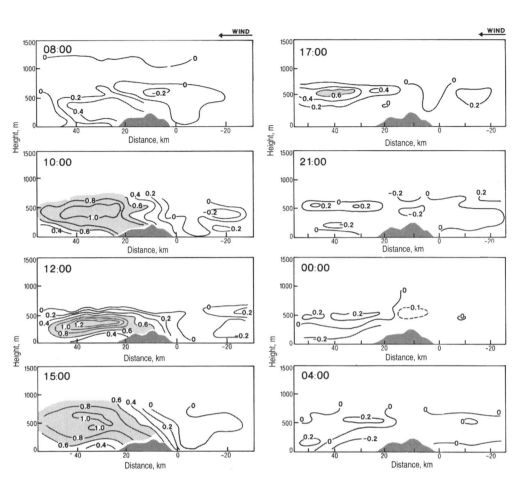

Fig. 14.15 The mean diurnal variation of the heat island and heat plume over Barbados in summer (after DeSouza, 1972).
Temperature deviations from the mean are given in °C.

away from the centre of the heat island in a downwind direction. In the Barbados case the zone of convergence and upward vertical motion moves to the edge of the island (Fig. 14.16, *left*). Similarly, over Columbus, Ohio, the zone of vertical velocity moves away from the central city such that over that part the net vertical motion is actually downward (Fig. 14.16, *upper right*).

The drifting of constant-level balloons over New York illustrates the generation of uplift within urban heat islands (Fig. 14.16, *lower right*). Such uplift is usually sufficient to enhance local rainfall by 10 to 15 per cent. Some examples of such heat island augmenta-

tion of rainfall are given in Figure 14.17. Particularly striking in the case of St Louis is that maximum enhancement of rainfall takes place in the downwind (south-eastern) sector of the city where urban-induced convergence prevails, whereas in the upwind (north-western) sector where the divergent limb of the urban cell is present rainfall is somewhat below normal. In the case of Barbados, in neutral or weakly disturbed conditions, rainfall occurs predominantly at night (when the cell shown in Figure 14.16 is reversed to give convergence and uplift over the island). When the atmosphere becomes strongly disturbed and synoptic influences predominate, the diurnal rainfall pattern is the

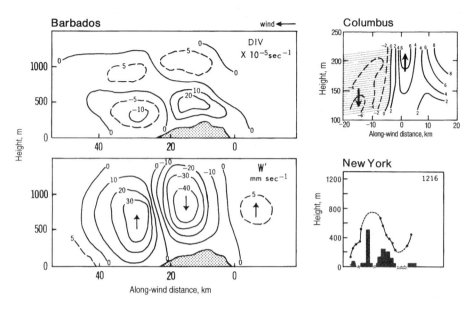

Fig. 14.16   Mean divergence and vertical velocity fields; *left:* over Barbados in summer (after DeSouza, 1972) and *upper right:* over Columbus, Ohio, before midnight in inversion conditions (after Angell et al., 1971); *lower right:* vertical displacement of a constant-level balloon released at 150 m over New York (after Hass et al., 1967).

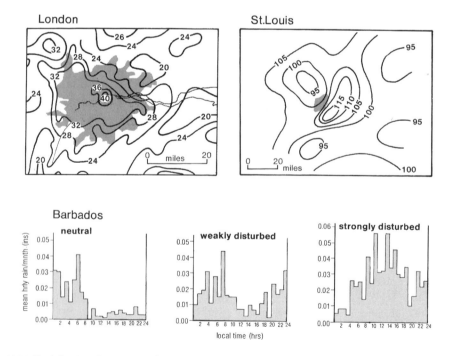

Fig. 14.17   Total thunder rain 1951–1960 (in inches) over London (after Atkinson, 1968) and the average rural–urban ratio of summer rainfall in the St Louis area 1949–1968 (after Changnon, 1972); the diurnal variation of hourly rainfall in neutral, weakly disturbed and strongly disturbed synoptic conditions over Barbados (after Garstang, 1965).

opposite and maximum hourly rainfall occurs during the day.

Urban heat islands may be modelled in a number of ways. Only one, a simple energy balance radiation model, will be considered here to illustrate how the prediction of surface urban temperature fields may be undertaken.

An urban-heat-island model

A one-dimensional (x-direction) model may be developed by assuming a flat surface and the movement of a uniform air mass over a city. The energy balance is given by

$$R_n = H + LE + G - F \qquad (14.11)$$

where the symbols are as defined earlier and where F is artificial heat generation by the city. The equation may be solved for a column of air above any point on the surface. If this is done over the urban area and over the adjacent rural area and the results are compared, then the urban heat island effect may be obtained. If the equation is solved for a series of grid squares, the model may be extended to two dimensions (x and y directions) and the spatial variation of urban-induced effects may be determined.

The procedure is to reduce, where possible, all the terms in equation (14.11) to mathematical expressions that are functions of surface temperatures and then to solve the resultant equation for this temperature. There are many ways of doing this. One possible approach will be illustrated here. Each term will be considered in turn.

Net radiation, R_n, is determined from the difference between net short-wave radiation, S, and net-long wave radiation, I. Thus

$$R_n = S - I \qquad (14.12)$$

where

$$S = \frac{S_0}{r^2}(1 - \alpha)\cos Z T_r^m \qquad (14.13)$$

in which S_0 is the solar constant, r is the earth's radius vector, α is surface albedo, T_r is atmospheric transmissivity and m is optical air mass. The zenith angle, Z, is determined from

$$\cos Z = \sin\phi\sin\delta + \cos\phi\cos\delta\cos h \qquad (14.14)$$

where ϕ is latitude, δ is solar declination and h is hour angle. Optical air mass may be approximated by

$$m = [\cos Z + 0.15(90 - Z + 3.885)^{-1.253}]^{-1} \qquad (14.15)$$

Net long-wave radiation is determined using equations such as those discussed in Chapter 7. The Brunt equation

$$I = \epsilon\sigma T_0^4 - \epsilon\sigma T_2^4 (0.66 + 0.39\sqrt{e}) \qquad (14.16)$$

in which T_0 is the surface temperature and T_2 is the temperature at the upper boundary level (a level at which the urban heat island may no longer be discerned) is often used. At this stage the surface temperature has been introduced for the first time into equation (14.11) via the specification of the amount of infrared cooling from the surface in equation (14.16). All the other equations thus far reduce to numbers.

Turbulent transfer within the boundary layer may be estimated as a bulk adiabatic transfer, K, from

$$K = \frac{k^2 \rho u_2 (1 - \gamma R_i)^p}{(\ln z_2/z_0)^2} \qquad (14.17)$$

where u_2 is wind speed at height z_2 at which the urban heat island ceases. It is assumed that eddy diffusivities of heat, matter and momentum are the same, that a logarithmic profile describes the variation of wind speed with height and that the atmospheric stability is neutral. Departures from neutral conditions may be handled by the correction factor $(1 - \gamma R_i)^p$ in equation (14.17) where $\gamma = 16$ and $p = 3/4$ for $R_i < 0$ and $\gamma = 5$ and $p = 2$ for $R_i > 0$. The Richardson number is estimated from

$$R_i = \frac{g(T_2 - T_0)\,(z_2 - z_0)}{u_2^2 T_2} \qquad (14.18)$$

Roughness length, z_0, may be estimated from

$$z_0 = \frac{\bar{h} A_e}{2 A_g} \qquad (14.19)$$

where $\bar{h}$ is the average height of the roughness elements (e.g., houses or large buildings), A_e is the silhouette area of the elements and A_g the area of the grid square over which z_0 is being determined.

Using the bulk adiabatic transfer, the *sensible heat flux* may be approximated by

$$H = K C_p\,(T_2 - T_0) \qquad (14.20)$$

Likewise, the *latent-heat flux* may be approximated by

$$LE = K L\,(q_2 - q_0) \qquad (14.21)$$

where the surface specific humidity is estimated from the saturated specific humidity and relative humidity. The latter is usually taken as the fraction, R, of the total area occupied by freely transpiring surfaces. Thus

$$q_0 = R[3.74 + 2.64\,(T_0/10)^2]10^{-3} \qquad (14.22)$$

in which T_0 again appears.

It remains to determine *substrate heat flux, G.* This may be done from

$$G = \frac{k_s\,(T_0 - T_z)}{z} \qquad (14.23)$$

where k_s is the thermal conductivity of the substrate and T_z is its temperature at depth z. The latter may be determined from the heat-conduction equation in one dimension in which

$$\frac{\partial T_z}{\partial t} = \frac{k_s}{\rho_s C_s}\frac{\partial^2 T_z}{\partial z^2} \qquad (14.24)$$

where ρ_s and C_s are density and specific heat of the substrate and $\partial^2 T_z/\partial z^2$ is the change with depth of the substrate temperature gradient.

Finally, artificial heat generation by the city, F, must be estimated from actual combustion of fossil fuels and electricity for each grid square. From a land-use appraisal for each grid square being considered, values of surface roughness, z_0, heat capacity, $\rho_s C_s$, albedo, α, thermal conductivity, k_s, transmissivity, T_r, and evaporating fraction, R, may be assigned. For different times of day different stabilities govern the values of γ and p in equation (14.17). Upper boundary conditions for wind speed, u_2, temperature, T_2, and specific humidity, q_2, at the height of the upper boundary, z_2, must be specified. In addition, initializing temperatures, vapour pressures and substrate temperatures must be specified and solar factors ϕ, δ and h defined. Once all this has been done, the model may be run to calculate the surface value of R_n, H, LE and G for each grid square. Finally, the spatial variation of the surface temperature, T_0, may be determined. In this way the urban-heat-island effect may be modelled.

Though crude and in some respects debatable, the model works and gives results that compare favourably with actual conditions, such as for Christchurch, New Zealand (Fig. 14.18).

AIR-POLLUTION CLIMATOLOGY

The prediction of what will happen to pollution when it is released into the atmosphere relies on the knowledge of the complex and individual nature of local climates and weather. From day to day, and at different times of the day, pollution concentration levels fluctuate in response to the changing state of atmospheric stability, to concomitant variations in mixing depth and to the effect of mesoscale and microscale wind systems on the transport and dispersion of air pollution.

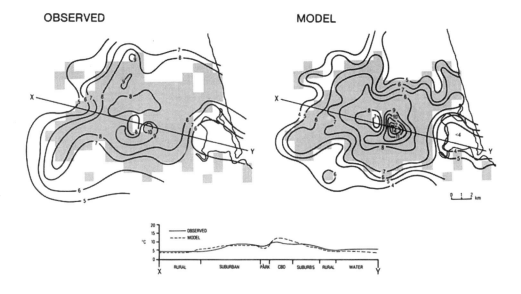

Fig. 14.18 Nocturnal, winter urban heat-island temperatures (°C) over Christchurch, New Zealand, and those predicted by an energy balance model (after Tapper et al., 1981).

Stability characteristics

The diffusion of atmospheric pollutants into a greater volume of atmosphere reduces the concentration of the polluting material. This occurs most effectively under conditions of free convection when the mixing layer is deep. Unstable conditions of this type occur most frequently in summer during the day. Conversely, diffusion is inhibited by stable conditions in the boundary layer. Most surface pollution is trapped in surface inversions. On occasions and at certain times of the day, however, the surface inversion may not be present and pollution will disperse, only to be prevented from diffusing freely upward by the presence of absolutely stable layers or elevated inversions. Both surface and elevated inversions over southern Africa need to be considered.

Surface inversions

Radiative inversions are particularly important in air-pollution meteorology because of their nightly occurrence under clear, calm and dry conditions (Chapter 7). They start developing shortly before sunset and by 21:00 may be several hundreds of metres deep. Acoustic radar allows continuous measurements of inversions to be made and traces show the degree to which surface inversions may be influenced by the flow above them. In Figure 14.19 (*upper left*) the coupling between a surface inversion and a weaker inversion layer above is evident (at both A and B). Sudden deepening of a surface inversion (such as at B) is not common, but may occur as a result of advection or following the onset of sudden, but slight, turbulent mixing. Often the flow near the top of the inversion is decoupled from that above the temperature discontinuity and ripples from the flow above may be imprinted on the top of the inversion. Such is the case in Figure 14.19 (*upper right*) where the effects of gravity waves with a period of about an hour are apparent. Superimposed on these are irregular saw-toothed fluctuations probably the result of high-frequency Kelvin waves consequent upon the shear between the lower-level stable layer

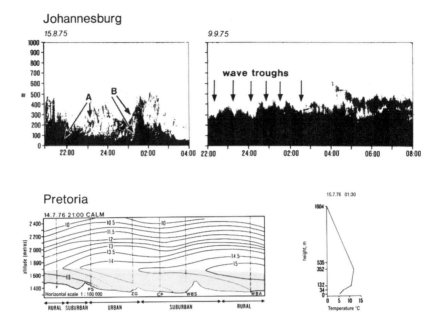

Fig. 14.19 *Upper left*: acoustic radar traces to show surface nocturnal inversions over central Johannesburg, the coupling between a double-layer structure (A) and a sudden deepening (B); *upper right*: a uniformly deep 300 m inversion over Johannesburg exhibiting gravity and Kelvin-wave ripples on its upper surface; *lower*: a 300 m deep surface inversion blanketing Pretoria on a calm winter evening together with a radiosonde sounding through the inversion (after Von Gogh, 1978).

and faster-moving less stable air above. Over the eastern Highveld a low-level wind maximum is frequently evident at the top of the inversion.

Surface inversions several hundreds of metres deep are common over Johannesburg and Pretoria in winter (Fig. 14.19, *lower*). Over southern Africa as a whole, the continuous nature of surface inversions is clearly illustrated in Figure 14.20. Low relative humidity and little cloud cover (particularly in winter) explain why annual frequencies of 01:30 surface inversions are greatest over the interior of Namibia and Botswana. The decrease of annual inversion frequency toward the south-east is the consequence of the increased interference in the process of inversion formation by weather associated with the large-scale, eastward-moving weather-producing systems. In contrast to the nocturnal situation, midday (13:30) surface inversions seldom

occur over southern Africa. They are present occasionally during the day over the Namib coast owing to the modification and stabilization of the lower atmosphere by the cold Benguela Current. Over southern Africa as a whole the winter nocturnal surface conversion has a depth of 400–600 m and a strength of 5–7 °C. Pollution released into stable inversion air is seldom able to rise through it and disperses slowly in clearly defined plumes (Fig. 14.21, *upper left*).

Surface inversions over flat terrain tend to be purely radiational. In hilly or mountainous terrain, however, cold-air drainage in katabatic flow causes the accumulation of cold air in valleys to form valley inversions (Fig. 14.21, upper right). These tend to be common everywhere in southern Africa, even in areas where only slightly sloping surfaces occur.

Annual frequencies

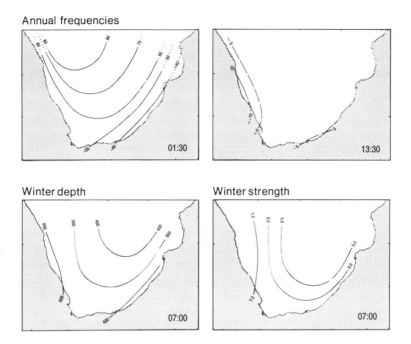

Fig. 14.20 Annual surface-inversion frequencies (percentage) at midnight and midday (*upper*) and generalized winter, early-morning inversion depth, *m*, and strength, °C, over southern Africa (*lower*).

Fig. 14.21 *Upper left:* a trapped pollution plume in a stable, surface radiation inversion over reasonably flat ground, Canterbury Plain, New Zealand; *upper right:* a valley inversion over Pietermaritzburg; *lower left:* a plume dispersing vertically in inversion-free ridge-top conditions, while pollution in the valleys is trapped over Pietermaritzburg; *lower right:* an elevated subsidence inversion over Bergville, KwaZulu-Natal (photos: P. D. Tyson).

Elevated inversions

Elevated inversions occur commonly in high-pressure areas. Sinking air warms adiabatically to temperatures in excess of those in the mixed boundary layer. The interface between upper, gently subsiding air and the mixed boundary-layer air is therefore marked by an absolutely stable layer or an elevated subsidence inversion (Figs. 14.21, *lower right* and 14.22, also see Figs. 6.5e and 10.33). Such types of elevated inversions are the most common over southern Africa. Other types do occur, however. The advection of cool sea-breeze air over warm land (and indeed any cool air over warm land), the advection of rural air over a city and cold air wedging beneath warm in cold-frontal discontinuities will produce elevated inversions as well.

A radiosonde ascent to show an elevated inversion over Pretoria as well as acoustic radar traces to show elevated inversions over Johannesburg are given in Figure 14.23. Like their surface counterparts, elevated inversions may be continuous over considerable distances, not only locally over cities and adjacent areas, but also on an extensive regional or even sub-continental scale. In Figure 14.23 it can be seen that the average winter base height of the first ele-

vated inversion above the ground varies from less than 1 000 m over the west coast to an average height of about 1 500 m over the eastern plateau. The strength of the temperature discontinuity is seldom greater than 2 °C. This is sufficient to trap all pollution below the inversion base and to ensure a highly adverse air-pollution climate.

AIR-POLLUTION DISPERSION

When pollution is emitted from a stack, the rate at which it is diluted is a function of the wind velocity and atmospheric stability. Wind speed governs the volume of air passing the stack so that concentration per unit volume decreases with increasing wind speed. As wind speed increases, so too do short-period turbulent fluctuations caused by surface roughness and internal shearing between air layers. These forced convection elements effect dilution by rapidly mixing the stack plume with clean air. Strong winds also transport pollutants with rapidly decreasing concentration downwind; weak winds have less of a diluting effect and may transport pollution considerable distances without much dispersion taking place.

Wind direction determines the path followed

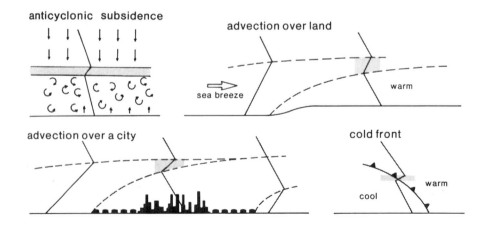

Fig. 14.22 Elevated inversions caused by subsidence, by the advection of cool air over warm, rough surfaces, and as a result of a cold front (adapted in part from Oke, 1978).

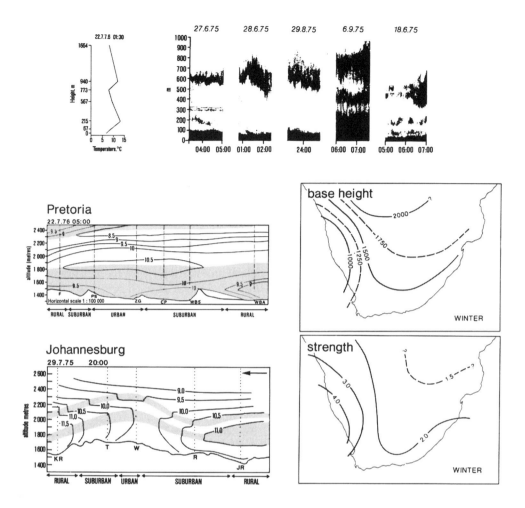

Fig. 14.23 *Upper:* Pretoria (Irene) radiosonde profile and acoustic radar traces over central Johannesburg to show elevated inversions; *centre* and *lower left:* extended regional elevated inversions over Pretoria (Von Gogh, 1978) and Johannesburg, the latter draping the topography; *centre* and *lower right:* the generalized base height (*m*) and strength (°C) of the first elevated inversion in winter over southern Africa as a whole).

by pollutants. Although the mean wind direction may remain approximately unchanged over a period of time, short-period direction and speed changes, due to turbulence, cause the plume to diffuse sideways. The degree to which lateral diffusion will take place depends on the time over which sampling is conducted. Horizontal spreading tends to have a conical form and within the cone the spread tends to conform to a Gaussian (normal) distribution away from the central line defined by the mean wind direction (Fig. 14.24).

Effective stack height

Pollution emission from a continuous point source is recorded at the ground at some point downwind from the stack. The level of concentration is reduced, however, by increasing the height of the stack, h_s, since this permits more effective eddy diffusion. If, in addition, the temperature of the stack emission is well above that of the environment and the stack exit velocity is high, the plume will rise due to buoyancy and momentum by an additional height, Δh (Fig.

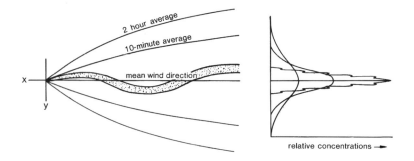

Fig. 14.24 Instantaneous, 10-minute and 24-hour average continuous air-pollution plumes (*left*), together with the corresponding cross-wind distributions of pollution (*right*).

14.25). The *effective stack height*, h, is therefore

$$h = h_s + \Delta h \tag{14.25}$$

A variety of formulae exist for the calculation of Δh. These must accommodate the characteristics of the rising plume under conditions of varying stability, turbulence and crosswind characteristics. The initial velocity of the rising plume relative to the ambient air causes shearing between the stack gas and the air. This leads to mechanical turbulence on the edges of the rising plume with mixing taking place between the air and the plume. This entrainment reduces the plume momentum generated by the stack exit velocity and in a stable environment will also reduce its buoyancy by mixing with cooler air. The effect of a crosswind is to produce horizontal momentum with subsequent shearing, turbulence and entrainment. In a stable environment the plume loses buoyancy by entrainment and rises by a finite amount to the height at which the plume temperature equals that of the ambient air.

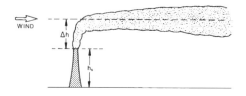

Fig. 14.25 Plume rise, Δh, from a stack of height h_s.

The Moses-Carson formula provides an example of an empirically derived, statistical plume-rise formula. For all stability classes

$$\Delta h = S\left(-0.029\,\frac{V_s d}{u} + 5.35\,\frac{h^{1/2}}{u} \right) \tag{14.26}$$

where S represents stability correction factors (2.65, 1.08, 0.68 for unstable, neutral and stable conditions respectively), u is the mean wind speed, V_s is the stack exit velocity, d is the stack diameter and h the heat emission rate.

The Briggs model offers a better, process-based way of modelling plume rise. In neutral and unstable conditions

$$\Delta h = \frac{7.4 h_s^{\,2/3} F^{1/3}}{u} \tag{14.27}$$

where h_s is stack height, u is mean wind speed at stack height and the buoyancy flux, F, is given by

$$F = g w_s \left(\frac{d}{2}\right)^2 \left(\frac{\rho_a - \rho_s}{\rho_a}\right) \tag{14.28}$$

where g is the force of gravity, w_s is the exit velocity of effluent, d is stack-top diameter, ρ_a is the density of air at stack height and ρ_s is the density of the stack gas. Under stable conditions

$$\Delta h = 2.9 \left(\frac{F}{uG}\right)^{1/3} \tag{14.29}$$

where G is a stability parameter given by

$$G = \frac{g}{\theta_s} \frac{\Delta\theta}{\Delta z} \qquad (14.30)$$

where θ_s is potential temperature at stack height and $\Delta\theta/\Delta z$ is the lapse rate of potential temperature.

Plume rise is effectively curtailed by temperature inversions, often leading to the development of spectacular plume types as the illustrations in Figure 14.26 show.

increasingly larger sinuous loops. High instantaneous ground-level concentrations may occur where the plume comes to ground. However, in general, ground-level concentrations decrease downwind. Small forced-convection eddies continuously break up the plume along its edges and cause further dilution.

• *Coning* occurs in an environment with neutral stability. Without free convection to expand the plume, or a stable environment to compress it, the plume grows vertically and laterally into a

Fig. 14.26 *Upper:* a slowly diffusing early-morning ribbon plume in the Witwatersrand area (photo: R. E. Pretorius); *lower left:* plume rise and fanning over Pietermaritzburg; *lower right:* the vertical inhibition of a strong, forest-fire plume by an inversion over the Canterbury Plains, New Zealand (photos: P. D. Tyson).

Plume behaviour

The form of a plume after emission is determined by stability conditions and the degree of turbulence and may be classified into five basic types (Fig. 14.27).

• *Looping* is characteristic of an unstable environment in which large free-convection eddies carry the plume up and down along a path with

cone shape through diffusion generated by forced convection and mechanical turbulence.

• *Fanning* requires a stable (inversion) environment. With vertical motion suppressed under these conditions the plume remains thin, but viewed from above may acquire a fan shape, or, alternatively, it may form a straight or meandering ribbon plume. Such plumes occur commonly at night under clear, calm conditions.

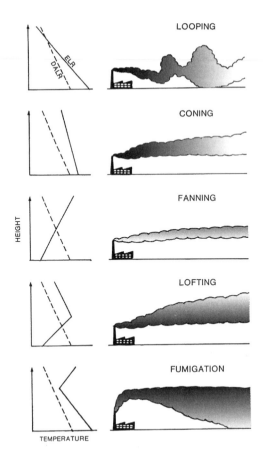

Fig. 14.27 Plume types and their relation to atmospheric stability (after Bierly and Hewson, 1962).

• *Lofting* usually occurs in the early evening while a surface radiative inversion develops. Within near-surface air the downward flux of matter and momentum is prevented by the stable conditions. Unstable air above the inversion permits the plume to disperse upward. This type of plume is usually transitory and as the inversion deepens is replaced by a fanning plume.

• *Fumigation* is the reverse of lofting. An elevated inversion prevents the upward growth of the plume, while the lapse temperature profile beneath permits the plume to disperse downward to the surface. Fumigations produce the most adverse pollution conditions and may

occur in a number of different ways (Fig. 14.28).

(a) *Thermal fumigations* occur when a radiation inversion is eroded by surface heating. In the presence of a fanning plume, the content of the plume is carried to the surface once the lapse conditions deepen to reach the plume. High ground-level concentrations result until such time as the elevated inversion dissipates.

(b) *Dynamic fumigations* are of a number of different types and may occur in association with the passage of a frontal system or, more usually, with the inland advance of a lake or sea breeze or the up-valley advance of a local valley wind. They may also occur in coastal areas in association with the subsidence ahead of a coastal low, followed by rapid mixing of trapped pollution with the passage of the low. Boundary-layer fumigations are most frequently associated with turbulent mixing within urban heat islands. Finally, dynamic fumigations may occur to the lee of hillslopes owing to the development of a standing eddy formed downwind of the crest. The same phenomenon is to be observed behind the lee edges of tall buildings.

The Gaussian Plume Model

The meandering behaviour of a plume and the random nature of atmospheric turbulence mix pollutants so that their concentration has a Gaussian (normal) distribution about the plume's central axis in both the horizontal, y, and vertical, z, planes (Fig. 14.29). This distribution permits the mathematical specification of the dispersion curves and the development of a plume dispersion model.

Consider a continuously emitted plume of pollution from a stack at height h above the ground. Near the stack no ground-level pollution will be measured. At a certain distance from the stack pollution from the elevated source will be encountered and will thereafter increase rapidly with distance x downwind, will reach a maximum and then will decrease steadily with further distance downwind. The under-

TYPES OF FUMIGATION

(a) thermal

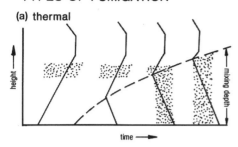

(b) dynamic

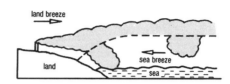

(c) dynamic / synoptic

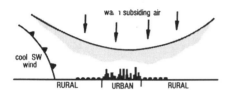

(d) urban boundary layer

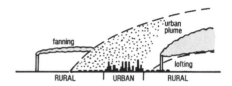

(e) lee slope

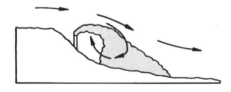

(f) local obstacle

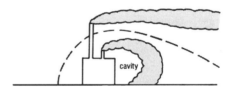

Fig. 14.28   Different air-pollution fumigation types.

lying assumption of the model is that, as the pollution disperses horizontally across the wind and vertically, half will move away from the centre line on each side. If, say, 1 000 units of pollution are emitted and dispersed horizontally, 500 will spread to the left and 500 to the right. Of the 500 units to the right, half (250) will go left, half right, and so on. The process continues so that at step 2 the distribution will be 250, 500 and 250 units across the spreading plume, at step 3 the distribution will be 125, 375, 375 and 125 units, and so on. At each step the spreading distribution of pollution is Gaussian (normal) with the standard deviation of the dispersion

across the plume, σ_y, increasing with increasing x. The same argument and distribution holds for the vertical. In each case the Gaussian equation of the general form in x–y coordinates describes the distribution:

$$y = \frac{1}{(2\pi)^{1/2}\sigma} \ \exp -\left(\frac{x}{2\sigma^2}\right) \qquad (14.31)$$

On the basis of the above rationale, the model is developed as follows. In the *downwind direction* (i.e., x-direction) the ground concentration of pollution χ is given by

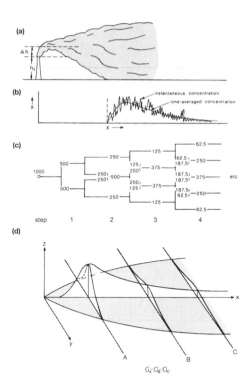

Fig. 14.29 (a) Plume rise and spread; (b) down-wind, along-axis concentrations from the plume; (c) a Gaussian spread of 1 000 units of pollution in the cross-wind direction; (d) the down-wind and cross-wind spreading of pollution.

$$\chi = \frac{Q}{x} = \frac{Q}{u\Delta t} = \frac{Q}{u} \tag{14.32}$$

for unit time ($\Delta t = 1$), where Q is the rate of emission, x is distance downwind that a stack-top wind of mean speed u moves pollution in unit time. In the *crosswind direction* (i.e., y-direction) the Gaussian spread is described by

$$\frac{1}{(2\pi)^{1/2}\sigma_y} \exp - \left(\frac{y^2}{2\sigma_y^2}\right) \tag{14.33}$$

In the *vertical direction* the spread is

$$\frac{1}{(2\pi)^{1/2}\sigma_z} \exp - \left(\frac{z^2}{2\sigma_z^2}\right) \tag{14.34}$$

at the ground. As a consequence of being emitted at an effective height, h, above the ground the concentration becomes

$$\frac{1}{(2\pi)^{1/2}\sigma_z} \exp - \left[\frac{(z+h)^2}{2\sigma_z^2}\right] \tag{14.35}$$

However, not all pollution is immediately absorbed at the ground. Reflection takes place. This is handled by assuming that a mirror-image virtual source at $-h$ is also producing pollution. The combined spread in the vertical as a result of both the real and virtual sources is then

$$\frac{1}{(2\pi)^{1/2}\sigma_z} \left\{ \exp - \left[\frac{(z+h)^2}{2\sigma_z^2}\right] + \exp - \left[\frac{(z-h)^2}{2\sigma_z^2}\right] \right\} \tag{14.36}$$

Multiplying the expressions (14.32), (14.33) and (14.36), gives the concentrations $\chi_{(x,y,z,h)}$ at distance x downwind, at distance y across the wind, at height z above ground, and from an effective stack top at height h above ground as

$$\chi_{(x,y,z,h)} = \frac{Q}{2\pi\sigma_y\sigma_z u} \exp - \left(\frac{y^2}{2\sigma_z^2}\right)$$

$$\left\{ \exp - \left[\frac{(z+h)^2}{2\sigma_z^2}\right] + \exp - \left[\frac{(z-h)^2}{2\sigma_z^2}\right] \right\} \tag{14.37}$$

Distance downwind is established in the model through the specification of the crosswind and vertical dispersion coefficients, σ_y and σ_z, by

$$\sigma_y = ax^b \tag{14.38}$$

and

$$\sigma_z = ax^b \tag{14.39}$$

where a and b are regression coefficients that are dependent on atmospheric stability.

Plume rise is modelled in neutral and unstable conditions using the Briggs equations given in equations (14.27 to 14.30).

Finally, it is necessary to model the wind speed at stack height from ground-level winds. This may be done using a power profile of the form

$$\frac{u_s}{u_g} = \left(\frac{z_s}{z_g}\right)^p \qquad (14.40)$$

where u_s is mean wind at stack height z_s and u_g is the ground-level wind at a standard recording level, z_g.

Depending on the state of atmospheric stability, it is possible to model pollution by determining, first, plume rise and hence effective stack height, secondly, the mean wind at stack height, and, thirdly, the actual concentrations for different time intervals (e.g., hourly, daily, seasonal, etc.)

Many modifications of the Gaussian Plume Model may be made to accommodate fumigations and other conditions; likewise more sophisticated models such as those of the integrated and sheared puff variety may be developed. The Gaussian Plume Model remains a useful one, however, particularly for illustrating the manner in which a simple model may be developed. It may be adapted to a variety of different uses and circumstances and is a useful first approximation to the highly complex task of modelling atmospheric pollution from a single continuous point source. The model may be applied easily and has modest data requirements. It does, however, suffer from weaknesses that limit its usefulness. These include the neglect of downwind absorption and deposition of pollutants, photochemical changes in the plume and the effect of friction on wind-direction changes with height (which causes the plume to spread in the cross-wind direction). The model also fails when meteorological conditions change over short time periods (the order of an hour) and cannot be used in regions of rough terrain where local winds are an important feature of mesoscale climates.

MESOSCALE AIR CIRCULATIONS

All local winds owe their origins to discontinuities in energy-balance fields or regimes. These discontinuities may occur across shorelines of lakes or seas, between sloping surfaces and the air beyond, or between mountains and valleys. In considering land and sea breezes and topographically induced winds on slopes and in valleys, a distinction must be made between local and regional winds.

Local winds on slopes and across shorelines

As a consequence of differences in the energy-balance regimes on slopes and in the free air adjacent to them, and the resultant differential heating (or cooling) of the air on the slope and in the free air, local baroclinic fields develop on slopes. Deformation of isobars and isotherms is opposite by day and night (Fig. 14.30, *upper*). Within the baroclinic zone, and to a depth H, solenoidal circulations are initiated. The circulation in each cell has a counteracting effect on its neighbours, except in the outer layer of cells. The consequence is the initiation of a low-level, up-slope *anabatic flow* of air up warmed slopes by day. At the top of the local baroclinic layer a compensating *return flow* completes the closed circulation. By night radiational cooling of the slopes and the reversal of the baroclinicity produces down-slope *katabatic flow* and its return flow. Katabatic flow may be concentrated in valleys running down the slopes, or may occur as shallow sheet flow (such as is illustrated in Figure 14.38).

Across shorelines similar localized baroclinic fields develop by day with land heating to produce *sea breezes* and their *return currents* in a closed circulation with convergence and ascent over the land and the opposite out to sea (Fig. 14.30, *lower*). As the system strengthens by day, so the sea-breeze front advances inland. Over flat country the daytime sea breeze may only arrive far inland at night. The sea breeze reaches Kalgoorlie in Western Australia 360 km

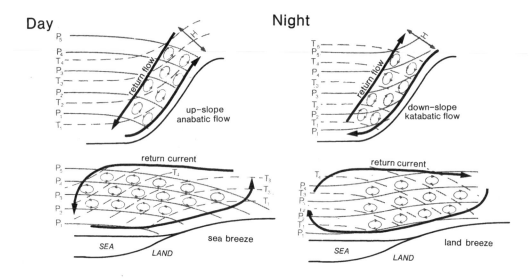

Fig. 14.30 Local baroclinic zones and solenoidal circulations producing anabatic and katabatic winds and their return flows, and land and sea breezes and their return currents.

inland from the sea at 21:00. Similarly on occasions a south-westerly sea breeze may be observed to begin blowing at the Gamsberg (on the Escarpment inland from Walvis Bay) around midnight. With late-afternoon weakening it is usual for sea breezes to be blowing inland, while they have ceased altogether along the coast. By night, with the cooling of the land, a *land breeze* and its *return current* develop. Land and sea breezes are common along the coastline of southern Africa. They are strongest along the west coast where the horizontal temperature gradient between the cold Benguela Current and the hot, vegetation-bare desert is greatest. The extent of the cold west-coast Benguela water is clearly evident in satellite photographs, as too is the warm water of the Agulhas Current to the east of KwaZulu-Natal and the deformation of the Current to the south of the Western Cape in the retroflection area (see earlier in Fig. 13.5). Gradients from the warm Agulhas Current to the well-vegetated east coastal areas are much smaller than those along the west coast, and the local winds are correspondingly weaker. On both coasts the sea breezes adjust to the Coriolis force so that, as they strengthen by day, they are deflected increasingly to the left to become north-easterly winds along the KwaZulu-Natal coast and south-westerly winds along the Namib shore. That the prevailing gradient winds on the two coasts are in the same direction as the sea breezes often makes the identification of the latter difficult.

Within valleys, local airflow is much more complicated than either on slopes or across shorelines. The degree of complexity of the airflow depends on the orientation of the valleys. East–west valleys (where sunrise and sunset heating and cooling will be reasonably uniform on each slope) behave differently from north–south valleys (where sunrise and sunset heating and cooling effects are in strong contrast) in many respects (Fig. 14.31).

If the rising sun warms both slopes approximately equally, then a double helical circulation may be initiated. Anabatic winds will develop up both slopes and subsidence will occur over the central valley. Later in the morning, as a strong axial temperature gradient begins to develop between the mouth and head of the valley, the up-slope winds begin to weaken as an up-valley *valley wind* is initiated. Above this an *anti-valley wind* will complete the circulation. The valley wind advances up the valley and by

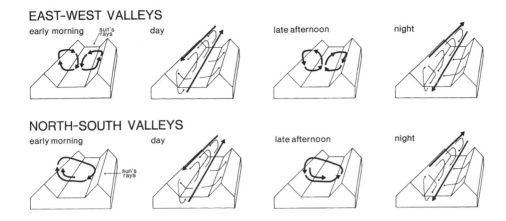

EAST-WEST VALLEYS

early morning   sun's rays   day   late afternoon   night

NORTH-SOUTH VALLEYS

early morning   day   sun's rays   late afternoon   night

Fig. 14.31 A model of the diurnal variation of local airflow in valleys.

early afternoon may be blowing strongly throughout the valley to a depth exceeding the flanking ridgelines. In the late afternoon, as the slopes begin to cool, the valley wind weakens near the slopes and along the valley floor. A reverse helical circulation to that of the early morning begins to blow with descending kata-batic flow down the valley slopes. Ascending motion occurs over the central valley. This phase is highly transitory (and may even be absent altogether) and is soon replaced by down-valley flow in the *mountain wind* which blows in response to the now-reversed temperature gradient between the head (cooler) and mouth (warmer) of the valley. To satisfy continuity the mountain wind has an *anti-mountain wind* as its compensating return current.

When the rising (setting) sun warms (cools) one slope, while the other is cooling (warming), uni-cellular overturning of air takes place (Fig. 14.31, *lower*). For the rest of the day and night the circulation within the valley will be the same. Since valleys seldom maintain fixed directions, the exact nature of the circulations they produce around sunrise and sunset will vary from place to place. During the day, flow will be up-valley; by night it will be down-valley provided the sky is cloudless (to allow energy-balance contrasts to be greatest) and the gradient winds are not too strong. As the air flows along

the valley, it will bank up and be stronger on the outer slope of a bend and be lower and weaker on the inner slope. Anti-mountain winds blow above the level of the flanking ridges and so are prone to obliteration by gradient winds to an extent far greater than the parent local winds in the valleys below. Valley winds dominate and are strongest in summer (when heating effects are greatest); mountain winds dominate and are strongest in winter (when cooling effects are most pronounced).

Over the eastern-plateau slopes of KwaZulu-Natal, along a 180 km transect between Giant's Castle and Durban, local winds conform close-ly to the above model. Examples of different stages in the development of the various regimes are given in Figure 14.32. Sections across the Bushman's valley at Giant's Castle show the development of the down-valley mountain wind, as well as the development of single-cell overturnings of air within the valley during the transition phase between the occur-rence of mountain and valley winds. Over Pietermaritzburg it can be seen clearly how the evening mountain wind advances down the val-ley and the early-morning valley wind advances up the valley as the systems deepen and develop by night and day. Similarly, at the coast at Durban the inland advance of the morning sea breeze and inland extension of the evening land

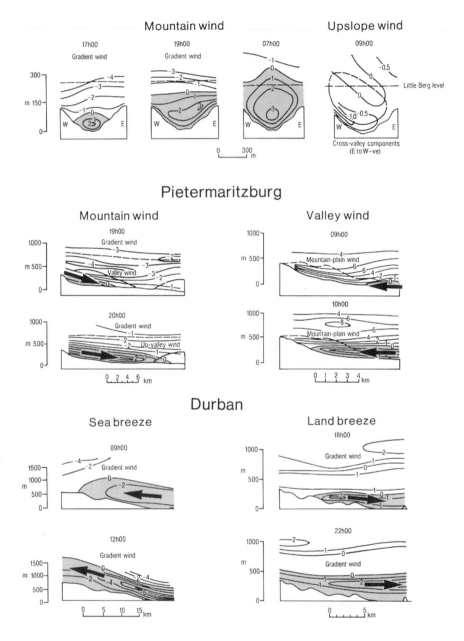

Fig. 14.32 *Upper:* cross sections through the down-valley *mountain wind* in the Bushman's valley at Giant's Castle; *centre:* the onset of the *mountain wind* in the Msinduzi valley, Pietermaritzburg; *lower:* the onset of the *sea breeze* over Durban. Velocities are given in m s⁻¹ as components parallel to valleys or normal to the coastline.

breeze are clearly evident. Simultaneously throughout KwaZulu-Natal on all slopes, in all valleys and at the coast, local winds undergo evolution and decay as one system changes to another, as one is overwhelmed by another, and as one becomes integrated with another. It becomes difficult to separate the various local effects, as for instance near the coast between an inland-advancing sea breeze moving up a valley and a valley wind developing and advancing inland up the same valley. To further complicate matters regional temperature gradients between coastal areas and the Drakensberg produce yet another suite of topographically induced winds.

Regional topographically induced winds

Just as the temperature gradient between slopes and adjacent free air and between the heads and mouths of valleys produce local pressure gradients driving local winds along slopes and within valleys, so the temperature gradients between mountains (escarpments) and plains

and/or the coast produce regional airflows on an appropriately larger scale. *Mountain–plain* winds produce large-scale regional airflow between cooler mountains and warmer plains by night. *Plain–mountain* winds produce an opposite flow between cooler plains and warmer mountains by day. Both winds may have anti-wind counterparts, though these are seldom observed, since they are weak and prone to being masked by gradient flow. Such regional winds have been reported in North America between the Rocky Mountains and plains to the east, in Europe between the Alps and plains to the north, in India, South America and elsewhere. In southern Africa they are particularly well developed seaward of the Escarpment over KwaZulu-Natal and over the Namib desert.

The diurnal sequence of local and regional topographically induced wind development over KwaZulu-Natal is presented in Figure 14.33. Around sunset mountain winds are initiated in individual valleys. At the coast the land breeze begins to blow. Overlying these local winds a deep plain–mountain wind prevails within the boundary layer to a depth of about 1 000 m. In

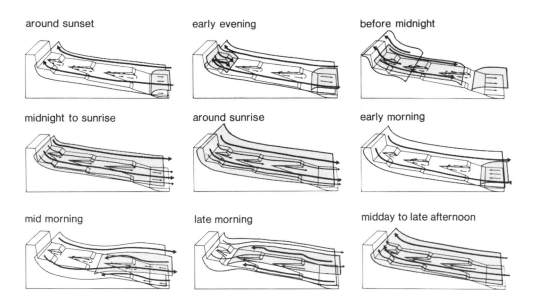

around sunset early evening before midnight

midnight to sunrise around sunrise early morning

mid morning late morning midday to late afternoon

Fig. 14.33 The diurnal variation of regional and local topographically induced winds between the Escarpment and sea over KwaZulu-Natal.

Drakensberg escarpment

By night

By day

Drakensberg foothills

By night

By day

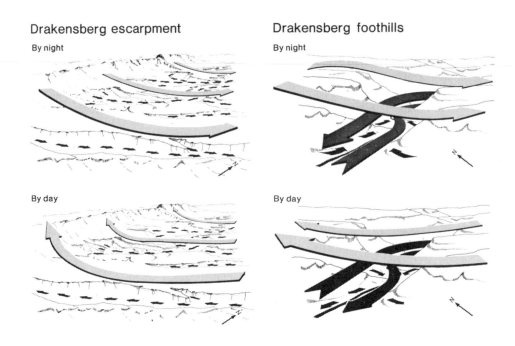

Fig. 14.34 Mountain–plain winds and plain–mountain winds overlying their mountain- and valley-wind counterparts in the Drakensberg escarpment zone and foothills region by night and day.

the early evening a mountain–plain flow begins off the Escarpment slopes and across the Drakensberg foothills of the Little Berg, the land breeze extends further inland at the coast, mountain winds strengthen and the plain–mountain wind begins to weaken. Sometime before midnight the plain–mountain wind decays completely over the coast and the KwaZulu-Natal Midlands, but continues blowing weakly near the Drakensberg. More importantly, by this time the mountain–plain wind has advanced to near the coast. Around midnight the mountain–plain wind and land breeze become completely integrated and the distinction between these two winds and mountain winds in individual valleys becomes difficult, if not impossible, to make. Between midnight and sunrise the mountain–plain flow deepens and strengthens so that around sunrise the flow is about 1 000 m deep. After sunrise the land breeze decays, the sea breeze is initiated and valley winds begin to advance up the valley. By

mid-morning the plain–mountain wind has been initiated and is advancing inland toward the Escarpment. Overlying it is a decaying mountain–plain wind even deeper than earlier. By late morning the plain–mountain wind has almost reached the Drakensberg and the mountain–plain wind is about to disappear. From midday to late afternoon the plain–mountain wind dominates and it is difficult, if not impossible, to distinguish between sea-breeze, valley-wind and plain–mountain-wind effects.

Near the Escarpment and in the Drakensberg foothills the layering of the diurnal airflow in and above the valleys is characteristic (Fig. 14.34). The picture is even more complicated with anti-winds developing on certain occasions and at certain times. These are by no means as regular as their generating counterparts and are difficult to observe, since they often blow in the same direction as the gradient wind above. The manner in which the local and regional winds dominate the near-ground airflow is well exemplified by 3-hourly wind roses for Colenso in the upper Tugela valley (Fig. 14.35). During the night and early morning, flow is down the valley and away from the Drakensberg; by day and in the early evening it is up the valley and toward the Escarpment. Equally clear is the fact that mountain/mountain–plain wind effects are dominant in winter and valley/plain–mountain winds are predominant in summer.

Time–height measurements of the mountain–plain wind on specific occasions are given in Figure 14.36, together with examples of observations of the wind advancing seaward over Drakensberg valleys, over Pietermaritzburg, Cato Ridge and Kloof and finally over Durban and out to sea over the Indian Ocean. As hollows fill with cold air by katabatic flow, as valleys fill with cold mountain-wind air and as the whole region is overlain by the cool mountain–plain wind drift, so a distinctive low-level temperature stratification develops over Kwa-Zulu-Natal. It is highly stable and inversions predominate in every valley and above them (Fig. 14.37). Within the low-level airflow pollution or fog act as clear tracers of the local and regional winds which are easily recognizable (Fig. 14.38).

In air-pollution and regional-airflow studies it is necessary to model local winds and particularly the variation of wind speed with height. Examples of characteristic wind profiles, some instantaneous, some hourly averages and some longer-period averages, are given in Figure 14.39. All are *Lagrangian* profiles, which are profiles obtained by taking observations of air particles

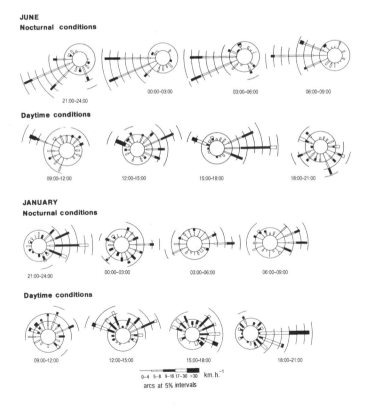

Fig. 14.35 The diurnal variation of wind at Colenso (data from Langenberg, 1974). Winds from a westerly quarter represent largely mountain and mountain–plain wind effects; those from an easterly quarter, valley and plain–mountain wind effects.

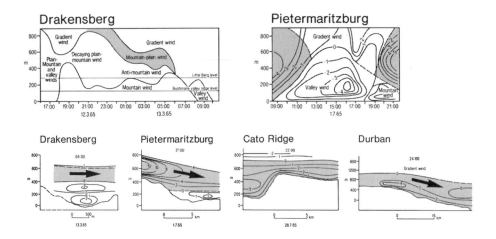

Fig. 14.36 *Upper:* time–height sections to show the mountain–plain wind (shaded); *lower:* examples of the mountain–plain wind along a transect from the Drakensberg to Durban. Wind speeds (m s^{-1}) are given as components parallel to valleys.

(balloons) moving with the wind at different heights, in contrast to *Eulerian* profiles, in which airflow is measured at a fixed reference point. The wind-profile equations (14.3) to (14.6) are for the latter types of airflow. The Lagrangian profiles of local thermodynamic winds are best described using a Prandtl profile of the form

$$u = u_m \frac{\exp(\pi/4)}{\sin(\pi/4)} \, \sin\left(\frac{z\pi}{H}\right) \exp -\left(\frac{z\pi}{H}\right) \quad (14.41)$$

where u_m is the maximum velocity observed at height h and H is the depth of the air stream. An alternative model is a parabolic variation of wind speed with height of the form

$$u = u_m \left\{ 1 - \frac{(z-h)^2}{h} \right\} \quad (14.42)$$

The Prandtl model best describes the mountain and valley wind over Pietermaritzburg and the valley winds in the Drakensberg. The parabolic profile is good for mountain and anti-mountain winds in Drakensberg valleys and for plain–mountain winds in the Drakensberg region. Rarely are mountain-wind profiles steady throughout the night. Instead the wind tends to surge periodically (every 60–75 minutes). As

adiabatic heating and local horizontal divergence occur in air accelerating down-valley in a surge, the local temperature (and hence pressure) gradient is weakened. Deceleration consequently occurs until radiational cooling again increases the pressure gradient and the cycle repeats itself. Across the continent, over the central Namib desert and between the Escarpment and the cold Benguela Current of the Atlantic Ocean coast, a similar suite of local and regional winds blow the year round, exhibiting some of the clearest boundary-layer oscillations to be found anywhere (Fig. 14.40). Like their east-coast counterparts, they seldom exceed about 1 000 m in depth, are vigorous, and blow constantly unless disturbed by strong synoptic systems. They show clear annual rhythms. Valley and plain–mountain winds predominate over the central Namib in summer and are less frequent in winter; the opposite is the case with mountain and mountain–plain winds, which show a near-inverse frequency distribution with a clear winter peak. On the Namib coast the sea breeze sets in at about 09:00. Thereafter it backs to south-west often blowing with speeds in excess of 50 kph. The sea-breeze front advances steadily inland so that by 15:00 it has penetrated about 60–70 km into the central Namib. At

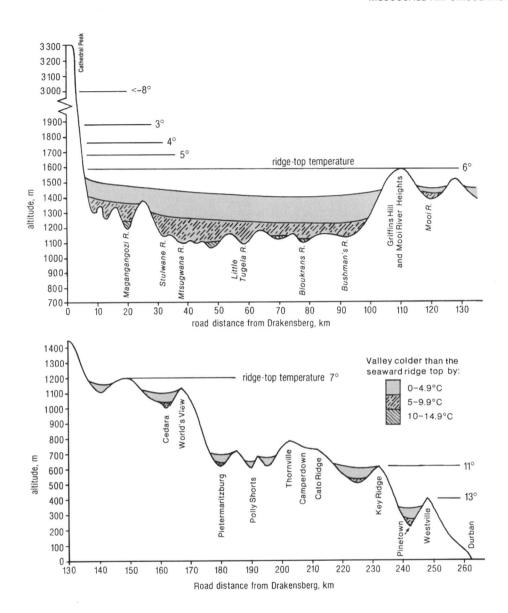

Fig. 14.37 A simplified temperature section across KwaZulu-Natal from the Drakensberg to Durban on a clear winter night from 00:00 to 06:00, to show the development of valley inversions. Valley cooling is given in the upper section as the difference between the Griffins Hill ridge-top temperature and valley temperatures; in the lower section it is the difference between the seaward ridge temperature and that in the adjacent valley. The decrease of ridge-top temperatures with height is shown. The analysis is based on fieldwork by R. Washington during June 1987. A similar traverse on a cloudy night showed that no valley inversion formed under such conditions.

Gobabeb it only blows for a few hours before being replaced by coastward-moving air. Sea breezes blow throughout the year on the Namib

coast, but with slight frequency maxima at the times of the equinoxes.

The regular rhythm of the daily oscillation of

Fig. 14.38 *Upper left:* sheet katabatic flow near Pietermaritzburg; *upper right:* the mountain wind down Town Bush valley, Pietermaritzburg; *centre left:* tributary mountain winds contributing to large-scale lower-valley airflow; *centre right:* large-scale mountain-wind airflow down the Msinduzi valley; *lower left:* mountain and mountain–plain wind advection of pollution over Pietermaritzburg (photos: P. D. Tyson); *lower right:* airflow out to sea in the mountain–plain wind over Durban (photo: R. A. Preston-Whyte).

the boundary layer over the central Namib is clearly evident in the regular occurrence of plain–mountain and mountain–plain winds in the time sections illustrated in Figure 14.41. The winter examples illustrate the dominance of the low-level mountain–plain winds, driven by cooling effects, at that time of year. In contrast, the summer examples show the dominance of plain–mountain winds, driven by heat-ing effects, during the hot season.

The spatial pattern of the ebb and flow of the regional winds over the Namib is particularly clear (Fig. 14.42). In winter the cool-air drainage in the mountain–plain wind is deeper than in summer. Conversely, the summer warm-air advection in the plain–mountain wind is deeper than in winter.

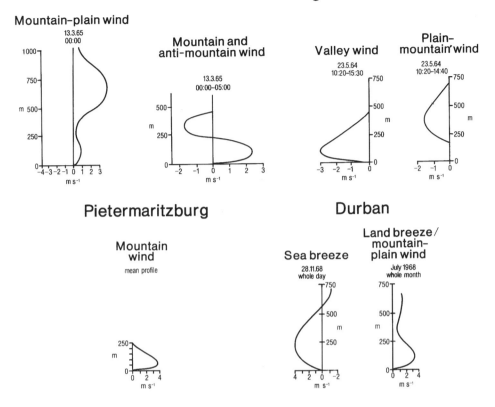

Fig. 14.39 Mean profiles of various local and regional topographically induced winds in KwaZulu-Natal.

Some environmental implications of local winds

Environmental local winds of all types are associated with highly stable air. They are the most non-turbulent winds of any kind and cause frost hollows and valley inversions wherever hollows and valleys are to be found. They are able to transport pollutants tens to hundreds of kilometres without much dispersion or dilution. Given their regularity of occurrence in suitable terrain, they represent highly adverse pollution conditions, particularly so when it is realized that, with early-morning decay of every nocturnal local wind, conditions are ideal for the occurrence of pollution fumigations (see Figs. 14.21 and 14.38). These may be associated with advancing sea breezes, valley winds or plain–mountain winds. Over the period of time covering the changeover from one regime to another it is possible for pollution to be recirculated (see Chapter 15 and Fig. 15.15).

Not all effects of local winds are adverse. The advection of cool sea-breeze air enhances the comfort of otherwise hot coastal areas. Cool plain–mountain winds bring relief on hot days. Valley and plain–mountain winds often produce mist in the mist belt of KwaZulu-Natal and frequently act as a local trigger of afternoon thunderstorms, as in the Drakensberg in KwaZulu-Natal and Mpumalanga.

By day

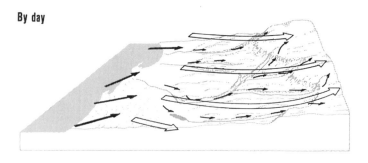

By night

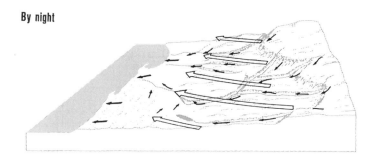

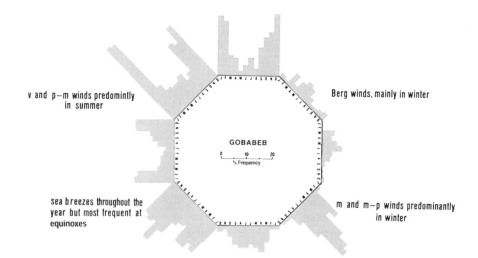

v and p—m winds predomintly
in summer

Berg winds, mainly in winter

GOBABEB

0   10   20
% Frequency

sea breezes throughout the
year but most frequent at
equinoxes

m and m—p winds predominantly
in winter

Fig. 14.40 *Upper:* a schematic model of the local components of airflow over the central Namib to show the occurrence of sea breezes, valley (v) winds and plain–mountain (p–m) winds by day in summer and land breezes, mountain (m) winds and mountain–plain (m–p) winds by night in winter; *lower:* monthly variation of wind direction frequencies at Gobabeb, which is located on the south-east- to south-west-trending Kuiseb river in the central Namib desert.

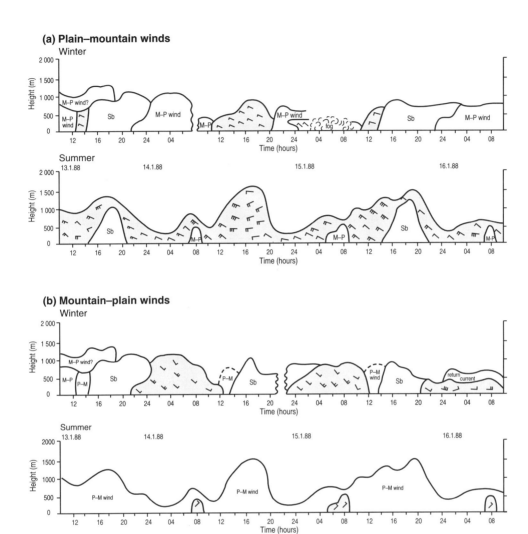

Fig. 14.41 Time–height sections of hourly winds determined from pibal ascents in the first 2 000 m above the surface at Gobabeb for 72-hour periods in winter and summer. In (a) plain–mountain winds are shaded; in (b) mountain–plain winds are shaded. Wind arrows fly with the wind: one feather represents wind speeds of 2.5–4.9 m s^{-1}, two feathers 5.0–9.9 m s^{-1} and three feathers 10.0–14.9 m s^{-1}. Sb denotes sea breeze, M–P mountain–plain wind.

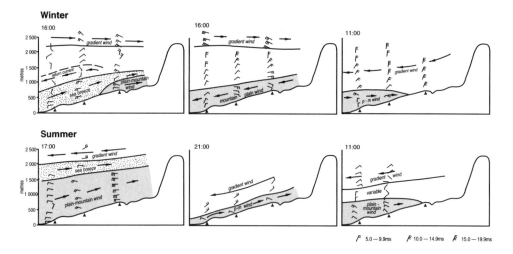

Fig. 14.42 Transects between the west coast and Escarpment over the central Namib to show the patterns of growth and decay of local winds in summer and winter. Wind arrows fly with the wind: one feather represents wind speeds of 2.5–4.9 m s^{-1}, two feathers 5.0–9.9 m s^{-1} and three feathers 10.0–14.9 m s^{-1}.

ADDITIONAL READING

SCORER, R. S. 1997. *Dynamics of Meteorology and Climate.* Chichester: Wiley.

SCHNEIDER, S. H. (ed.). 1996. *Encyclopaedia of Weather and Climate,* Vols. 1 and 2. Oxford: Oxford University Press.

GARRATT, J. R. 1992. *The Atmospheric Boundary Layer.* Cambridge: Cambridge University Press.

STULL, R. B. 1988. *An Introduction to Boundary Layer Meteorology.* Dordrecht: Kluwer Academic.

ATKINSON, B. W. 1981. *Meso-scale Atmospheric Circulations.* London: Academic Press.

OKE, T. R. 1978. *Boundary Layer Climates.* London: Methuen.

MONTEITH, J. L. 1973. *Principles of Environmental Physics.* London: Edward Arnold.

FLOHN, H. 1969. 'Local wind systems' in *General Climatology,* 2. Landsberg, H. E. (ed.). World Survey of Climatology, Vol. 2. Amsterdam: Elsevier.

MUNN, R. E. 1966. *Descriptive Micrometeorology.* New York: Academic Press.

SELLERS, W. D. 1965. *Physical Climatology.* Chicago: University of Chicago Press.

15

THE TRANSPORT OF AEROSOLS
AND TRACE GASES

The atmosphere over any region is never completely calm. It is always in motion and circulating on a variety of scales ranging from the local and mesoscale to the synoptic and larger. Any aerosols or trace gases suspended within the atmosphere, particularly those near the surface, will be transported on the same scales by the air in circulation. Vertical transport will be controlled by the stability structure of the atmosphere and horizontal transport by the local thermo-topographic winds near the surface or by larger-scale circulation in changing synoptic flow fields.

THE PROCESS OF TRANSPORT

Transport in the atmosphere is brought about by the circulation prevailing at any one time and then continued by the circulation occurring later. Thus at a synoptic scale, transport of a particular atmospheric constituent may be initiated by, say, an anticyclonic condition on *day 1*, continued with the passage of a trough on *day 2*, and further continued with the development of a ridging anticyclonic condition on *day 3* behind the front associated with the trough, and so on. The horizontal transport of air (and what is contained therein) may be determined by trajectory analysis. The principle of such analysis is demonstrated in Figure 15.1. On *day 1* the material is transported from A to B along trajectory AB. By the time it has arrived at B the pressure pattern and associated circulation has changed from anticyclonic to cyclonic and transport is continued along trajectory BC on *day 2* to give the two-day trajectory ABC. By the time the material has reached C the synoptic situation has become affected by a ridging anticyclone and the material continues along a

Day 1   **Day 2**   **Day 3**

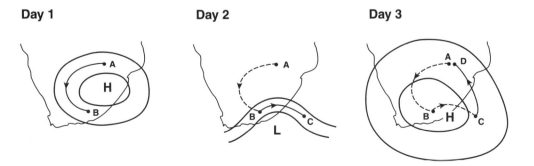

Fig. 15.1 Daily transport trajectories associated with different circulation systems over South Africa.

new trajectory CD to give the complete three-day trajectory ABCD. In the process the air has been recirculated back to the point from which the transport was initiated. Thus, whereas the transport is produced by the sequence of changing circulation types, the resulting transport pattern bears little apparent relationship to the individual types.

TRAJECTORY MODELLING

Two types of trajectory analysis may be used. In *isentropic trajectory modelling* the parcel of air being followed moves along isentropic surfaces; that is, along planes of constant potential temperature. Alternatively, in *kinematic trajectory modelling* no specific surface is specified and the parcel follows the path dictated by the prevailing relative strengths of the *u, v* and *w* components of the wind. In southern African transport studies the movement of air, trace gases or aerosols has usually been determined by kinematic analysis.

In kinematic trajectory modelling individual back-and-forward trajectory pathways from desired points of origin are determined from six-hourly, three-dimensional wind fields at different pressure levels. Vertical velocities are determined from adiabatic non-linear mode initialization, which permits the diabatic as well as the adiabatic processes that influence vertical motion to be considered. Usually trajectories are followed for seven days or longer and are more complicated than the simple situation illustrated in Figure 15.1. To minimize errors, clusters of trajectories are determined for a variety of points close to the source area from or to which transport is being determined. The centroid trajectory is then taken to be representative. In all transport studies, local and mesoscale transport needs to be distinguished from that occurring on a synoptic scale.

LOCAL AND MESOSCALE TRANSPORT

Transport near the surface in the lowest section of the boundary layer is almost always induced by horizontal spatial discontinuities in temperature, pressure and density fields, like those caused by land–water or urban-rural transitions, or by topographically induced local wind systems of the kind considered in Chapter 14. Nocturnal katabatic flow and down-valley mountain winds in highly stable air within a surface inversion frequently transport pollution trapped within the inversion long distances down slopes or valleys (Fig. 15.2, *upper*). On a larger scale, between a plateau and plains or a mountain range and plains, mountain–plain winds, blowing by night and in the early morning, may transport aerosols and trace gases in stable conditions over distances up to 200 km or more. This frequently happens in South Africa between the Escarpment and lower-lying land sloping down to coastal areas. It is particularly well developed in KwaZulu-Natal during winter between the Drakensberg, the Midlands beyond and the sea.

By day, the reverse transport may occur, provided that the unstable conditions in which valley and plain–mountain winds occur are capped with absolutely stable layers or inversions. In the absence of such capping, dispersion may be effected within a few hours, as happens in the closing stages of fumigation conditions (see Chapter 14). Large topographical contrasts are not needed to produce the nocturnal and day-time regional winds that effect local and mesoscale transports. For instance, over the Highveld of central South Africa, providing synoptic flow fields are weak, material produced in the major urban and industrial areas may be transported by day in shallow plain–mountain winds towards the plateau edge to the east beneath the frequently-occurring ~700 hPa stable discontinuity (Fig. 15.3). The next night, the material that has accumulated over the escarpment area is transported out over the Lowveld by the nocturnal drainage of air off the

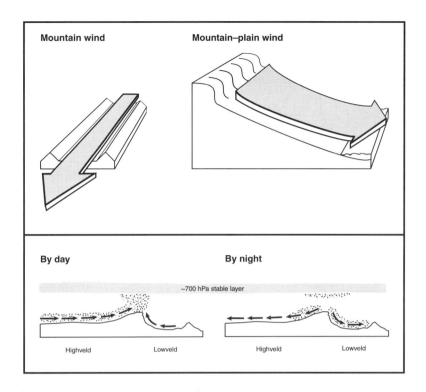

Fig. 15.2 Mesoscale transport associated with nocturnal mountain and plain–mountain wind systems (*upper*) and with the diurnal variation of low-level regional winds over the Highveld and Lowveld of South Africa (*lower*).

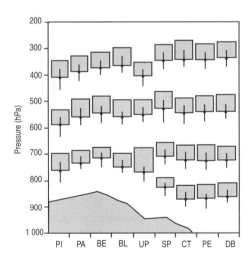

Fig. 15.3 The stable stratification of the atmosphere over South Africa on no-rain days. Absolutely stable layers are indicated. Of these the ~500 hPa layer is the most persistent. All occur throughout the year on no-rain days.

Escarpment in mountain and mountain–plain winds (Fig. 15.2, *lower*). Material initially not carried to the Escarpment will recirculate back towards the source area over the Highveld.

Another type of nocturnal mesoscale transport may occur in surface-inversion conditions over flat ground in the absence of local winds as flow decouples at the top of an inversion layer. With such decoupling, flow begins to accelerate along the top of the inversion and a low-level jet forms as the roughness of the ground surface is replaced by the relative smoothness of the inversion top (Fig. 15.4, *upper*). The jet rises during the night as the inversion deepens. Any pollution trapped near the top of the inversion layer is transported away from the source area in the region of the low-level wind maximum over distances of 300–500 km per night. Such mesoscale jets may be spatially continuous over large areas and considerable distances, as an

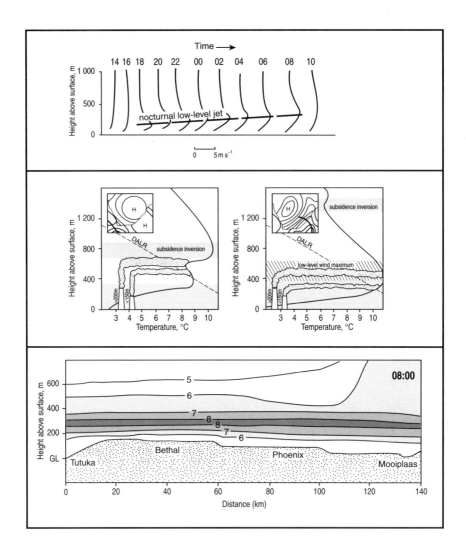

Fig. 15.4 *Upper:* the development of a nocturnal low-level jet at the top of an inversion layer; *middle:* the trapping of pollution plumes beneath and within inversions and stable layers over the Highveld of South Africa; and *lower:* the horizontal extent of a low-level, inversion-top jet over the Highveld (after Tosen, 1997). Wind speeds are in m s^{-1}.

illustration from the Highveld of South Africa shows (Fig. 15.4, *lower*).

SYNOPTIC-SCALE TRANSPORT

With most of the different types of synoptic circulation fields that occur in the lowest layers of the atmosphere over southern Africa, transport of air, and whatever is carried with it, is predominantly towards the east coast and the Indian Ocean (Fig 15.5). Recirculation back over the subcontinent is brought about by anticyclonic circulation fields, mainly those associated with transient ridging highs that follow the passage of westerly waves and cold fronts.

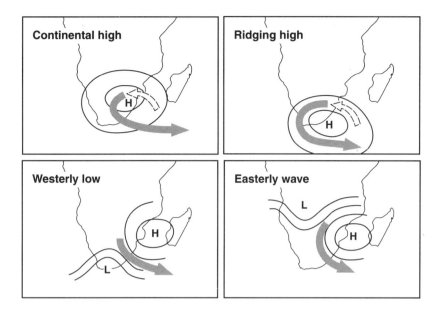

Fig. 15.5 Major transport pathways associated with different circulation types over southern Africa. Shaded arrows indicate direct transport; broken arrows show recirculation.

THE SOUTHERN AFRICAN HAZE LAYER

Over much of the year, aerosols are trapped in a dust pall beneath the ~500 hPa stable discontinuity to form a pronounced haze layer. The layer blankets the whole subcontinent from South Africa to northern Zambia and beyond. The top of the haze layer is usually clearly defined at altitudes of 4–6 km and represents a sharp discontinuity between aerosol-laden and clean air (Fig. 15.6). It occurs throughout the year on no-rain days and consequently most frequently in the rainless winter conditions of the summer-rainfall region of southern Africa. The top of the layer oscillates in height from day to day with a mean at about the 500 hPa level in the middle troposphere. The top coincides with the ~500 hPa absolutely stable layer, which is often highly persistent. On one occasion over the Highveld of South Africa, the stable layer was observed to persist for 40 days without a break during spring.

The primary constituents of the haze layer are surface-derived dust, urban-industrial pollution, biomass-burning products and marine aerosols. Generally, aeolian (wind-borne) dust is the most significant contributor to the total aerosol loading. However, in the vicinity of and downwind from urban and industrial areas (and often significant distances downwind) anthropogenically derived aerosols assume a greater significance, as do biomass-burning products at certain times of the year near the source fires that produce them. Over southern Africa, marine aerosols are significant only over coastal and adjacent inland areas and are seldom observed far inland.

A section through an October haze layer extending from Johannesburg to north of Lusaka is illustrated in Figure 15.7 (*upper*). The discontinuity in aerosol number density, aerosol mass and extinction at the top of the haze layer is marked (Fig. 15.7, *lower*). Trace gases, such as carbon monoxide, and organic hydrocarbons, such as butane, acetylene and

Fig. 15.6 Examples of the surface to ~500 hPa aerosol pall in the haze layer capped by the ~500 hPa absolutely stable layer; *left:* over eastern Botswana in winter; and *right:* over Gauteng, South Africa, in early summer. In both cases the top of the layer occurs at an altitude of ~5.5 km (photos: P. D. Tyson).

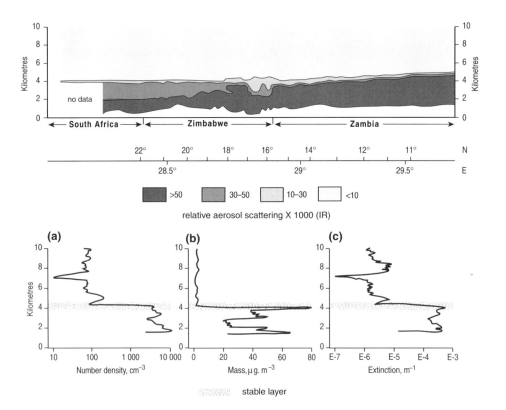

Fig. 15.7 The aerosol pall in the African haze layer extending from South Africa to northern Zambia and beyond on one occasion in October (*upper*), together with vertical profiles of aerosol (a) number density, (b) mass and (c) extinction in and above the layer (*lower*).

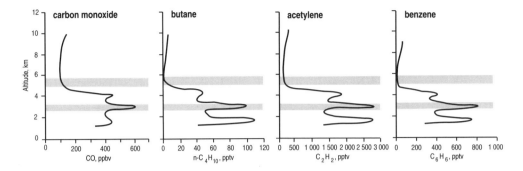

Fig. 15.8 Vertical profiles of selected trace gases over southern Africa during TRACE-A/SAFARI measured during October 1992 (after Blake et al., 1996). The climatological heights of the ~700 and ~500 hPa absolutely stable layers are shown.

benzene, are likewise trapped within the haze layer (Fig. 15.8). Near source regions, concentrations of aeolian dust, urban and industrial aerosols, biomass-burning products and marine aerosols may be high. However, with time, transport and mixing, equilibrium background loadings are achieved. These exemplify regional conditions. Background conditions within the aerosol haze layer are best sampled routinely at remote, high-altitude sites, such as the one located just below the summit of Ben MacDhui (altitude of 3 001 m) in the Drakensberg mountains of South Africa near Barkly East. The Ben MacDhui site allows sampling to be carried out at about the height of the ~700 hPa absolutely stable layer. That is, it enables sampling to be effected at roughly the middle of the 4–6-km-deep aerosol haze

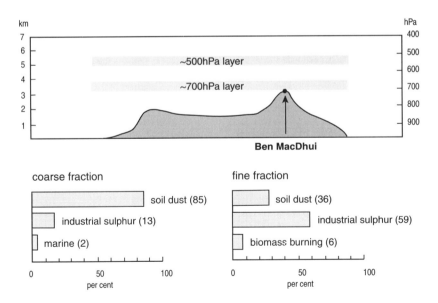

Fig. 15.9 *Upper:* the location of the high-altitude, remote, background-aerosol Ben MacDhui site in relation to the ~700 and ~500 hPa absolutely stable layers over South Africa; *lower:* the annual percentage source apportionment of coarse- and fine-fraction aerosols at Ben MacDhui.

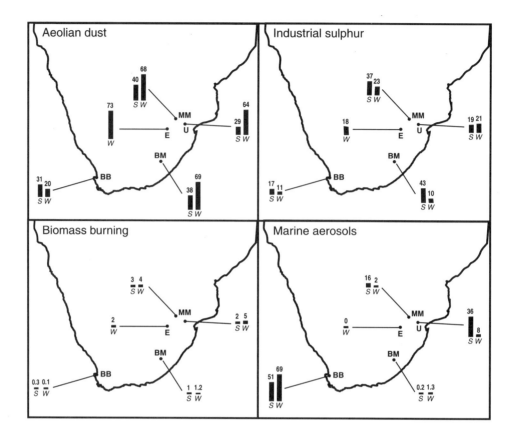

Fig. 15.10   Summer (s) and winter (w) percentage contributions of aeolian dust, industrial sulphur, biomass burning and marine aerosols to the total detected background aerosol loading over South Africa. BB denotes Brand se Baai on the west coast, BM the Ben MacDhui high-altitude site on the edge of the Lesotho massif in the Drakensberg mountains, E Elandsfontein on the central Highveld plateau, MM Misty Mountain on the Escarpment slopes and U Ulusaba in the Lowveld.

layer that extends to the ~500 hPa level over South Africa (Fig. 15.9).

Of the annual total background aerosol loading observed in the lower troposphere at Ben MacDhui, aeolian dust and urban and industrial sulphur products (emitted at distances up to 1 000 km or more along the transport pathways to the sampling site) are the primary ingredients. In the coarse fraction of the aerosols (2.5–10 μm), aeolian dust constitutes 85 per cent of the annual loading; industrial sulphur makes up 13 per cent (Fig. 15.9). In the fine fraction (< 2.5 μm) the situation is reversed and industrial sulphur constitutes 59

per cent of the total aerosol loading and aeolian dust only makes up 36 per cent. In both fractions, biomass-burning products and marine aerosols together constitute only 4–5 per cent of the total loading.

Seasonally, in both summer and winter, at the surface on the South African plateau and slopes below the Escarpment, aeolian dust is the major constituent of the background aerosol loading of the atmosphere, followed by industrial sulphur (Fig. 15.10). Biomass-burning produces large numbers of aerosols only in the vicinity of the burning, not in the background loading. In general, in both seasons, biomass burning products and marine aerosols are negligible constituents of the haze layer over inland South Africa. Within the lower troposphere, the contribution by aeolian dust to the total loading is not always greatest. In summer, in the lower troposphere at Ben MacDhui, the contribution of industrial sulphur to the total loading exceeds that of the aeolian dust. The high contribution in the upper portion of the 4–6-km-deep haze layer at this time of year is due to enhanced oxidation, during long-distance transport in the moist summer atmosphere, of sulphur dioxide emitted at distant sources. In the dry winter atmosphere, such oxidation does not occur to the same extent and the particulate sulphur loading is less.

TRANSPORT PATTERNS

The nature of the anticyclonic transport that occurs over southern Africa with such frequency is shown in two different cases in Figure 15.11. In the first instance (Fig. 15.11, *upper*), two days before reaching the vertical section showing the aerosol distribution between South Africa and Zambia, air entered South African airspace from the west and Atlantic Ocean. It was then transported anticyclonically to the section wall and thence westward over southern Africa, from where it moved south and recurved back to the east over South Africa before exiting to the Indian Ocean in the vicinity of Durban (to read positions on the ground, perpendiculars must be dropped from the trajectory to the surface in Fig. 15.11). The period of transport over land was 8–9 days. In the second example given, trapping of aerosols beneath the ~500 hPa stable layer was likewise evident, the transport was again anticyclonic and the residence time of the air being transported over southern Africa was up to two weeks (Fig. 15.11, *lower*).

The kinematic trajectory analysis of horizontal transport associated with continental anticyclonic conditions over a five-year period over southern Africa reveals that from source regions all over the subcontinent the predominant transport pathway is towards the Indian Ocean. In Figure 15.12 the percentage numbers of trajectories following the pattern indicated by mean centroid trajectories show how little transport takes place to the Atlantic Ocean and how long it takes to reach the indicated positions along the most frequently occurring trajectory. Not only does more transport take place to the Indian Ocean in a well-defined plume, but it takes place more rapidly. On average it takes 4–5 days to reach about longitude 40° E over the Indian Ocean. With ridging anticyclones increased flow to the Atlantic Ocean occurs, but the westerly plume to the Indian Ocean remains the dominant feature of the transport field.

Transport across the Indian Ocean

The plume to the Indian Ocean has been shown to retain its identity across the whole ocean basin before passing south of Australia in both winter (July) and summer (January) (Fig. 15.13). In winter, about 30 per cent of the air from southern Africa reaches 140° E. In summer, with weaker westerlies, the African plume is less frequent in the Australasian sector of the southern hemisphere and more air recirculates out of it and towards the equator to the east of southern Africa. Satellite observations reveal that, on occasions, southern African dust that accumulates in the haze layer may be transported across the entire Indian Ocean to Australia in 8–9 days (Fig. 15.14).

Likewise, lower-tropospheric ozone has been observed to be transported across the Indian Ocean in a clear plume from southern Africa. The trace gas radon from South Africa has been measured in rainfall on Amsterdam Island midway across the ocean.

It is possible to model transports of trace gases across the Indian Ocean using general circulation models (GCMs). Chemical-transport models are able to replicate well the plumes coming off southern Africa.

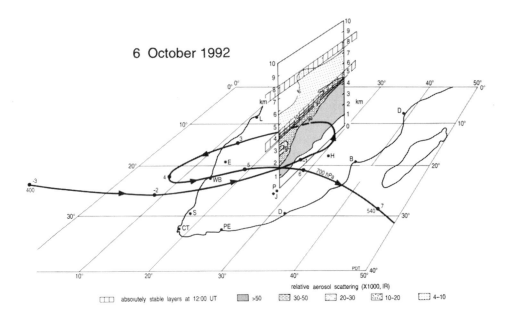

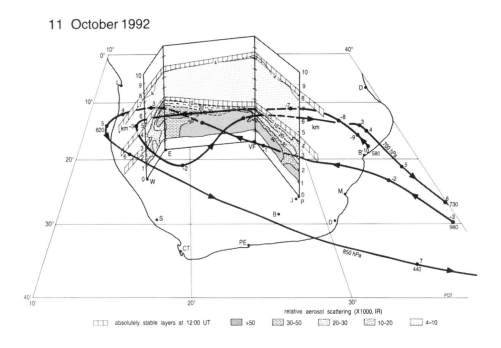

Fig. 15.11 Trapping of aerosols beneath absolutely stable layers over southern Africa in October. Trajectories through the sections give examples of transport pathways that were observed on the two occasions. Pressure levels along the trajectories, in hPa, are at daily intervals (negative indicates days before reaching the section; positive indicates days after passing through the section).

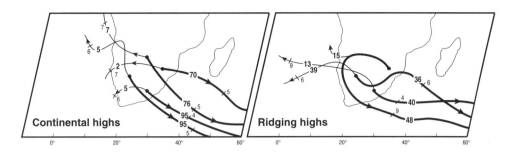

Fig. 15.12 Transport of air in the surface-800 hPa layer; *left*: from four widely spaced locations over southern Africa with the occurrence of continental high-pressure systems; *right*: from three localities with ridging highs. In each case only the most frequently occurring transport pathways are shown. Bold figures on the pathways give the total percentage transport across the meridian; figures at the designated marks on the pathways indicate the time in days taken to reach those positions from the points of origin.

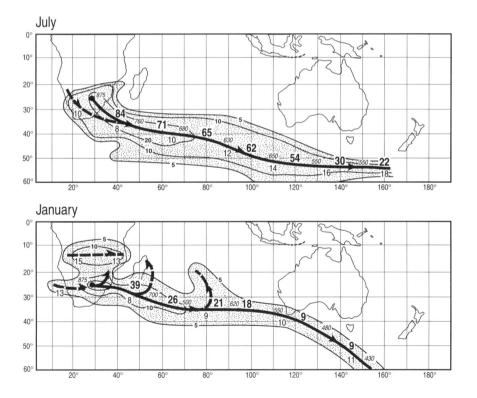

Fig. 15.13 Mean 850–800 hPa trajectory transport fields for July and January from a source area on the industrial Highveld of South Africa. Contours give the mean percentage frequency of transport of air parcels. The heavy line indicates the locus of most frequent transport. Mean percentage transports within the plume to given longitudes are indicated in heavy numbers. Light numbers along the most frequent pathway indicate days of travel to the point, geopotential height of which is given in italic.

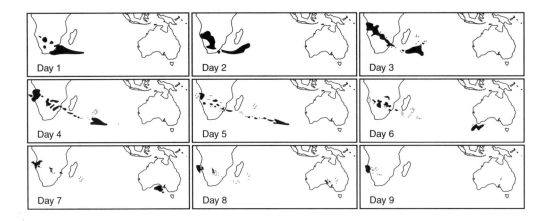

Fig. 15.14 TOMS satellite aerosol determination of dust transport across the Indian Ocean over a 9-day period in September (after Herman et al., 1997).

Simulations of carbon dioxide transport using four different GCMs are illustrated in Fig. 15.15.

Mean transport pathways

With the occurrence of continental anticyclones over southern Africa, within the surface to 500 hPa layer, 75 per cent of all air circulating over southern Africa exits to 35° E at an average height of 675 hPa at a mean latitude of 31° S with a mean transit time of 3.4 days (Fig. 15.16). Only 5 per cent of air circulating over the subcontinent is transported by semi-permanent continental anticyclones to the Atlantic Ocean. These anticyclones prevail

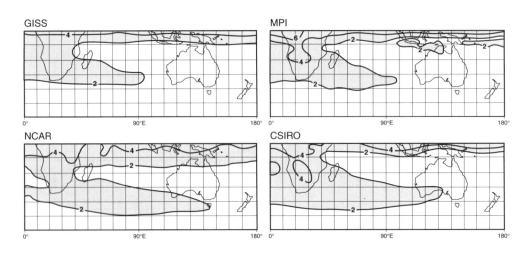

Fig. 15.15 GCM transport modelling of carbon dioxide transport in the South Africa-Australasian sector of the southern hemisphere. Peak-to-peak amplitudes, ppmv, at 500 hPa are given for the Goddard Institute of Space Science (GISS), the National Centre for Atmospheric Research (NCAR), the Max Planck Institute (MPI) and the Commonwealth Scientific and Industrial Research Organization (CSIRO) models (Rayner and Law, 1995).

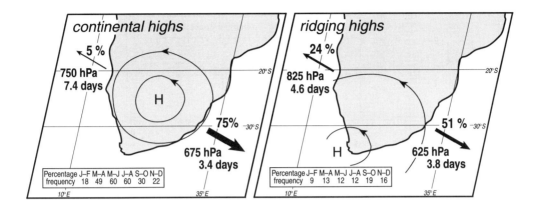

Fig. 15.16 Mean vertically integrated transport of air from the surface to 500 hPa with the occurrence of continental and ridging anticyclonic conditions. The mean position and geopotential height (hPa) of the maximum frequency pathways at 10° E and 35° W, and the percentage transport and times of transit (days) along them, are shown. Bimonthly transport frequencies of occurrence are given in the insets.

most frequently in autumn, winter and spring, but are by no means absent in summer. Transient ridging anticyclones transport more aerosols and trace gases to the Atlantic Ocean (24 per cent) than do their stationary counterparts. Nonetheless, the greatest transport – 51 per cent – remains to the Indian Ocean in the southern African plume.

As air is transported over southern Africa from the Atlantic Ocean it becomes trapped in the stable layers that are so prevalent over the subcontinent at preferred levels and which prevail so frequently, not only in autumn, winter and spring, but also on no-rain days in summer. During the 1992 international Southern African Fire Research Initiative (SAFARI) research programme, from August to October 1992 only about 20 per cent of the air trapped beneath the ~500 hPa layer actually penetrated upward through the layer (Fig 15.17). Trapping beneath the ~700 hPa layer was likewise considerable. When it is realized that the ~500 hPa stable layer may prevail unbroken for spells of 40 days at a time between May and October, the potential for the accumulation of aerosols and trace gases in the atmosphere over southern Africa is high. On average during the 3-month field observation period of SAFARI, air circulated for about a week over the subcontinent

before exiting in the transport plume to the east. Subsequent research has revealed that the residence time over the subcontinent may be a lot longer at times because of recirculation.

RECIRCULATION

It has long been known that the recirculation of air, aerosols and trace gases may occur with local winds, for instance when pollution is transported out to sea in a land breeze only to be returned to land the next day in the sea breeze. Similarly, aerosols or trace gases transported by night from the KwaZulu-Natal interior in the cool, stable drainage of air in the ~300 m deep mountain–plain wind may be carried out to sea (Fig. 15.18). Thereafter they may be entrained into north-easterly gradient winds associated with the South Indian Anticyclone over the Agulhas Current before being recirculated onshore in the sea breeze further south several hours later.

Changing synoptic-scale circulation fields produce extensive recirculation in the manner illustrated in Figure 15.1. Such recirculation occurs on a variety of spatial and time scales over southern Africa, ranging from the local to the subcontinental (Fig. 15.19). It is possible to find air from Namibia being transported south and

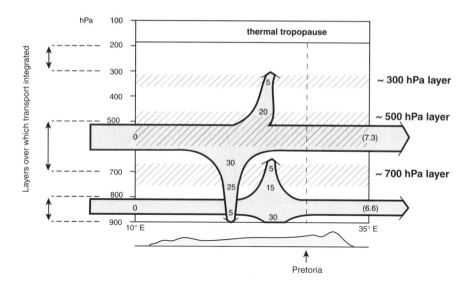

Fig. 15.17 Trapping of transport streams beneath and within the mean ~500 hPa and ~700 hPa stable layers observed during August–October of SAFARI-92. Percentage penetration of air parcels through the layers is indicated, as well as the days taken in transit over South Africa (in brackets).

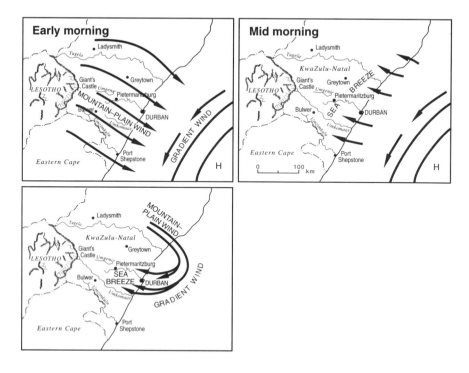

Fig. 15.18 *Upper:* early and mid-morning surface local and regional winds and potential transport pathways over KwaZulu-Natal; *lower:* low-level recirculation that may occur under such conditions.

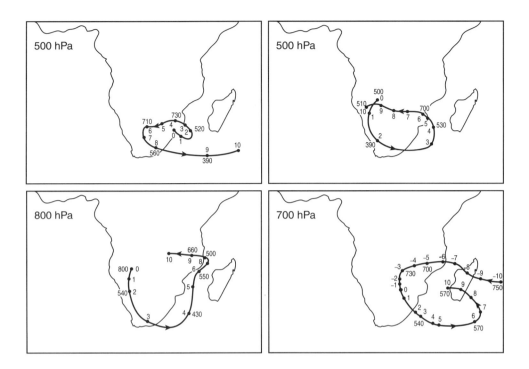

Fig. 15.19 Forward trajectories to show recirculation at different scales over southern Africa. The starting height is specified in each case. Geopotential height, hPa, is specified at designated days along the trajectories. Positive days indicate time after leaving the source point. In the lower right diagram both backward and forward trajectories are given from a point over the central subcontinent. In this case, negative days indicate the time before reaching the point, positive days the time after leaving it.

east over South Africa, to the Indian Ocean, and north over the Mozambique Channel, before being recirculated back inland over Mozambique, Zimbabwe, and Botswana, and before returning to Namibia 10 days later (Fig. 15.19, *upper right*). Large-scale recirculation over a period of nearly three weeks has been observed from Madagascar to Botswana and back across Mozambique, Malawi, Zambia, Zimbabwe and South Africa (Fig. 15.19, *lower right*).

During the process of recirculation over southern Africa air sandwiched between different stable layers may become decoupled from that circulating in the layers above and below. Such an occurrence is illustrated in Figure 15.20 where simple anticyclonic transport over periods of 12 and 9 days is evident in trajectories crossing the vertical aerosol section over Gauteng, South Africa,

at both the 850 and 500 hPa levels respectively. By comparison, at the 700 hPa level, air recirculated twice over the region in a period exceeding 20 days before exiting to the east in the westerly transport plume from southern Africa.

MEAN MASS TRANSPORT

On the basis of a five-year analysis of trajectory patterns associated with transport in semi-permanent continental anticyclones over southern Africa, it is possible to determine not only the likely amount of air transported, but also the extent of recirculation under such conditions. At the central meridian roughly bisecting southern Africa, the total annual mass transport of aerosols beneath the ~500 hPa stable layer on the

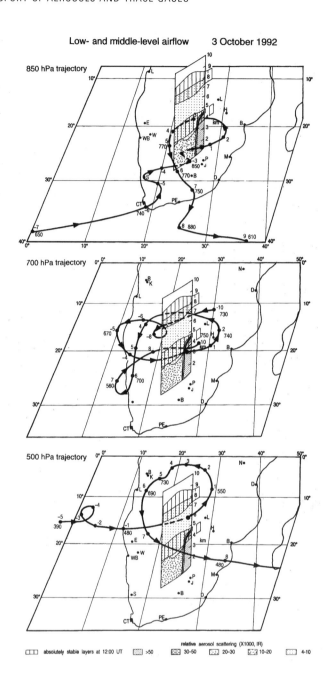

Fig. 15.20 Decoupling of flow at around 700 hPa. Aerosol trapping over Gauteng, South Africa, is given for an occasion during October. Back and forward trajectories at 850, 700 and 500 hPa are shown. Absolutely stable layers are indicated by vertical shading. Days of travel and daily geopotential heights are included along the trajectories (negative days indicate the time before reaching the section, positive days the time after leaving it).

southern, poleward side of the anticyclonic transport cell is estimated to be 39.1 Mt to the east, of which 17.3 Mt, or 44 per cent, is recirculated or recirculating aerosols (Fig. 15.21). On the northern, equatorward side of the vortex, the total transport is 11.5 Mt to the west, of which 5.1 Mt, also 44 per cent, is recirculated or recirculating particulate material.

INTER-HEMISPHERIC TRANSPORT FROM SOUTH AFRICA

Recirculation promotes both the accumulation of material to be transported and its further transport. The transport of dust from southern Africa to Australia has been mentioned above. Transport over the western Indian Ocean off southern Africa is frequent. Much less frequent is transequatorial flow from the southern subcontinent to the northern hemisphere in the African sector of the southern hemisphere. However, it does occur. Aerosols from South Africa and neighbouring countries have been traced to Kenya on the equator and beyond. On one such occasion,

air was transported from Mozambique over Zimbabwe, Botswana and South Africa to Madagascar and Kenya at levels from about 450 to 500 hPa (Fig. 15.22), which is at the mean height of the top of the African haze layer. During the 3–4 days during which the air circulated over southern Africa it acquired a high aerosol loading (in which silicon, sulphur, iron, potassium, and chlorine accounted for 95 per cent of the elemental composition). After accumulation, the aerosols were transported over Madagascar and the western tropical Indian Ocean and thence recirculated back to land over Kenya. After some stagnating in the equatorial trough, further onward transport from Kenya took place towards India. In this case, the transport period from the time relatively clean air entered southern African airspace until the aerosol material neared India was about 9 days. The high silicon loading of the aerosol would have accumulated during the transport of air over all the southern African countries traversed. In contrast, the high sulphur loading would have been acquired, in the main, in the passage of air over the industrial Highveld regions of South Africa during the first recirculation.

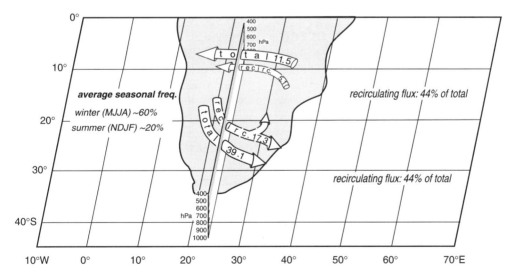

Fig. 15.21   Total and recirculating transport fluxes, Mt y^{-1}, associated with continental anticyclonic circulation systems. Fluxes are given for the vicinity of the central meridian over South Africa within the frequently occurring haze layer capped by the ~500 hPa absolutely stable discontinuity.

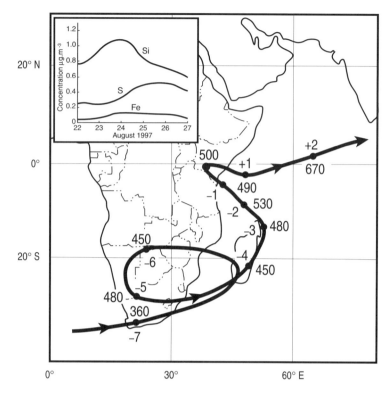

Fig. 15.22 Transport of silicon, sulphur and iron from southern Africa to Mount Kenya and beyond on an occasion in August. The height of the transport plume (hPa) is given at specified times (days).

THE PLUME TO THE INDIAN OCEAN

The atmospheric transport climatology of southern Africa is distinctive. In the horizontal, it is dominated by the anticyclonic transport of air, trace gases and aerosols in a large cell or gyre that is conceptualized in Figure 15.23. Transport to the Indian Ocean in a massive plume is the major feature of the system. More than 75 per cent of all transport out of southern Africa occurs in the plume. Considerable recirculation occurs over the land before this happens; much landward recirculation from the ocean reaches of the plume is also evident.

The vertical structure of the plume is as distinctive as is its horizontal counterpart. Aircraft flights through plumes allow cross-sections to be determined. These reveal that off both the west and east coasts of southern Africa plumes trans-porting air, and what is carried therein, are usually discrete features located above the boundary layer. They have a complex structure, are usually 3–4 km deep from the point they are recognizable above the boundary layer and may have a width of 1 000 km or more. An example of a distinctive vertical section through a pronounced dust plume off the east coast of South Africa is given in Figure 15.24. On this occasion, the plume was separated from the boundary-layer air below by relatively clean air. The plume was 3.5 km deep at its thickest. The direct offshore stream of the plume was directed towards Australia and was about 500 km wide. Two cores, A and B, could be identified in the offshore stream over Durban and south of St Lucia. Immediately alongside, a stream of aerosols recirculated inland in the opposite direction. The core of the recirculating stream occurred over the

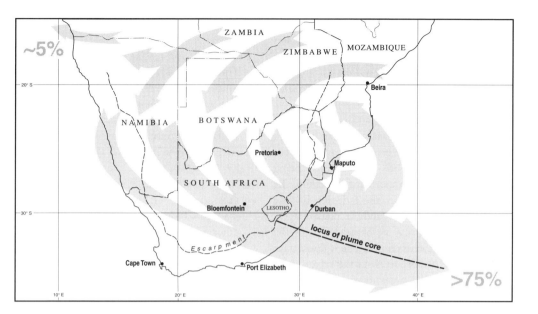

Fig. 15.23 Generalized atmospheric transport pathways over southern Africa, the southern African plume to the Indian Ocean and recirculation pathways.

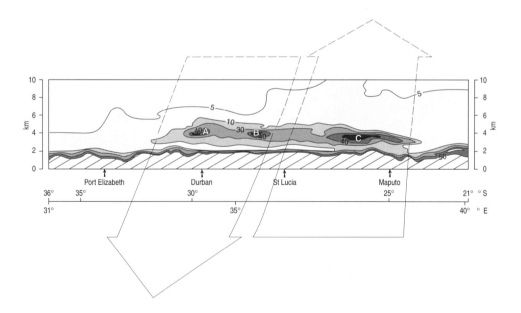

Fig. 15.24 An October transect from south of Port Elizabeth to north of Maputo. A clear, elevated, complex, multiple plume of aerosols extending approximately 1 000 km from south to north and vertically from about 2 to 5 km is evident. It is separated from and occurs above the marine boundary layer (hatched) and consists of a direct offshore transport plume with cores at A and B and a recirculating onshore plume with a core at C.

South Africa-Mozambique border region at C. The combined elevated offshore-streaming and onshore-recirculating haze layer off southern Africa extended from south of Port Elizabeth in South Africa to north of Maputo in Mozambique.

The large-scale transport and recirculation of aerosols and trace gases over and from southern Africa to the Indian Ocean occurs to an extent previously unrecognized. The material being transported exerts an important feedback on the regional climate. It alters the radiation balance and the microphysics of clouds; it alters temperature regimes and may be associated with the possible diminution of rainfall. The regional trace gases, and their transport from the region, contribute to increasing greenhouse-gas concentrations and enhanced global warming. The regional aerosols, and their transport from the region, contribute to a direct ameliorating effect on this global warming and play an important indirect role in the precipitation process.

ADDITIONAL READING

TYSON, P. D. and D'ABRETON, P. C. 1998. Transport and recirculation of aerosols off southern Africa: macroscale plume structure. *Atmospheric Environment*, 32, 1 511–1 524.

TYSON, P. D. 1997. Atmospheric transport of aerosols and trace gases over southern Africa. *Progress in Physical Geography*, 21, 79–101.

GARSTANG, M., TYSON, P. D., SWAP, R., EDWARDS, M., KALLBERG, P. and LINDESAY, J. A. 1996. Horizontal and vertical transport of air over southern Africa. *Journal of Geophysical Research*, 101, 23 721–23 736.

TYSON, P. D., GARSTANG, M., SWAP, R., KALLBERG, P. and EDWARDS, M. 1996. An air transport climatology for subtropical southern Africa. *International Journal of Climatology*, 16, 265–291.

TYSON, P. D., GARSTANG, M. and SWAP, R. 1996. Large-scale recirculation of air over southern Africa. *Journal of Applied Meteorology*, 35, 2 218–2 236.

16

CLIMATIC CHANGE AND
VARIABILITY

Climatic change is as old as the atmosphere itself. That climates have changed radically in the past in southern Africa is indisputable; that they will change again in the future is certain. Climate and its variability were always major environmental determinants with which the subcontinent's prehistoric inhabitants had to contend. So it was in historic times, still is at present and will be in the future. The adaptation by the people of southern Africa to droughts in the 1980s and 1990s is a case in point. Periods of warmth and cold, drought and flood, famine and plenty have occurred repeatedly in the past. All too often climate is thought to be that which constitutes the normal as determined from conditions occurring during part of the twentieth century. Even the climate of the twentieth century has varied. To gain an insight into the normality of changing climates it is necessary to go back in time many millions of years.

PRE-QUATERNARY CHANGES
IN CLIMATE

At least three major ice ages are known to have affected southern Africa; each lasted several million years. Between these, considerably warmer conditions prevailed. Evidence of the first major event may be identified in the Chuos, Numees, Griquatown and Daspoort tillites of the early pre-Cambrian occurring 2 000 million years (my) ago or more. The posi-

tion of Cambrian and pre-Cambrian times is given in the geological time scale shown in Figure 16.1. During the late pre-Cambrian, a second ice age gave rise to the Nama and other tillites about 600 million years ago. The causes of these early ice ages are not well understood. The third major glacial episode, the massive continental Dwyka glaciation, was experienced in the late Palaeozoic, during Permo-Carboniferous times about 320 million years ago, when conditions were more severe than at any time since (Fig. 16.2). Much of Gondwanaland, and the whole of Africa south of the Tropic of Capricorn, was covered by an extensive ice sheet as the ancient supercontinent drifted across the south polar region. Glacial conditions prevailed for a long period of time. Clear glacial pavements and striations, such as those near Niewoudtville in the Northern Cape (Fig. 16.3, *left*), are found at many places over southern Africa. During the Permian cool, temperate, wet conditions obtained and were associated with the laying down of the Ecca coals. Forests of *Dadoxylon* trees occurred in many places. Only their fossil remains are seen at present, such as in the eastern Free State (Fig. 16.3, *right*). Thereafter, in Beaufort times (Permo-Triassic), a lengthy period of warm, equable climates prevailed before an interval of intense aridity followed in the Triassic, as evident in the Clarens (Cave Sandstone) Formation. Polar ice caps did not exist and humid and arid environments were appreciably different from those of the present. During the

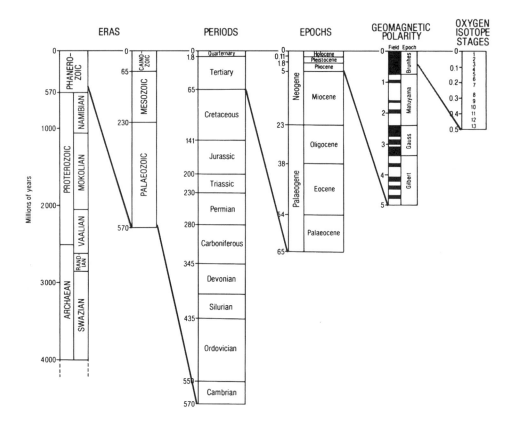

Fig. 16.1 Geological time scale, geomagnetic polarity scale and oxygen isotope stages.

Jurassic, the extensive volcanic outpourings would have significantly altered the regional climate, as did the Cretaceous fragmentation of Gondwanaland and the introduction of contiguous water bodies around the subcontinent with the formation of the Southern, Indian and Atlantic Oceans (Fig. 16.4). By 80 my the proto-oceans were clearly recognizable and southern Africa was bounded by water on three sides. The climatic implications and effects were profound. After a Cretaceous temperature maximum (Fig. 16.5), cooling set in. From late-Cretaceous times onward temperatures declined everywhere. Palaeotemperatures for sub-Antarctic surface water indicate what might have been happening further north over southern Africa. A prominent, abrupt decrease

occurred in the temperature of this water near the Eocene-Oligocene boundary at 38 my, with a drop of more than 5 °C in one hundred thousand years. Similar events appear to have been widespread throughout the world. As Antarctic snow and ice accumulated, shallow glaciation occurred on the continent and substantial sea ice formed. A permanent ice cap formed around 14 my and reached a maximum at about 6.5 my with attendant strong global cooling. With these changes in climate, vegetation began to change from tropical to temperate forest and eventually to fynbos in the south-western parts of the Western Cape.

The Namib acquired its great aridity early in the Miocene and by the mid-Tertiary the Kalahari was established as an arid to semi-arid

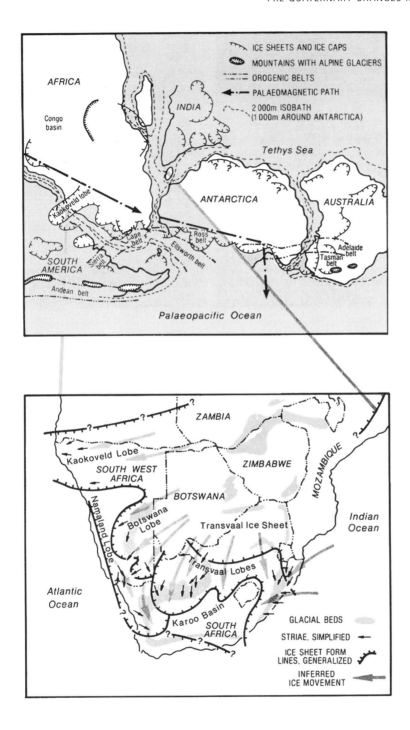

Fig. 16.2 Reconstructed Gondwanaland during the late Palaeozoic showing ice sheets, mountain glaciations and major orogenic belts, together with the corresponding glacial palaeogeography of southern Africa (modified after Crowell and Frakes, 1975).

Fig. 16.3 Illustrations of the manifestations of extreme shifts in past climates in South Africa, *left*: Permo-Carboniferous glacial striations cut into Table Mountain Sandstone near Niewoudtville, Northern Cape, during the Dwyka glaciation ~300 my ago; and *right*, the fossil trunk of a ~25-m-tall *Dadoxylon* tree from a Permo-Triassic temperate forest in the Harrismith area of the Free State (photos: P. D. Tyson).

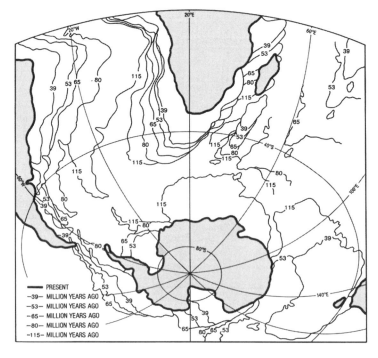

Fig. 16.4 The fragmentation of Gondwanaland (after Norton and Sclater, 1979; De Wit et al., 1986).

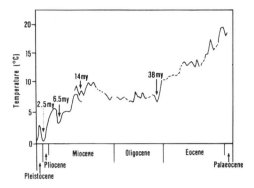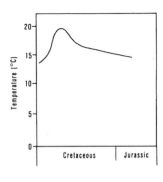

Fig. 16.5 Oxygen-isotope palaeotemperatures for the sub-Antarctic from the Palaeocene to Pleistocene (after Shackelton and Kennett, 1975) and for the New Zealand Region during the Jurassic and Cretaceous (after Deveraux, 1967).

environment. An event of major significance was the establishment of the Benguela Current and strong upwelling on the west coast by late-Miocene times, with the warm Agulhas Current showing evidence of its modern-day characteristics on the eastern margin of the subcontinent. Tectonic uplift of the subcontinent by several hundred metres from time to time during the Tertiary and Quaternary had direct palaeoclimatic consequences as surface temperatures decreased by about 0.6–0.7 °C for each 100 m of uplift. A major climatic shift took place across the Mio-Pliocene boundary, with west-coast conditions becoming cooler and more arid, while tropical conditions developed on the east coast. The east–west gradient of climate similar to that of modern times and local regional differences characterized the Pliocene. About 4 my of widespread environmental changes were initiated during a general phase of cooling. Stronger cooling occurred around 2.5 my ago and was responsible for mass extinctions of flora and fauna.

QUATERNARY CHANGES IN CLIMATE

Some of the best records of Quaternary climatic change are to be found in oxygen-isotope records obtained from deep-sea sediment cores

and cave deposits. Oxygen isotope stages are assigned to successive alternating interglacial and glacial episodes. Stages given odd numbers (starting with stage 1, the unfinished stage containing recent sediments) refer to interglacials (less continental ice), while stages given even numbers refer to glacials (more continental ice). Ice sheets have played an important role in determining the oxygen isotope record. During glacial periods ice sheets of isotopically light ice accumulate and oceans diminish in volume, becoming slightly more saline and enriched in the isotope ^{18}O. The opposite occurs during interglacials. The precise dating of the stages may vary from one sediment core to another. Both individual and globally averaged oxygen-isotope records from deep-sea cores indicate the extent to which climate has fluctuated over the last two million years. Alternating glacial and interglacial conditions occurred frequently with a quasi-periodicity of about 100 000 years.

All the essential features of the global curve are evident in the record for a core taken off the KwaZulu-Natal coast and which gives a good idea of what conditions around southern Africa were over the last half million years (Fig. 16.6). It would appear that glacial stages 2 and 6 were somewhat less severe in this locality than globally, but that stage 12 at about 430 000 BP (years Before Present) was severe. Interglacial 5 (the Last Interglacial or Eem), centred at

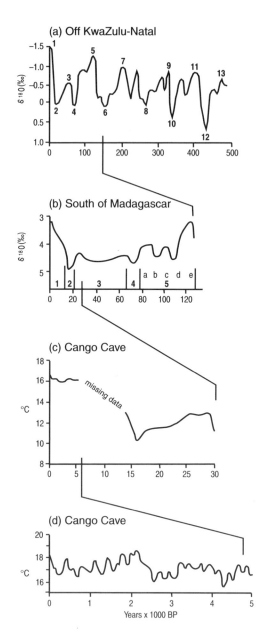

125 000 BP, was the warmest episode during the Pleistocene. Changes taking place over the last 130 000 years in the southern African sector of the southern hemisphere are illustrated by oxygen-isotope records determined from benthic (ocean-bottom) foraminifera in a core taken some 1 000 km south of Madagascar in the Indian Ocean at about 44° S, 51° E. The warmth at the maximum of the Last Interglacial (stage 5e) at 125 000 BP is clearly evident in Figure 16.6. During the Pleniglacial (stage 4) at about 75 000 BP a short-lived period of enhanced cooling occurred. The succeeding Inter-Pleniglacial (stage 3), from about 65 000 to about 25 000 BP, was a prolonged interval of cold conditions that ended with a slight amelioration just before the abrupt onset of the severest cooling. The Last Glacial Maximum (stage 2) was a significant event in southern Africa and was a phenomenon experienced throughout the southern hemisphere. The coldest conditions prevailed at 16 000 BP in southern South Africa at Cango Cave and thereafter temperatures began to rise rapidly. Throughout South Africa it would appear that ocean temperatures during the Last Glacial Maximum were about 5 °C lower than those of today. Climatic extremes would have increased considerably. However, the climate was never sufficiently severe to become more than periglacial. Thereafter, conditions ameliorated rapidly as the regional effect of global warming rose to a maximum during the Holocene Altithermal. It is necessary to examine these events in greater detail.

During the late Quaternary, Milankovitch forcing of climatic change in the southern African sector of the southern hemisphere was clearly discernible, not only through the ~100 000-year effect of changes in the eccentricity of the earth's orbit, but also in the ~23 000-year forcing exerted by changes brought about by the precession of the equinoxes (see Chapter 7 for more information on Milankovitch forcing).

Fig. 16.6 Different scales of climatic change in the southern African sector of the southern hemisphere as evidenced by: (a) oxygen-isotope analysis of an ocean sediment core taken off KwaZulu-Natal (after Prell et al., 1979), (b) oxygen-isotope analysis of an ocean sediment core taken south of Madagascar (after Shackelton, 1977), (c) and (d) stalagmite temperatures from Cango Cave (after Talma and Vogel, 1992).

PRECESSIONAL FORCING OF RAINFALL

With precessional forcing, the times of perihelion (when the earth is nearest to the sun) and aphelion (when the earth is furthest from the sun) change regularly. Currently, perihelion occurs in early January in the southern-hemisphere summer, and aphelion in early July in winter (Fig. 16.7). This causes both present-day southern-hemisphere summers and winters to be more severe than their northern-hemisphere counterparts. In 11 500 years the opposite will be the case and southern hemisphere summers and winters will be milder than those in the northern hemisphere. After a further 11 500 years the present situation will again prevail, completing the 23 000-year cycle.

At times when perihelion occurs in the middle of the southern-hemisphere summer, as at present, the meridional temperature gradient between equator and South Pole will strengthen, the southern-hemisphere summer ITCZ will do likewise and tropical forcing of southern African summer climates will be enhanced. Wetter conditions should prevail (Fig. 16.7). At the same time drier conditions may be expected to prevail over Africa north of the equator. At times when aphelion occurs in the middle of the southern-hemisphere summer, the meridional temperature gradient between equator and South Pole will weaken, the southern-hemi-

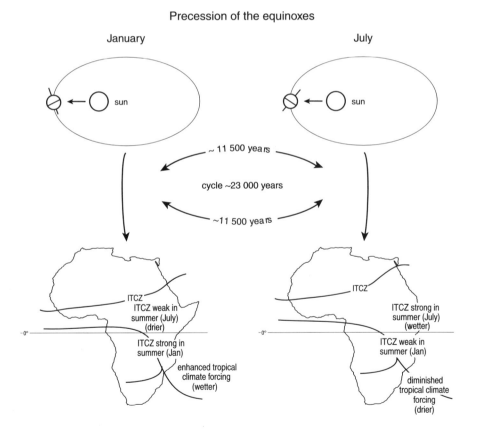

Fig. 16.7 The forcing of climatic change over Africa as a result of the 23 000-year cycle of the precession of the equinoxes.

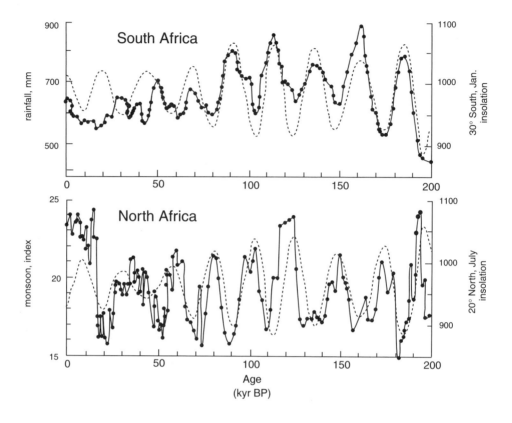

Fig. 16.8 The variation of rainfall at the Tswaing Crater north of Pretoria, South Africa, and a monsoon rainfall index for North Africa and changes in solar irradiance at 30° N and S consequent upon the 23 000-year cycle of the precession of the equinoxes over the last 200 000 years (after Partridge et al., 1997).

sphere summer ITCZ will do likewise and tropical forcing of southern African summer climates will diminish. Drier conditions will prevail. Climatic changes forced by precessional changes in this manner should be 180° out of phase in Africa south and north of the equator. Proxy rainfall series determined from sediments taken from the Tswaing Crater (formerly the Pretoria Saltpan), a 200 000-year-old meteorite impact crater, show the correspondence between variations in rainfall and the Milankovitch precessional solar radiational curve for 30° S and the out-of-phase relationship with conditions over northern Africa (Fig. 16.8).

THE LAST GLACIAL MAXIMUM TO THE HOLOCENE ALTITHERMAL

The period from 125 000 to around 16 000 BP in the southern hemisphere was one characterized at high latitudes by a series of rapid pronounced warmings followed by slow variable declines to progressively lower minima. The deuterium isotopic record from the Antarctic Vostok ice core illustrates how each maximum and each minimum was successively lower until the Last Glacial Minimum was reached at around 20 000 BP (Fig. 16.9). In Figure 16.9 a correspondence is shown to exist between the

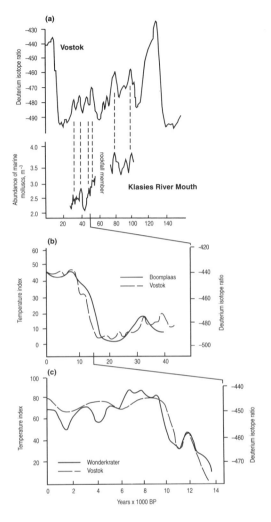

Fig. 16.9 The correspondence between deuterium-isotope ratios for the Vostok ice core, Antarctica, and: (a) the abundance of marine mollusc samples taken from different cultural stratigraphic horizons in the Klasies River Mouth cave (after Thackeray, 1992); (b) a principal-components pollen-assemblage temperature index determined by Thackeray (1987) for Boomplaas (after Thackeray, 1990); (c) a principal components pollen assemblage temperature index determined by Scott and Thackeray (1987) for Wonderkrater (after Thackeray, 1990).

decline to the most severe glacial conditions in Antarctica is replicated in many respects in the mollusc record. In each cultural horizon the greatest relative abundance of molluscs corresponds to periods of regional warming. Likewise, proxy temperature series developed from pollen sequences for Boomplaas, near Cango Cave, in the all-seasons rainfall region of southern South Africa and for Wonderkrater in the Bushveld of the summer rainfall region of the central plateau interior of the country show a close correspondence with the Vostok record.

Distinctive spatial palaeoclimatic gradients prevailed over southern African between 21 000 and 18 000 BP (Fig. 16.10). Rainfall appears to have been lower over the entire subcontinent at the time. A minimum of less than 40 per cent of the present mean annual total was experienced over the Kalahari region. The smallest diminution in rainfall occurred over eastern regions of South Africa. The reconstructed palaeotemperature field shows a clear south-to-north meridional gradient. Stable isotopic analyses of artesian groundwaters in the Uitenhage aquifer and from Stampriet, Namibia, and of the Cango Cave stalagmite show that temperatures were depressed by about 5 °C at the Last Glacial Maximum over large areas of southern Africa.

Widespread evidence is available to demonstrate that rapid warming took place in southern Africa after 16 000 BP. The sudden cooling associated with the Younger Dryas, which at around 11 000 BP punctuated the general amelioration of conditions after the Last Glacial Maximum, is evident in the Wonderkrater pollen record and in stable isotope data for Atlantic Ocean-seaboard molluscs. Thereafter, temperatures rose over much of the subcontinent. From the Boomplaas and Wonderkrater records it would appear that the Holocene Altithermal was manifest in South Africa between about 7 000 and 4 500 BP (Fig. 16.9). At this time temperatures were higher than those prevailing at present and were at their highest values since the Last Interglacial at 125 000 BP. Rainfall increased in some areas as

abundance of marine molluscs found in different cultural stratigraphic units in the Klasies River Mouth cave on the southern coast of South Africa and the Vostok sequence. The

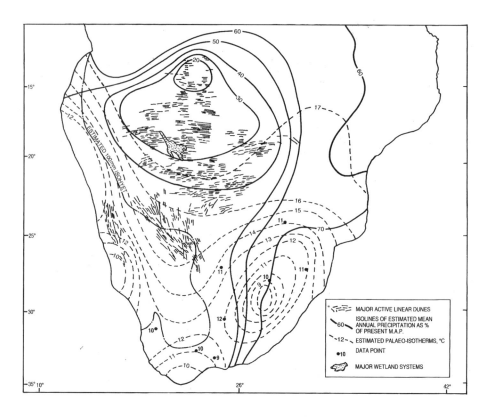

Fig. 16.10 A palaeoclimatic reconstruction of rainfall and temperature conditions at the time of the Last Glacial Maximum at 21 000–18 000 BP (after Partridge, 1997).

temperatures rose, with maximum receipt occurring at different times in different regions. Contemporaneously, the proportion of rain falling in summer increased as that in winter diminished. A climate reconstruction for the altithermal at around 7 000 BP is given in Figure 16.11. Considerable regional variation is evident. The semi-arid western interior appears to have received more rainfall, while the region to the east (including the area of the Tswaing Crater) was drier on the evidence of pollen and other proxy evidence.

CONDITIONS AFTER THE HOLOCENE ALTITHERMAL

The 5 000-year, high-resolution Cango Cave isotope record given in Figure 16.6 was for

some time the longest continuous terrestrial climatic series available for the post-Holocene Altithermal conditions in southern Africa. From this it appeared that two major cool events occurred in the third to fifth millennia before the present in the southern part of South Africa. The first of these took place from about 4 700 BP to 4 200 BP. The second was the so-called neoglacial period (reported extensively elsewhere in the world), which in South Africa occurred in the 800 years from 2 500 to 3 200 BP. The intervening period between the two cool events was mild. The end of the neoglacial was marked by the rapid warming after 2 500 BP, which marked a step change to higher mean conditions than those prevailing in the earlier part of the record.

The longest high-resolution Holocene palaeoclimatic proxy series in southern Africa is

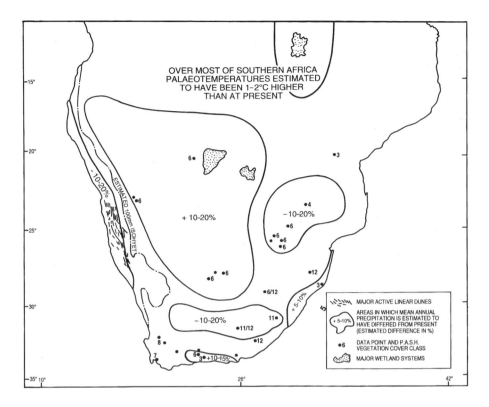

Fig. 16.11 A palaeoclimatic reconstruction of rainfall and temperature conditions at the time of the Holocene altithermal at about 7 000 BP (after Partridge, 1997).

the decadal-resolution 6 600-year record from a stalagmite obtained from Cold Air Cave in the Makapansgat Valley, south-west of Pietersburg, in the summer rainfall region of the central plateau. Oxygen and carbon stable isotope variations, as well as variations in the colour (greyness) of the stalagmite growth rings, are given in Figure 16.12. Increasing values of $\delta^{18}O$ and $\delta^{13}C$ represent warmer, wetter conditions and decreasing values a cooler, drier environment. A comparison of colour banding in the stalagmite with tree-ring widths reveals that individual growth layers, when evident, are annual. Growth-layer widths and rainfall are positively correlated, as are colour intensity and temperature.

A high degree of decade-by-decade variability occurred throughout the last 6 600 years. At times the variability was pronounced, as for instance between 1180 and 1220 AD, when $\delta^{18}O$ changed by ~2 ‰ in 40 years. The most rapid period of growth of the stalagmite occurred at around 4 000 BP and was followed by a period in which $\delta^{18}O$ decreased rapidly, $\delta^{13}C$ increased and a pronounced change in the colour composition of the speleothem occurred as the environment cooled to a minimum from 3 800 to 3 400 BP. This was followed later by the neoglacial cooling from 2 700 to 3 100 BP. Another period of cooler conditions prevailed in the central interior of South Africa in the centuries before 2 000 BP. The most pronounced, most sustained and longest perturbation in the 6 600-year record for the Makapansgat Valley was the five centuries of the Little Ice Age between 1300 and 1800 AD. The Little Ice Age in South Africa corresponds in time to the decline in solar irradiance that occurred during

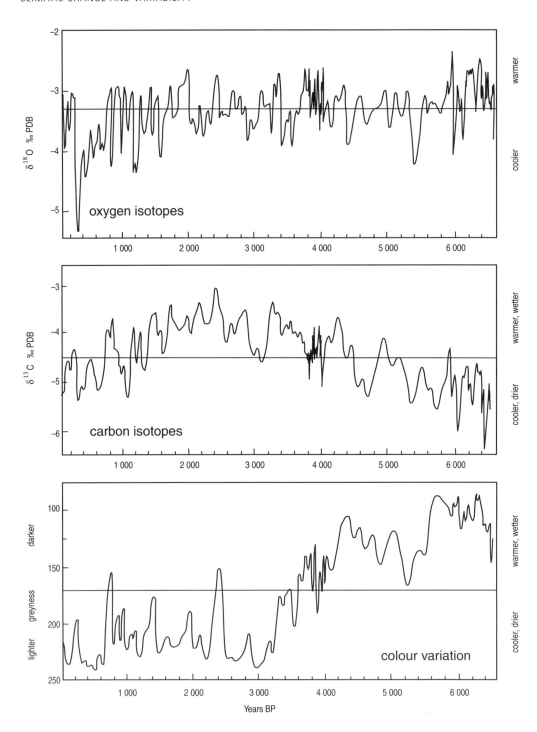

Fig. 16.12 A 6 600-year record of $\delta^{18}O$, $\delta^{13}C$ and colour density (grey-level) changes in a stalagmite taken from Cold Air Cave in the Makapansgat Valley south-west of Pietersburg. The raw data have been filtered with a 5-term binomial filter.

the Maunder Minimum from 1645 to 1715, the Sporer Minimum around 1500 and an unnamed minimum around 1350.

The most sustained period of warming occurred in the earliest part of the record until just after 6 000 BP and represents the last phase of the altithermal. Conditions were highly variable. Warmer conditions prevailed again before 4 000 BP and at around 2 400 BP. The four centuries from 900 to 1300 were generally warmer and highly variable, with warm periods punctuated by short cool periods, and are the regional expression of the Mediaeval Warm Epoch observed elsewhere in the world. The most prolonged and consistently warm period in the last three millennia was from 40 to 440 AD.

Spectral analysis of the last three millennia of the record reveals that the $\delta^{18}O$, $\delta^{13}C$ and grey-level (colour density) series for the Makapansgat Valley exhibit a number of quasi-regular variations. The first occurs at ~120 years. Oscillations also occur in the range 200–300 and 500–600 years and at around 800 years. In the oxygen series, the major peak is at ~800 years, in the carbon series it is the 200–300-year peak, while in the grey-level series it is at 500–600 years. Similar oscillations have been reported elsewhere. The ~120-year oscillation has been observed in South African tree-ring series.

Quasi-periodic oscillations of a similar period have been observed in several other records elsewhere in the world, including those from low latitudes in Kenya, from high latitudes in the northern hemisphere and in California. A ~200-year oscillation has been reported in Californian dendrochronological studies, from marine cores drilled on the western side of the Antarctic Peninsula, from Tibetan ice cores and from peat deposits in Denmark and Scotland. Possible forcing mechanisms for these oscillations include solar variability and lunar tidal effects. Fairly regular changes of temperature may have occurred in the area around Cango Cave every 300–400 years. A significant peak at ~450 years does occur in the stalagmite oxygen-isotope record from the Makapansgat Valley,

but not in either the carbon isotope or grey-level data. An oscillatory mode of about 420 years has been reported in Californian tree-ring studies.

REGIONAL COHERENCE OF LATE-HOLOCENE CLIMATIC CHANGES

In order to synthesize the findings for the Makapansgat Valley into a regional palaeoclimatology for the summer rainfall region of South Africa and beyond, it is necessary to compare findings for a multiplicity of sites and for different times with those of the Makapansgat Valley stalagmite record (Fig. 16.11). Few high-resolution series are available for comparison, but many dated climatic events have been observed from different regions. These cover mainly the earlier part of the record and corroborate individual features of the Makapansgat record. The evidence for coherent climatic changes over a large area of the subcontinent during the last two millennia or so is compelling.

The Little Ice Age cooling, with its associated aridification and accompanying increase in the frequency of major flood events, is documented in tree-ring analyses from widely separated sites in South Africa. Likewise, it is found in evidence from foraminifera in diatomaceous sediment cores taken off the west coast of Namibia, in the Cango Cave stable isotope record, in river sediment analyses in Namibia, in changing levels of Lakes Malawi and Chilwa and in high-resolution palynological analyses made in the Kuiseb Valley of Namibia. Declining sea-surface temperatures at this time are also evident in mollusc isotope records from coastal archaeological sites in the Western Cape Province and from recent sea-level fluctuations along the southern coast at Knysna. Palaeoflood hydrology of the lower Orange River valley points to greatly increased flooding in the river catchment at about the same time. The corroborating evidence comes from sites separated by

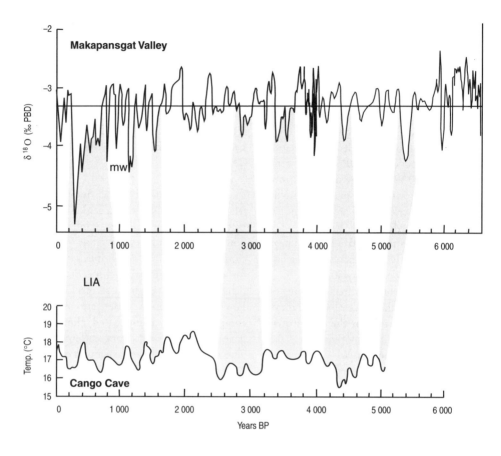

Fig. 16.13   A comparison of stalagmite oxygen-isotope records for the Makapansgat Valley and Cango Cave over the last six millennia. Shading indicates cool periods common to both sites. LIA denotes Little Ice Age, mw medieval warming.

thousands of kilometres across the entire sub-continent of southern Africa.

An increasing amount of evidence exists to support a highly variable, but generally warmer and wetter, mediaeval period in southern Africa. The generally warm interval from 880 to 1320 and its variable character has been identified in the coastal mollusc isotope record, in the Namibian river sedimentary record and from increased speleothem and tufa formation in the interior of southern Africa. Fluctuations in the earlier period of record are more difficult to corroborate in southern Africa owing to a paucity of time-series data with the required high resolution. The brief cool, dry interval in the mid fifth century is supported by the palynological

record, the Cango Cave isotope record and the palaeoflood record of the lower Orange River. The warm, wet period around 0–200 AD is evident in the offshore Namibian sediment analyses and the Cango Cave record. The cooler period from 300 to 0 BC is likewise supported by the Cango Cave record, as well as the palaeoflood record of the lower Orange River. It appears that the Makapansgat Valley record is representative of conditions over a large part of southern Africa during much of the Holocene.

The two longest, best-resolved Holocene palaeoclimatic records for South Africa are those from the Makapansgat Valley and Cango Cave. It is instructive to compare them (Fig. 16.13). All the major cool events at the northern site are replicated in the southern, except for that occurring before 2 000 BP at Makapansgat Valley, which had a warm counterpart at Cango Cave. The cool periods were almost always of longer duration in the south, whereas warm periods were almost always of longer duration in the north.

The longer periods of warming in the more tropical northern latitude near the Tropic of Capricorn and longer periods of cooling in the more temperate southern latitude at 33.5° S are consistent with the notion of an expanding and contracting circumpolar vortex of westerly winds (and the cooler, drier conditions these winds bring) and a coeval inverse contraction and expansion of the tropical easterlies (and warmer, wetter conditions). General circulation modelling (GCM modelling) shows that the expanding-contracting vortex model is driven by changing meridional temperature gradients between the poles and the equator. The times when peaks and troughs in the Makapansgat and Cango Cave series were 180° out of phase, such as at 700 BP and 1 300 BP, represent conditions when the tropical warming had reached the more northerly site with contraction of the circumpolar vortex, but had not penetrated sufficiently far south to reach southern South Africa.

In general, changes in the climates of the winter- and summer-rainfall regions of South Africa are 180° out of phase. Along the southern coastal area of South Africa and the adjacent interior, the changes in the transitional, all-seasons-rainfall region may at times be difficult to interpret. The region is influenced by the atmospheric circulation and weather processes affecting both the summer- and winter-rainfall regions. As such it is highly sensitive to small changes in the position of the boundary between the two. The consequence may be that at certain times the same relationships will naturally obtain between the all-seasons- and summer-rainfall regions. At other times different or even inverse relationships will prevail. When changes driven primarily by tropical circulations prevail sufficiently far south, then the conditions experienced in the two regions will correlate positively. The same will happen when temperate or sub-polar westerly circulations expand sufficiently far north. If neither happens, then changes in the all-seasons region are likely to correlate imperfectly, not at all, or even inversely with changes occurring in the summer-rainfall region.

GLOBAL TELECONNECTIONS

GCM modelling suggests that it should be possible to find teleconnections between southern African palaeoclimates and those in some areas of the globe, but not in others. Conditions in parts of Greenland and Scandinavia are likely to correspond with those in southern Africa. Clear correspondence between the Greenland GISP 2 ice-core $\delta^{18}O$ record and the Makapansgat Valley stalagmite isotope series is evident over the last three millennia (Fig. 16.14). Century-scale warming and cooling episodes between 800 AD and the present and earlier than 400 BC were approximately synchronous at the two localities in the northern and southern hemispheres. Between 800 AD and 400 BC the peaks and troughs appear to have been displaced on average by about two centuries. It is unlikely that the displacement is real; it is much more probable that it is an artifact of inaccuracies in one or both of the chronologies. Extreme events in the Makapansgat Valley and GISP 2 records often correspond. Cold events occurred in both site regions at around 1700, 1350, 850 AD and 1000 BC. Coeval warm events are observed at about 1000 AD, 100 AD–100 BC, 400 BC and 600 BC. Likewise, the anomalously low deuterium excess observed in the GRIP central Greenland ice core from about 1600 to 1800 (and reflecting lower temperatures in the North Atlantic Ocean during the Little Ice Age) corresponds to the lowering of temperature at Makapansgat that culminated in the minimum at around 1700.

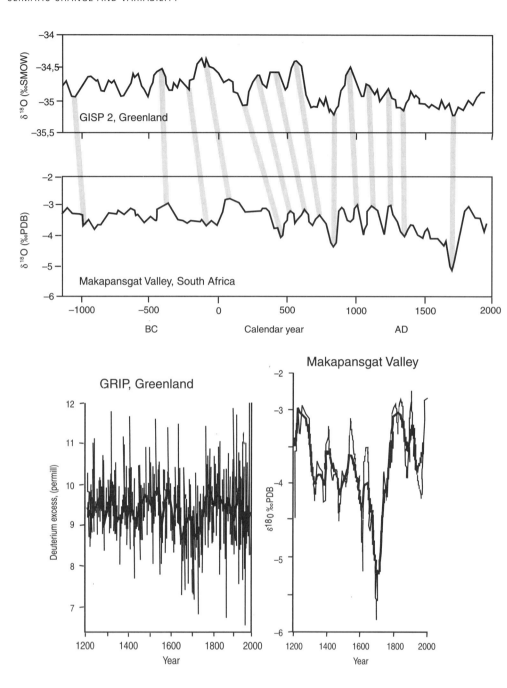

Fig. 16.14 *Upper:* a comparison of extreme events in the δ¹⁸O stalagmite record from the Makapansgat Valley with those in the δ¹⁸O GISP 2 ice-core record from Greenland (Meese et al., 1994; Alley et al., 1997) over the period 1150 BC to the present. Both series have been smoothed with 5-term running means; *lower left:* deuterium excess measured in the GRIP ice core, central Greenland (after Jouzel, 1998); and *lower right* δ¹⁸O over the last few centuries in the Makapansgat Valley. In all three cases the Maunder Minimum of the Little Ice Age at 1700 is present. All three series are characterized by abrupt changes and high variability.

THE PERIOD OF HISTORICAL RECORD

Late eighteenth- and nineteenth-century climate conditions may be inferred from travellers' journals, settlers' diaries and other historical sources. However, historical indicators are frequently difficult to interpret, often ambiguous, and must be used with caution to infer large-scale changes in climate.

Annual climate-anomaly maps may be constructed from the references to weather and climate in historical source material. This has been done for the Cape Colony in the nineteenth century (Fig. 16.15). The period 1825–1829 was characterized by a predominance of reports of droughts and desiccation, whereas during 1830–1833 flood and good-rain reports predominated. Other periods during which reports of droughts predominated were 1834–1843, 1849–1851, 1872–1878 and 1881–1885. Wetter years appear to have occurred between 1844 and 1848 and between 1852 and 1860, notably in eastern areas. Reports for 1862–1870 and 1886–1896 showed no clear predominance of either wet or dry conditions. Meteorological records show that the early part of the period 1886–1896 was dry, whereas the second half of the 1890s was wet. When an average of rainfall observations is taken over the period 1885–1896, wetter-than-normal conditions are seen to have prevailed. Reports of droughts between 1881 and 1885 correspond well with the actual rainfall record. Earlier than 1880 the paucity of measured rainfall data makes comparison difficult. From all accounts it would appear that the climate of the Cape during the nineteenth century was much like that of the twentieth. Not only was there considerable inter-annual rainfall variability, but also a tendency for spells of wet and dry years.

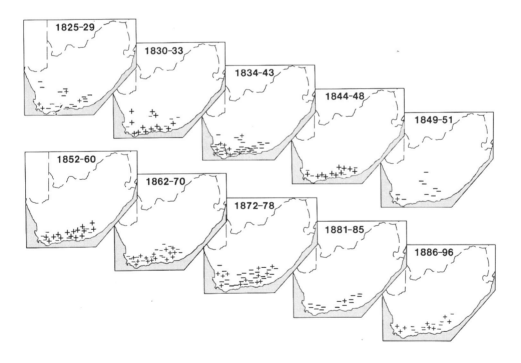

Fig. 16.15 Precipitation-anomaly maps for the nineteenth century determined by analysis of historical documents (after Vogel, 1987). Predominantly wetter (+) or drier (–) conditions are indicated in bold. Indeterminate situations occurred in the periods 1862–1870 and 1886–1896.

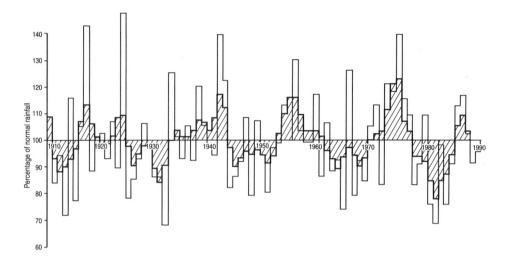

Fig. 16.16 An areally averaged regional rainfall series for the summer-rainfall region of South Africa for the period 1910–1990. Dates on the horizontal axis give the year in which the October–September rainfall years began.

THE PERIOD OF METEOROLOGICAL RECORD

During the twentieth century the climate of subtropical southern Africa has been character- ized by a high degree of both temporal and spa- tial variation. Much of the variability is random. However, within the record, real and significant non-random components are clearly identifi- able. A number of quasi-periodic rainfall oscil- lations have been reported. The most note- worthy is that with an average period of about 18 years, which affects the summer rainfall region of north-eastern South Africa to the greatest extent. An areally averaged rainfall series for this region is given in Figure 16.16.

During the runs of years included in a wet spell, years receiving above-normal rainfall have predominated, but have not occurred exclusive- ly; during dry spells, years of below-normal rainfall have predominated. In both wet- and dry-spell cases, integration over the duration of the spells results in above- and below-normal rainfall respectively. Contained within the gen- erally random year-to-year rainfall variability is an underlying non-random component which

has varied systematically for over eighty years. The quasi-periodic underlying oscillation is weak and seldom accounts for more than 20 to 30 per cent of the variance of local rainfall. Nonetheless, it is strong enough to have impart- ed a degree of regularity to rainfall variations over the last eight decades that cannot be ignored (Fig. 16.17).

Since the turn of the century eight approxi- mately 9-year spells of either predominantly wet years, which average to show above-normal rainfall during the spell, or predominantly dry years, showing below-normal rainfall, have occurred. These spells have affected most of southern Africa, not always at exactly the same time and not always affecting all regions equal- ly. What happened over South Africa is repre- sentative of much of the subcontinent. The period 1981/82–1989/90 was dry over most of the country. During the 1971/72–1980/81 wet spell high positive rainfall anomalies were expe- rienced almost everywhere. The period 1962/63–1970/71 was uniformly dry, whereas the 1953/54–1961/62 spell was generally wet except for parts of the central interior and one eastern area. The 1944/45–1952/53 dry spell

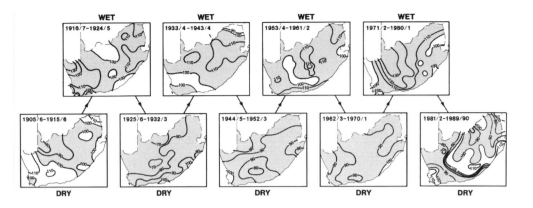

Fig. 16.17 Percentages of mean annual rainfall for designated wet and dry spells for the period 1905/6 to 1989/90.

was exceedingly dry over the whole country. Since the turn of the century, over the summer-rainfall region of north-eastern South Africa taken as a whole, the most consistently dry spell has been that of 1925/26–1932/33, when the regionally averaged rainfall for every year except one was below normal. The dry spells 1944/45–1952/53 and 1962/63–1970/71 had two and three years out of nine respectively when rainfall was below normal. The driest spell on average was that of 1925/26–1932/33. Between 1905 and 1981 the highest space-mean rainfall in any one year was experienced in 1924/25 (142.5 per cent of normal); the lowest in 1932/33 (66 per cent of normal). In comparison, during the great drought of 1982/83 the space-mean rainfall was 68 per cent of normal. The most persistently wet spell has been that of 1971/72–1980/81 with six consecutive years experiencing above-normal rainfall. In general, the dry spells have been more persistently dry than the wet spells wet. Not only is the relative variability (ratio of mean deviation to the mean) of rainfall greater during above-normal years, but so is the spatial variability within a particular wet spell. Dry spells have had a greater areal extent and spatial homogeneity than wet spells. With each successive wet spell since the turn of the century the areal extent of the excess-rain areas has increased.

TEMPERATURE CHANGES

Extensive long-series temperature records are not available for southern Africa as a whole. Some are for South Africa. The best of these are for maximum daily temperature. Areally averaged mean annual maximum temperature changes during the twentieth century over South Africa show that around 1920 the warmest conditions prevailed, whereas the 1960s and 1970s were cooler (Fig. 16.18). From 1970 onward the region has warmed steadily in the same way as the southern hemisphere as a whole. Decadal surface rock temperatures derived by inverse modelling of the temperature profile measured in deep boreholes in solid rock indicate that the surface-temperature increases over South Africa since 1850 have followed the hemispheric trends and that the peak warming of the air in the 1920s was followed a decade or two later by a surface rock-temperature response. Since 1980, trends in the regional air temperature, the regional surface temperature and the hemispheric temperature have been similar.

Over the period 1861–1980 the southern-hemisphere, hemispherically averaged, combined land-marine temperatures rose by 0.48 °C as a result of global warming. Regionally averaged South African surface rock temperatures

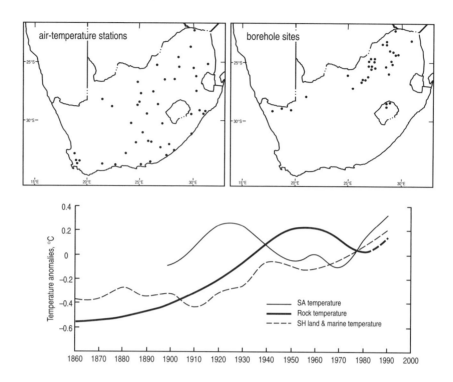

Fig. 16.18 Decadal southern-hemisphere combined land–sea temperatures (after IPCC, 1996) compared with South African regional maximum annual air temperature and borehole-derived surface rock-temperature anomalies, °C, for the sites shown. The 1990 rock temperatures are best estimates.

rose by 0.67 °C. Over the period for which the South African annual regional maximum air temperature series is available (1900 to 1990), the hemispheric land-marine temperatures rose by 0.52 °C, the South African regional air temperatures by 0.37 °C, and the borehole-derived surface temperatures by 0.48 °C. Given the uncertainties in the various data sets, there is an encouraging degree of agreement between the various measures of the regional manifestation of the global warming effect in the southern part of Africa.

The extent to which regional temperature changes in southern Africa have been similar to those experienced elsewhere may be assessed by comparing the changes that occurred between the successive twenty-year periods 1955–1974 and 1975–1994 (Fig. 16.19). Southern Africa, excluding the Angolan west coast and the area

to the north, warmed together with the southern Atlantic and Indian Oceans. Some cooling took place in the central South Atlantic Ocean. Warming dominated the entire region.

LINKS WITH RAINFALL ELSEWHERE IN AFRICA

At times teleconnections become evident between rainfall in southern Africa and that occurring in the rest of Africa. Rainfall in the decade 1950–1959, which includes six years of the 1953/54–1961/62 southern African wet spell, showed positive departures over most of Africa except in an equatorial belt where negative anomalies occurred (Fig. 16.20). In contrast, over the years 1968–1973 (i.e., during the 1962/63–1970/71 dry spell over southern

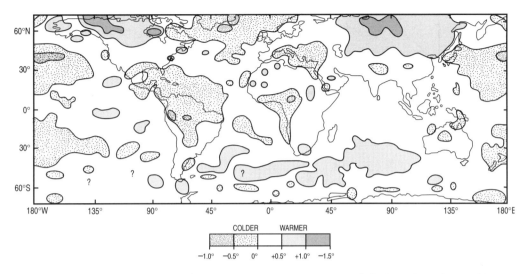

Fig. 16.19 Global temperature changes (°C) between the successive twenty-year periods 1955–1974 and 1975–1994 (after IPCC, 1996).

Africa), the departure field reversed to show generally negative anomalies over most of the continent and positive departures over the equatorial belt. The extent to which equatorial and subtropical regions south of the equator are inversely correlated is also evident in Figure 16.20.

ENVIRONMENTAL IMPLICATIONS OF DROUGHTS

The living habits of countless millions of people have become adjusted to the conditions of the last century and particularly the past few decades. This is especially so in regions where great pressures are placed upon natural resources of food, water and energy, the balance between the supply and demand of which may seriously be affected by small changes in climate. Such regions are often on the margins of deserts, or in the subtropical semi-arid parts of the world where the occurrence of droughts is endemic. Much of southern Africa can be so described and much of the subcontinent is prone to the desertification that may be triggered or exacerbated by drought. The process of desertification involves the impoverishment of arid, semi-arid and some subhumid ecosystems by the combined impact of human activity and drought. Desertification and the attendant problems of declining biological productivity, deterioration of the physical environment and increasing hazards for human settlement and life affect much of Africa. Southern Africa is no exception, as the region prone to the hazard of desertification shows (Fig. 16.21). The whole of Namibia and Botswana and more than half of South Africa are rated as potential desert, with large areas of central and northern South Africa at more than high risk.

The desertification hazard has developed in the past as the result of land vulnerability, human land-use pressures and climatic factors. Desert encroachment is seldom uniform and progressive. Instead, it takes place through the coalescence of islands of degradation, as fragile dryland ecosystems are allowed to degenerate through misapplied technology, bad management or other human controls, usually at times of severe climatic stress in prolonged droughts. It is under such conditions that the Karoo and desert conditions have been advancing to the east and north-east in South Africa (Fig.

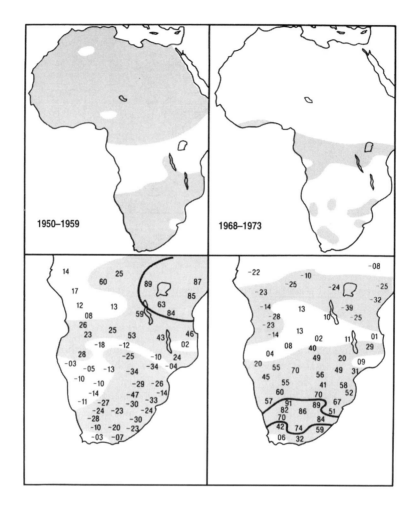

Fig. 16.20 *Upper:* continental-scale rainfall anomalies over Africa (positive shaded; negative unshaded) during the generally wet conditions of 1950–1959 and the generally dry conditions of 1968–1973 (after Nicholson and Entekhabi, 1986); *lower:* rainfall correlations over the period 1910–1970 between the regions enclosed by the heavy lines and the rest of Africa. The decimal point has been omitted (thus 52 reads as 0.52); shaded areas are significant at better than the 95 per cent level (after Nicholson, 1986).

16.22). It has been suggested that the eastward invasion of Karoo species of vegetation has been progressing at about 1.6 km per year. Undoubtedly the rash-like advance is retarded during extended wet spells and accelerated during runs of dry years. It has frequently been shown that, if given the chance, threatened or even severely damaged ecosystems may recover once climate ameliorates. It is crucial that the land must be allowed to recover. In South Africa this has not always happened and after each successive dry period veld recovery may have been insufficient to reverse the long-term tendency towards desertification in many parts.

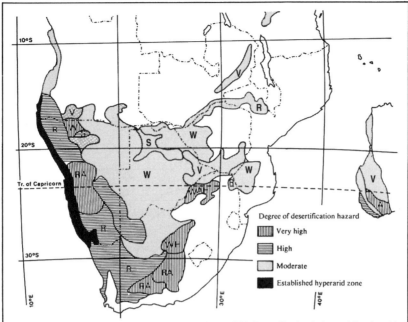

Desertification map for Southern Africa prepared for UNCOD. Desertification is here defined as 'the intensification or extension of desert conditions; it is a process leading to reduced biological activity, with consequent reduction in plant biomass, in the land's carrying capacity for livestock, in crop yields and human well-being.' Legend: W, surfaces subject to sand movement; R, stony or rocky surfaces subject to aerial stripping by deflation or sheet wash; V, alluvial or residual surfaces subject to stripping of topsoil and accelerated runoff, gully erosion on slopes and/or sheet erosion or deposition on flat lands; S, sufaces subject to salinization or alkalinization; H, subject to human pressure; A, subject to animal pressure.

Fig. 16.21   Degrees of desertification hazard for southern Africa (after UNCOD, 1977).

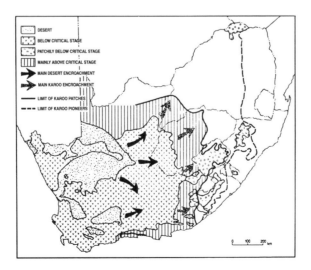

Fig. 16.22   Karoo encroachment over South Africa during the first half of the twentieth century (after Acocks, 1953).

VARIATIONS IN ATMOSPHERIC CIRCULATION

It has long been known that on a scale of a few rain days surface pressures tend to be somewhat below normal and that during runs of dry days they tend to be above normal over southern Africa. Likewise, on the scale of about a month during a wet spell, positive surface-pressure anomalies develop in the Gough Island region and negative anomalies tend to occur over Marion Island and over land; the opposite tendencies develop during the dry period (Fig. 16.23). At the 500 hPa level the development of a quasi-stationary trough in the westerlies over the west-coast region is frequently associated with wetter

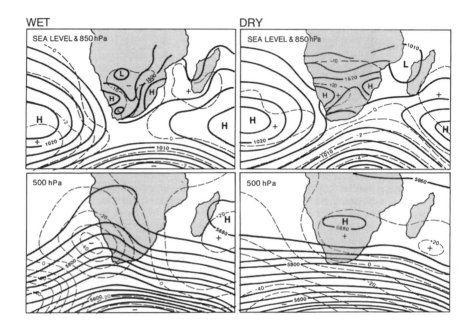

Fig. 16.23 Mean sea-level pressures (hPa), contours of the 850 hPa and 500 hPa surfaces (gpm, solid lines) and departures from mean conditions (broken lines) for midsummer 20–day wet and dry periods (after Triegaardt and Kits, 1963).

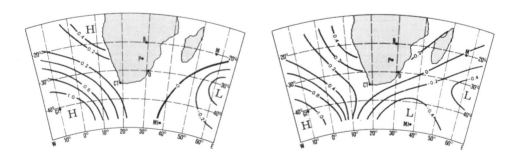

Fig. 16.24 Surface-pressure differences (hPa) existing between the 1954–6 wet and 1963–72 dry spells. *Left*: mean pressure of all summers (October–March) in the wet spell minus the mean pressure of all summers in the dry spell; and *right*: the mean of annual pressures during the wet spell minus the mean annual pressures of the dry spell.

conditions over the summer-rainfall region.

Climatological pressure anomaly fields are a manifestation of changes in the general circulation of the atmosphere and the frequency of specific perturbations in the atmospheric field of motion. If the sea-level pressure during the wet spell of 1954–1962 is compared with that of the 1963–1972 dry spell over the oceans surrounding southern Africa, then it is clear that, both for the summers and years as a whole, surface pres-

sures increase over the Gough Island region and fall to the south-east of the subcontinent during extended wet spells (Fig. 16.24). The same pattern holds for a comparison of the dry spell of the 1960s with the wet spell of the 1970s. At the 500 hPa level the pressure anomaly fields show distinctive regional gradients and opposite patterns for wettest and driest months in the two near-decadal spells, for wettest and driest seasons and for the spells as a whole (Fig. 16.25).

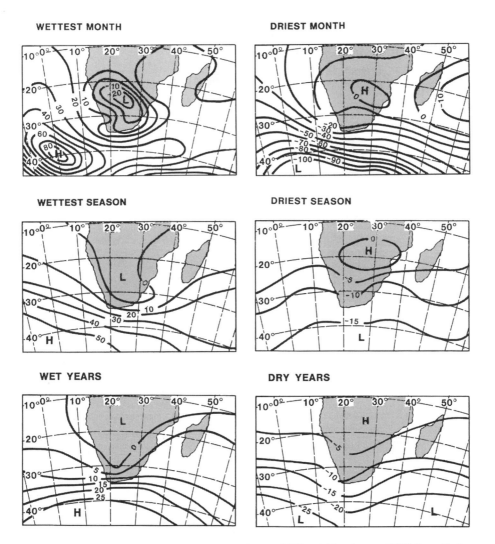

Fig. 16.25 Mean 500 hPa deviations (gpm) for: the wettest January (1978) and driest January (1969); the wettest summer (1975/76) and the driest summer (1965/66); all the wet months in the extended wet spell of 1972/73 to 1978/79 and all the dry months in the extended dry spell of 1963/64 to 1970/71. The symbols H and L indicate relative states only.

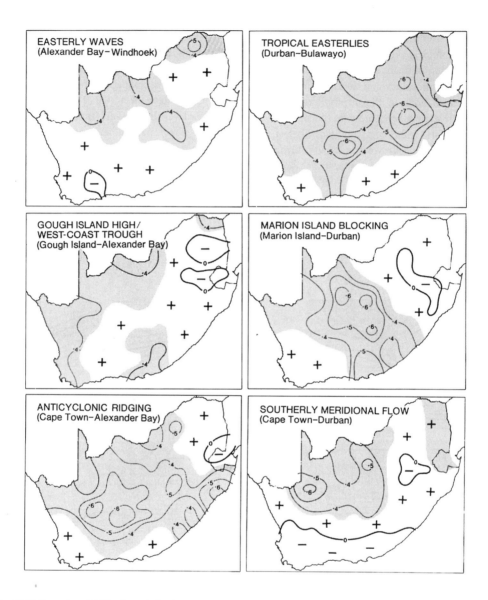

Fig. 16.26 Examples of fields of correlation between individual station-pair pressure indices (and hence geostrophic winds) and rainfall series. Shading indicates regions in which correlation coefficients are locally significant at the 90 per cent confidence limit. Positive signs denote regions of positive correlation, negative signs regions of negative correlation.

Thus in the middle troposphere, during wet conditions, negative pressure deviations occurred over parts of the interior and the strongest positive deviations over the south-western ocean area in the region of Gough Island. Weaker positive deviations occurred over the adjacent southern and south-eastern ocean areas.

During dry conditions the gradients reversed.

The pressure changes modulating the occurrence of extended wet and dry spells are naturally accompanied by changes in the wind field. By determining the time changes in pressure gradients between specific stations in the southern African region, changes in geostrophic flow

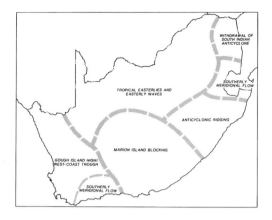

Fig. 16.27 The regional distribution of predominant *annual* atmospheric circulation controls of wet and dry spells. Each region encompasses an area within which like circulation variables at either the 850 hPa or 500 hPa pressure level correlate most strongly with annual rainfall totals. In all cases, except that of the far north-eastern region affected most by the South Indian Anticyclone, the 500 hPa control is dominant.

patterns may be linked to rainfall changes. Correlation fields for the association of 500 hPa wind anomalies with rainfall over the period 1958 to 1978 are given in Figure 16.26. The occurrence of tropical easterly flow and easterly waves may be inferred from the Durban-Bulawayo pressure index and Alexander Bay-Windhoek index respectively. Both correlate positively with rainfall and show that at an *annual* scale, as perturbed 500 hPa tropical easterly flow increases so does rainfall over northern parts of the country. Likewise, as pressure rises over the Gough Island region and an upper trough forms over the west coast, so rainfall increases over western regions. Marion Island blocking is linked to both blocking and the occurrence of tropical-temperate cloud bands and affects a north-west to south-east region of central Africa. Ridging anticyclones produce a distinctive correlation field, as too does southerly meridional flow. That these annual fields are plausible and reasonable is confirmed by daily and monthly synoptic experience. If the contribution to annual rainfall of the various flow

types at both 850 and 500 hPa is analysed and the dominant control is extracted, then it appears that what is happening at the 500 hPa level is more important than what is happening at the surface. The predominant *annual* circulation controls of southern Africa's rainfall are all at the 500 hPa level, except in the far north-east of South Africa where a localized Indian Ocean Anticyclone control is important (Fig. 16.27). The regions given in Figure 16.27 are those within which the specific circulation type contributes most to annual rainfall variability over the year as a whole.

The effect of the Walker Circulation and the Southern Oscillation on climatic variability in southern Africa has been discussed previously in Chapter 13. It is substantial. Events initiated in the South Pacific Ocean undoubtedly play an important role in determining pressure, temperature and wind anomalies over southern Africa and hence rainfall (see Fig. 13.20). The mechanisms sustaining the teleconnection are discussed in Chapter 13 (see Fig. 13.15) and act primarily by controlling the occurrence of upper-level tropical easterly flow and the location of preferential regions for cloud-band formation.

Cloud bands form in the manner discussed in Chapter 12 and join major regions of tropical latent-heat release and disturbances in the tropical easterlies with the westerly perturbations to the south of Africa. The South Atlantic Convergence Zone is often linked to the African Cloud Band Convergence Zone (not to be confused with the ITCZ) in a coherent wave structure (Fig. 16.28), with the divergent parts of the upper wave (the north-west to south-east sections) being associated with the cloud bands and the convergence part (the south-west to north-east section) producing subsidence and clear skies below. During wet months, cloud bands tend to form and precipitate preferentially over South Africa. In dry months within the rainfall season, the bands show little tendency for a preferential locus and they occur over a wider belt and are dispersed towards the east.

On the basis of an analysis of a 20-year

Cloud vectors

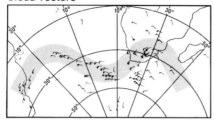

Cloud bands: wet conditions

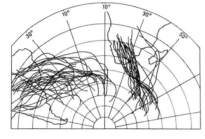

Cloud bands: dry conditions

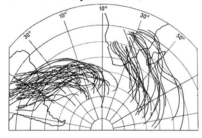

Fig. 16.28 Satellite cloud (wind) vectors on a particular day at about 300 hPa to illustrate the Atlantic wave, together with positions of major cloud bands during a wet January and a dry January (after Harrison, 1986).

period, including the dry spell of the 1960s and the wet spell of the 1970s, it is possible to suggest an explanation for the recent extended wet and dry spells of the twentieth century.

The wet-spell situation

During the approximately 9-year wet spells the ascending limb of the African Walker cell is situated over tropical Africa (Fig. 16.29). Upper tropospheric zonal flow at 20° S becomes anomalously easterly; lower in the atmosphere flow

becomes anomalously westerly. At 30° S the zonal anomaly is westerly at upper levels and easterly at 500 hPa and below. Zonal flow in the subtropical jet to the south of the continent is strengthened; less so in the Gough Island region, more so in the vicinity of Marion Island. During dry spells the pattern of zonal wind anomalies is the opposite in all respects.

Along the west coast of southern Africa the upper-level meridional flow is poleward during wet spells; in the lower troposphere it is equatorward in the Hadley-cell mode. In contrast, over the eastern central parts of the subcontinent, the meridional structure is far more complicated (Fig. 16.30). At 20° S a convergence

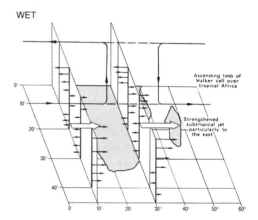

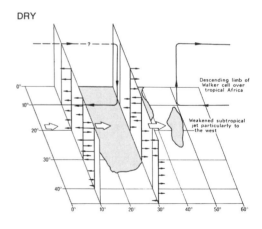

Fig. 16.29 A model of zonal wind-component anomalies over southern Africa during wet and dry spells of about nine years' duration each.

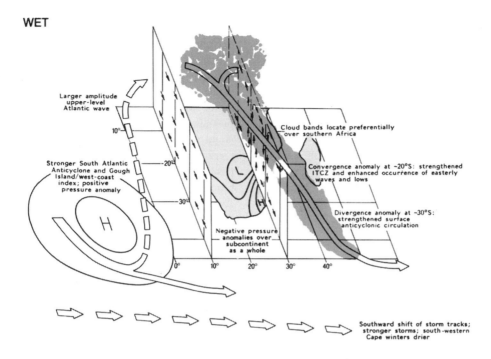

WET

Larger amplitude
upper-level
Atlantic wave

Stronger South Atlantic
Anticyclone and Gough
Island/west-coast
index; positive
pressure anomaly

Cloud bands locate preferentially
over southern Africa

Convergence anomaly at ~20°S: strengthened
ITCZ and enhanced occurrence of easterly
waves and lows

Divergence anomaly at ~30°S:
strengthened surface
anticyclonic circulation

Negative pressure
anomalies over
subcontinent
as a whole

Southward shift of storm tracks;
stronger storms; south-western
Cape winters drier

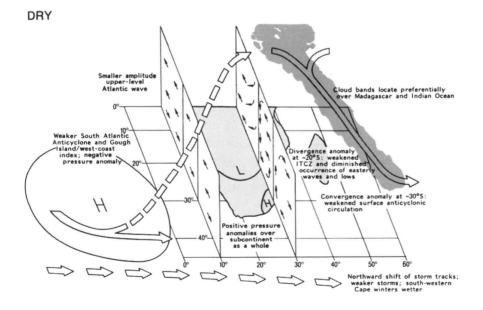

DRY

Smaller amplitude
upper-level
Atlantic wave

Weaker South Atlantic
Anticyclone and Gough
Island/west-coast
index; negative
pressure anomaly

Cloud bands locate preferentially
over Madagascar and Indian Ocean

Divergence anomaly
at ~20°S: weakened
ITCZ and diminished
occurrence of easterly
waves and lows

Convergence anomaly at ~30°S:
weakened surface anticyclonic
circulation

Positive pressure
anomalies over
subcontinent
as a whole

Northward shift of storm tracks;
weaker storms; south-western
Cape winters wetter

Fig. 16.30 A model of the anomalous meridional circulations over southern Africa during wet and dry spells. The relative positions of the upper-tropospheric Atlantic wave, preferred zones for cloud-band formation, the surface manifestations of the South Atlantic Anticyclone and locations of cyclonic storm tracks are also shown.

anomaly occurs, inter-tropical convergence and the Inter-Tropical Convergence Zone will be strengthened and the formation of easterly wave perturbations and tropical lows will be facilitated. At the same time at 30° S, a divergence anomaly occurs, strengthening the localized high occurring over eastern areas and strengthening the north–south pressure gradient. Enhanced easterly airflow and moisture advection from the Indian Ocean occur in the middle to lower troposphere. Over the subcontinent as a whole, negative pressure anomalies will prevail. To the south-west, over the Gough Island region, pressures will rise to above normal with the strengthening of the South Atlantic Anticyclone. The Gough Island/west-coast circulation index will increase, the Atlantic wave will deepen and the zone for the preferential occurrence of major cloud bands will be located over central and eastern southern Africa, with the northern point of inflection of the wave occurring in the region of the ascending limb of the African Walker cell and in the region of maximum tropical latent-heat release. Enhanced zonal easterlies at 20° S will be associated with an enhanced flux of westerly momentum into the atmosphere; this, in turn, will be transported poleward along the cloud-band conduit to be injected into the upper westerlies around Marion Island where the subtropical jet will be strengthened accordingly. Enhanced meridional energy fluxes diminish the tropical-temperate temperature gradient and a poleward shift of storm tracks occurs to the south of the continent resulting in a diminution of winter rainfall in the south-western part of the Western Cape. Enhanced advection of thermal vorticity causes storms to become somewhat stronger along their more southerly tracks.

The dry-spell situation

During the extended dry spells of about nine years the situation, on average, is reversed. Along the west coast, upper-tropospheric flow becomes equatorward; low-level flow becomes poleward in the Ferrel-cell mode. At 30° E a divergence anomaly at 20° S weakens the Inter-Tropical Convergence Zone and diminishes the occurrence of easterly waves. An opposite convergence anomaly at 30° S weakens the surface anticyclone and diminishes the north–south pressure gradient and moisture advection, which is more from the Atlantic than the Indian Ocean. The overall pressure anomaly becomes positive over the subcontinent, negative over Gough Island, and the South Atlantic Anticyclone and the Gough Island/west-coast index weaken. The configuration of the Atlantic wave changes and both the area of maximum tropical heat release and the zone for the preferred occurrence of major cloud bands move eastward. The injection of westerly momentum into the subtropical jet in the vicinity of Marion Island is diminished and the jet weakens commensurately, but to a lesser degree than it does in the region of Gough Island. Meridional fluxes of energy diminish, the tropical-temperate temperature gradient strengthens and westerly storm tracks move equatorward bringing more winter rain to the south-western part of the Western Cape from somewhat weaker storms. Beyond the confines of the winter-rainfall region the frequency of winter rainfall increases in the summer-rainfall region during dry spells. The model does not imply that mirror-image medium-term variations will take place over Africa north of the equator. The role played by the generation of major cloud bands, the importance of the changes in the South Atlantic Anticyclone (and, synergistically, the ridge of the southern-hemisphere zonal wave 1) and the interaction of the South Atlantic and African cloud bands via the Atlantic wave are reasons why the same thing does not occur north of the equator. It is not surprising that near-decadal extended wet and dry spells of the kind observed over southern Africa have no counterpart to the north. In contrast, oscillations with frequencies similar to those of the Southern Oscillation are found throughout Africa. The distinctive wet and dry spells of southern Africa appear to be modulated mainly via the anticyc-

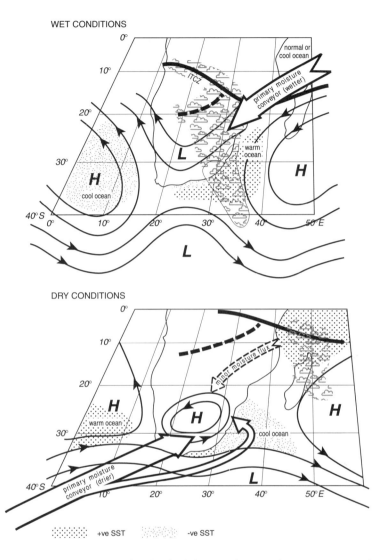

Fig. 16.31 A conceptual model to illustrate changing circulation controls, sea-surface temperatures, moisture-transport conveyors and loci of tropical convection in extended wet and dry spells over southern Africa.

lonic circulation variations occurring in the region of Gough Island and appear to be linked to the Atlantic wave. Consequently, rainfall teleconnections may be expected between eastern South America, and particularly the centres of tropical heat release over Brazil, and southern Africa. Though not strong, they are indeed present.

Figure 16.30 needs to have the effects of sea-surface temperature changes and moisture transport added to it. To avoid unnecessary complication of already complicated diagrams, this has been done separately (Fig. 16.31). The qualitative model given in Figures 16.30 and 16.31 hold equally well for centennial and millennial scales.

PALAEOCLIMATIC CIRCULATION CHANGES

At the time of the Last Glacial Maximum, the circulation over most of South Africa and adjacent countries was dominated by equatorward expansion of the circumpolar westerlies and the transient weather disturbances therein. Cyclonic depression tracks were displaced northward. The proportion of winter rainfall received in the summer-rainfall region would have been greater than today. However, it is unlikely that the seasonal rainfall cycle would have been inverted, except in southern areas now near the boundary between summer- and winter-rainfall regions. Mean annual rainfall would have been much less than at present owing to the inability of the increasing winter-rainfall component to compensate for the diminishing summer rains. In contrast, the circulation prevailing at the time of the altithermal was likely to have been characterized by poleward contraction of the circumpolar vortex with synchronous expansion southwards of the tropical easterlies and the tropically induced disturbances therein. Summer rainfall would have increased and wetter conditions would have prevailed over many areas. Winter rainfall in the summer-rainfall region would have been at about the same negligible level as that experienced at present.

According to this model, warmer and wetter conditions would have been initiated first in the more tropical northern parts of South Africa and adjacent countries and would have advanced south to reach the southern coast of South Africa last. Evidence for such a north–south gradient of change at about the time of the altithermal suggests that the poleward advance of increasing annual moisture receipt toward higher latitudes was quasi-linear, commenced at around 7 500 BP at the Tropic of Capricorn and reached the southern coast of South Africa at about 3 500 BP. The altithermal conditions appear to have advanced poleward at a rate of $3°$ to $4°$ of latitude per millennium. By the same token, cooler, drier conditions would have been initiated first in southern South Africa with the onset of a cool period and would then have advanced northward. A comparison of the Cango Cave and Makapansgat records shows that this was indeed the case during the Holocene. The fact that cool episodes at Cango Cave prevailed for a century or two longer than at Makapansgat on most occasions when they occurred during the last six millennia, and that warm episodes were of longer duration at Makapansgat, suggests the model holds for scales of decades to millennia.

However, whenever meridional adjustments of the circumpolar westerly and tropical easterly circulations occurred, the situation would have been complicated by contemporaneous east–west displacements, notably of upper-tropospheric standing waves and tropical-temperate cloud-band convergence zones. Zonal gradients of palaeoclimatic conditions would have been superimposed on the meridional to produce complex regional responses, as happens at present. The circulation changes of the atmosphere over southern Africa, since the Last Glacial Maximum at least, appear to have been analogous to those producing the modern decadal-scale, extended wet and dry spells over the subcontinent. The circulation model describing present-day changes over southern Africa appears to do equally well in accounting for those of the past in the region.

ABRUPT CLIMATIC CHANGES

Just as weather may change rapidly from day to day, so may climate change abruptly from decade to decade and even from year to year. Abrupt climatic changes have been illustrated in the record from the Vostok ice core (Fig. 16.9), that of the Greenland GISP and GRIP ice cores (Fig. 16.14) and in the Makapansgat Valley stalagmite record (Fig. 16.12). As the resolution of the palaeoclimatic record improves, so it becomes apparent that the time scale of climatic variability decreases. It is now clear from annually-resolved Greenland ice-core records that major changes may take place within as little as three years. More usually the changes take place over longer periods. The changes over the last 50 000 years in the North Atlantic Ocean region illustrate the point (Fig. 16.32). Periods of relatively long-term cooling were followed a number of times by abrupt warming during which ice-flow debris was deposited on the ocean floor during a Heinrich event, producing a sediment layer rich in polar foraminifera. The coupling between the atmosphere, ocean and cryosphere systems in high northern latitudes is clear and has been linked to abrupt reversals of the thermohaline circulation and the Broecker subsurface oceanic conveyor belt shown previously in Figure 13.12. In Chapter 13 it was shown that the thermohaline production and circulation of North Atlantic Deep Water was produced by the mixing of cool, fresh water of polar origin and warm, saline water of tropical origin and that only small increases in the density of the resultant

mixed water are needed to produce deep convection and sinking to initiate the conveyor belt. Within the Atlantic Ocean, mechanisms likely to affect the thermohaline circulation on various time scales are illustrated in Figure 16.33. In the coupled atmosphere–ocean–cryosphere system changes in the hydrological cycle, expressed in terms of the water fluxes precipitation-evaporation ($P_{rain}-P_{ev}$) for the ocean and snowfall-ablation ($P_{snow}-P_{ablation}$) for snow and ice, are due to changes in ocean temperature. The formation of North Atlantic Deep Water (NADW) is affected by changes in ice volume and extent (V) and regulates the intensity of the thermohaline circulation (C). Changes in Antarctic Bottom Water (AABW) formation are likewise affected by the thermohaline process and act to regulate the overall thermohaline circulation of the system. The thermohaline circulation affects the temperature of the overall system (T) and is also affected by it.

Changes in the Atlantic Ocean thermohaline

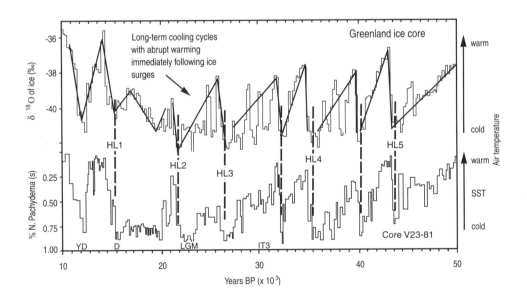

Fig. 16.32 A comparison of the GRIP Greenland ice-core oxygen-isotope record with the relative abundance of polar foraminifera in the North Atlantic sediment core V23–81 (after Bond et al., 1993). YD denotes Younger Dryas, D deglaciation, LGM the Last Glacial Maximum, IT3 Interstadial 3 and HL Heinrich Layer.

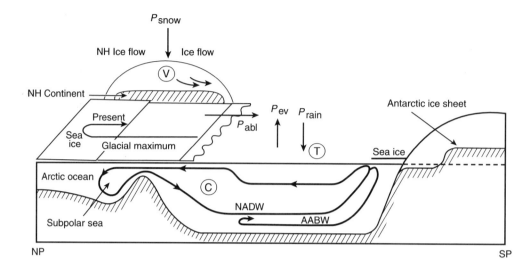

Fig. 16.33 A section through the Atlantic Ocean to show the mechanisms likely to affect the thermohaline circulation on various time scales (after Ghil and McWilliams, 1994).

system do affect southern African climate. The teleconnections that were shown earlier between Klasies River Mouth and Antarctica (Fig. 16.9) and between the Makapansgat Valley and Greenland (Fig. 16.14) illustrate the nature of the relationship.

ADDITIONAL READING

PARTRIDGE, T. C. and MAUD, R. R. 1999. *The Cenozoic in Southern Africa*. New York: Oxford University Press.

DEACON, J. and LANCASTER, N. 1988. *Late Quaternary Palaeoenvironments of Southern Africa*. Oxford: Clarendon Press.

TYSON, P. D. 1986. *Climatic Change and Variability in Southern Africa*. Cape Town: Oxford University Press.

VOGEL, J. C. (ed.). 1984. *Late Cainozoic Palaeoclimates of the Southern Hemisphere*. Rotterdam: Balkema.

17

THE PREDICTION OF FUTURE CONDITIONS

Prediction depends on the existence of models suitable for the task in hand. Models in meteorology and climatology are basically of two main types, the *empirical* and the *theoretical*. Empirical models are based on observational data and are usually descriptive or statistical. While offering useful and often essential insights into the nature of the phenomena being studied, they seldom provide a sound basis for prediction. Nonetheless, they are used widely in applied meteorology and climatology, for instance in agroclimatology for the estimation of crop yields. Empirical models are generally the least powerful of the various modelling approaches available (Fig. 17.1). Theoretical models are much more powerful, are deductive and are formulated from the fundamental equations governing the physical processes under consideration. The models always have to be simplifications of the real atmosphere, and often the constituent processes require *parameterization*, the representation of physical effects in terms of oversimplified parameters. The laws governing complex processes are not always fully understood. *Feedback* mechanisms produce complications in which a positive feedback modifies a particular parameter in the same sense as the perturbation being considered, whereas a negative feedback produces an opposite modification.

A second set of simplifications relates to the time and space resolution of the models. Generally speaking, the finer the spatial resolution,

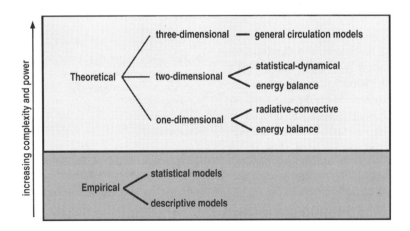

Fig. 17.1 Some types of meteorological and climatological models.

the more accurate the model. However, the finer the resolution the greater the data and computational problems encountered. If too coarse a resolution is used, then sub-grid-scale processes may be excluded from the model. Similar considerations govern the choice of the time resolution. Most models incorporate some form of time-step in the calculations, such that after processes have been allowed to act for a certain time new conditions are determined. These then form the new initial conditions for further computations and so on until the model has run for the required amount of time.

It is beyond the scope of this book to discuss all the various approaches to modelling and prediction in meteorology and climatology. Instead, some examples from numerical weather prediction and climate modelling will be considered.

NUMERICAL FORECASTING OF WEATHER

For many decades attempts have been made to increase the objectivity of weather forecasting by the use of models. Progress in forecasting has always been notoriously difficult owing to the fact that so much of the motion of the atmosphere is made up of perturbations (cyclones, anticyclones, troughs and ridges) all progressing through various stages of development and decay. In addition, the atmosphere is characterized by frequent instability of flow and much randomness. The movement and development of the perturbations may be extrapolated from a series of synoptic charts with some success by experienced forecasters. However, accelerations and intensifications of systems are difficult to predict. Net divergence and deepening are particularly difficult to assess, since they are so often the result of small differences between large quantities, themselves difficult to determine with any accuracy.

The basis for all modern numerical forecasting and climate modelling of the general circulation of the atmosphere are the so-called *primitive equations*, the basic non-derived equa-

tions describing the state of the atmosphere. The primitive equations appear in many slightly different forms and describe the processes commonly included in general circulation models (Fig. 17.2). The equations given here are in component form, where all time derivatives refer to changes at a fixed point in space in the *x, y, p* coordinate system. The Coriolis parameter is taken as positive in the northern and negative in the southern hemisphere.

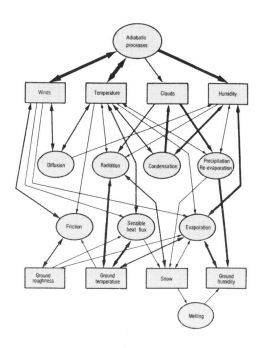

Fig. 17.2 An illustration of the processes commonly included in atmospheric general circulation models. The thickness of the arrow gives a qualitative indication of the importance of the interaction represented (after Simmons and Bengtsson, 1984).

For motion:

$$\frac{\partial u}{\partial t} = -u\,\frac{\partial u}{\partial x} - v\,\frac{\partial u}{\partial y} - w\,\frac{\partial u}{\partial p} - \frac{\partial \phi}{\partial x} + fv + F_x \tag{17.1}$$

$$\frac{\partial v}{\partial t} = -u\,\frac{\partial v}{\partial x} - v\,\frac{\partial v}{\partial y} - w\,\frac{\partial v}{\partial p} - \frac{\partial \phi}{\partial y} - fu + F_y \tag{17.2}$$

The hydrostatic equilibrium:

$$\frac{\partial \phi}{\partial p} = -\frac{RT}{p} \qquad (17.3)$$

For conservation of energy (first law of thermodynamics):

$$\frac{\partial T}{\partial t} = -u\frac{\partial T}{\partial x} - v\frac{\partial T}{\partial y} + w\left(\frac{R}{C_p}\frac{T}{p} - \frac{\partial T}{\partial p}\right) + \frac{H}{C_p}$$
$$(17.4)$$

For continuity:

$$\frac{\partial w}{\partial t} = -\left(\frac{\partial u}{\partial x} + \frac{\partial v}{\partial y}\right) \qquad (17.5)$$

For moisture conservation:

$$\frac{\partial q}{\partial t} = w\frac{\partial q}{\partial p} + \left(u\frac{\partial q}{\partial x} + v\frac{\partial q}{\partial y}\right) + C = 0 \quad (17.6)$$

Symbols are as defined previously. F_x and F_y are accelerations arising from friction and turbulence, H represents diabatic heating and C incorporates the influence of condensation and evaporation on water-vapour content.

Instead of using pressure as the vertical coordinate, it has become customary to use sigma coordinates in which

$$\sigma = \frac{p}{p_s} \qquad (17.7)$$

where p is pressure at any level and p_s is surface pressure. The effect of topography is thereby eliminated and the earth's surface coincides with the surface $\sigma = 0$. Other coordinates follow the terrain, becoming more nearly horizontal with increasing height (Fig. 17.3).

In the motion equations the left-hand terms are the unknowns, which, when computed for the many grid points that cover the forecast area or globe and combined with the other solutions, will produce the predicted flow patterns. The first and second terms on the right-hand side of the motion equations (17.1) and (17.2) represent the advection of the velocity field. The third terms give the local change of u and v components of velocity with upward movement through the σ surfaces. The fourth term gives the rate of change of geopotential in zonal and meridional directions along a σ surface. The fifth term is the component of the Coriolis force acting in the zonal and meridional directions. Finally, the last terms represent the surface

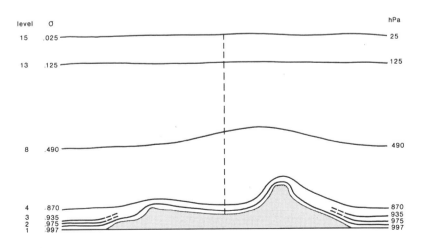

Fig. 17.3 The distribution of sigma and isobaric surfaces in a 15-layer model.

frictional stress and effects of eddy viscosity.

Nothing needs to be said about the hydrostatic equilibrium as it has been considered fully in earlier chapters. Equation (17.4) for the conservation of energy establishes local changes in temperature as comprising the horizontal advection of temperature (the first two terms on the right-hand side) and vertical changes through ascent plus heat added from the surface and through diabatic latent-heat release in the condensation process. The continuity equation has been discussed fully before and relates convergence and divergence to the shrinking and stretching of air columns.

Lastly, changes in vapour content of the air in the moisture conservation equation are related to convection (first term on the right) and advection (second term) plus phase changes associated with condensation. In the computation q is not allowed to exceed its saturation value; after saturation the latent heat released is added to H in equation (17.4).

A minimum of two levels is required to describe the baroclinic growth of mid-latitude disturbances. An early primitive-equation weather-prediction model of the early 1970s used six layers. These consisted of the boundary layer, three layers in the troposphere and two in the stratosphere. A horizontal polar stereographic grid of 53×57 points at six levels, that is a total 18 125 points, was used and daily computations of the changes with time of the variables used in the above equations were made for all these points. As was the case with the application of the barotropic forecast models, the best forecasts resulted from a combination of the pure numerical forecast and the forecast based on the subjective skill of the forecaster.

With the advent of more powerful computers, it has become common to use more than six levels. A variety of higher resolutions allows a more accurate dynamical representation and facilitates parameterization. In tropospheric modelling it is common to now use 18 levels or more. Mesospheric models may use 30 levels or more; stratospheric models use around 50 levels. Some parameterizations of the lower boundary

layer require that at least two or three levels lie within the first few kilometres of the surface.

A model used by the South African Weather Bureau is the primitive-equation unified model developed by the United Kingdom Meteorological Office and the Hadley Centre. It uses the primitive equations in a form similar to that constituted by equations (17.1) to (17.6); in detail it incorporates a number of sophisticated refinements and improvements. Rapidly moving inertial-gravity waves and lateral diffusion are incorporated in the model, as too are sub-grid-scale processes and external forcing. Ascent within the model and cooling through radiation may cause supersaturation. Excess moisture is removed as precipitation (either as rain or snow) and the latent heat released is fed back into the convective process, which is based upon the parcel theory and detrainment. Within each grid box, buoyant plumes with different temperatures, humidities, areas and heights are assumed to exist. These entrain air from their surroundings and rise until no longer buoyant. Precipitation may be evaporated into the layer immediately below that in which condensation occurs, in which case extracted latent heat alters the temperature of the model atmosphere accordingly.

Incoming radiation is modelled in a manner similar to that used in the heat-island model (see Chapter 14). Seasonal and diurnal effects, atmospheric temperature structures, cloud and humidity are incorporated. Albedo is defined as a function of latitude and surface type. Sea-surface temperatures are held constant throughout the forecast period. Long-wave radiation is modelled using empirical emissivities. In addition, radiative exchange is allowed to take place within the model boundary layer. Boundary-layer diffusion assumes different eddy diffusivities for heat and momentum and uses classical Fickian equations. Surface fluxes are assumed proportional to the near-surface gradients of temperature, moisture and momentum. The surface temperature is determined from the energy balance. Bulk Richardson numbers are used (as in the urban-heat-island model dis-

ANALYSIS DAY   **24-HOUR FORECAST**   **48-HOUR FORECAST**

Streamline fields

Fig. 17.4 Observed streamline fields and those forecast 24 hours and 48 hours ahead using a 1980s 15-level version of the primitive equation model of the United Kingdom Meteorological Office, together with derived divergence fields for the same days (by courtesy of G. C. Schulze, 1987). Units of divergence are 10^{-7} s^{-1}.

cussed in Chapter 14) and within the boundary layer fluxes are determined using a mixing-length approach. The model uses a sub-grid-scale penetrative convection scheme and a simple cloud model. Land-surface processes are parameterized, and a model of the vegetation cover is included in which different types of vegetation may be specified.

In its weather forecasting mode, the model is initialized with global synoptic data at a standard time and then run to produce 24-, 48-, 72- and 96-hour forecasts. In an early version of the model, using a coarse-mesh version and a global grid of 129×67 points at 15 levels, totalling 129 645 points, it was possible to assimilate data

and to produce a 24-hour forecast in less than 12 minutes. Now the model is even faster. The input data are constantly changed as new synoptic observations are reviewed and the model is run for standard forecast times using, in each case, the most up-to-date data available. An example of the actual situation over southern Africa on a particular early-winter day and the 24- and 48-hour numerical forecasts and derived divergence fields from a mid-1980s version of the model are given in Figure 17.4. Streamlines at 850 and 500 hPa show the passage of a westerly trough across South Africa, followed rapidly by a second trough. The divergence fields at the two levels show the

Fig. 17.5 Meteosat photographs for the days of the 24-hour and 48-hour forecasts given in Figure 17.4. On both occasions the two low-pressure troughs and associated cloud and weather were correctly forecast by the numerical general circulation model.

superimposed surface convergence (divergence) and upper-level divergence (convergence) responsible for the uplift, cloud and precipitation ahead of the trough, and subsidence and clearing of the weather behind. In Figure 17.5 the actual satellite photographs for the three days in question illustrate the extent to which the observed cloud patterns conformed to the passage of the forecast disturbances.

The extent to which 24-hour numerical rainfall forecasts reflected the actually observed occurrence of precipitation over parts of KwaZulu-Natal, the Western Cape, North West Province and the Northern Cape was close (Fig. 17.6). In contrast, the actual rainfall amounts forecasted are not nearly as accurate. Likewise, forecasts of rainfall for 36 and 48 hours ahead of time may be quite good if only

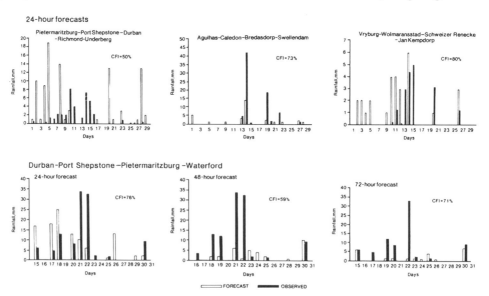

Fig. 17.6 Rainfall forecasts using the version of the United Kingdom Meteorological Office model quoted in Figure 17.4. *Upper:* 24-hour forecasts for various areas in South Africa during April 1987; *lower:* 24-, 48- and 72-hour forecasts for a coastal region in KwaZulu-Natal for part of March 1987 (after Schulze, 1987). A correct forecast index (CFI) is defined as the percentage of days in the forecast period with a qualitatively correct rain/no-rain forecast.

rain or no rain is predicted; again the actual amounts forecast leave much to be desired. Notwithstanding the considerable improvements in numerical forecasting models over the last few years, it remains difficult to predict actual rainfall amounts with any accuracy.

CLIMATE MODELLING AND PREDICTION

Climate involves variations in which the atmosphere is forced by factors external to the climate system and is influenced by and interacts with internal components of the system. The internal interactive components in the climate system include the atmosphere, the oceans, the cryosphere (sea ice, land ice, which includes the ice sheets of Greenland and the Antarctic and mountain glaciers, and snow cover), the land and its features (including vegetation, albedo,

biomass and ecosystems) and hydrology (including rivers, lakes and surface and underground water) (Fig. 17.7). The full three-dimensional character of the climate may be determined using general circulation models similar to those used in numerical weather forecasting. The vast amount of input data necessary for climate modelling and the computational time required to simulate climates over long time scales used to place major practical constraints on this type of modelling. Today such modelling is done routinely.

The principles of climate modelling are similar to global-forecast modelling, only time scales differ. The climate system is a dynamic system in transient balance. Fluxes (time changes into and out of the system) achieve different budgets over different time and space scales. The most important fluxes are those of solar and infrared radiant energy, water (particularly vapour) and mass (matter). The climate

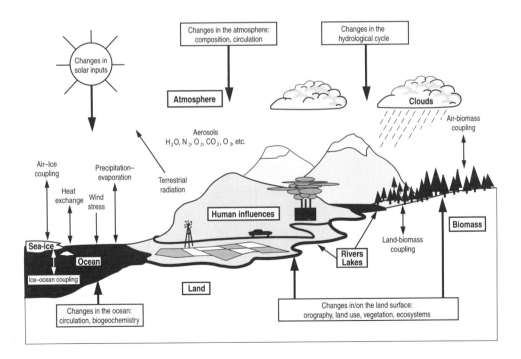

Fig. 17.7 The components of the global climate system (bold), their processes and interactions (thin arrows) and some aspects that may change (bold arrows) (after IPCC, 1996).

changes either because the forcings imposed on the planet change or because the dynamics of the system cause internal variations in fluxes and budgets. Forcings are caused by variations in agents outside the climate system, such as solar variations (e.g., Milankovitch forcing), volcanic eruptions or anthropogenic activity. A feedback occurs when a portion of the output from the action of the system is added to the input to subsequently change the output. The result can be an amplification of the process (a positive feedback) or a dampening of the original process (a negative feedback).

In the earliest atmospheric general-circulation-model (GCM) simulations of future climates, the ocean was treated as a stagnant swamp with zero heat capacity and no circulation. Subsequent GCMs were coupled to a slab ocean (usually about 50 m thick) which was well mixed and in which ocean heat transport was described, but no currents occurred. In early numerical schemes, the land surface was merely a reflector of solar radiation and an emitter of infrared radiation. Since then, various land-surface and hydrological parameterization schemes have allowed the effect of changing vegetation cover and rainfall-runoff processes to be incorporated in increasingly sophisticated ways. Likewise, interactive thermodynamic sea-ice

formulations and more sophisticated cryospheric processes have progressively been introduced into the models. Mixed-layer models allowed an annual seasonal cycle to be introduced into the simulations. Usually the models were run in equilibrium mode in which the CO_2 content of the atmosphere was doubled and the model run until equilibrium was reached. Mixed-layer models are little used now, except in diagnostic studies, where their relative ease of use and comparative computational cheapness ensures their continued acceptability.

Modern models all work in the transient mode wherein greenhouse gases and other atmospheric constituents are allowed to increase at fixed rates (usually specified by one of the IPCC emission scenarios) until a required condition (doubling, for example) has been achieved. In modern climate models a multi-layer atmospheric model is coupled to a multi-layer circulating-ocean model, allowing feedbacks from one to the other. The numbers of layers in the atmosphere may exceed thirty; in the oceans more than twenty are common. The various atmospheric, land-surface, hydrological and cryospheric parameterization schemes have all been improved over their original counterparts and the models are major improvements over their ancestral varieties. However, they

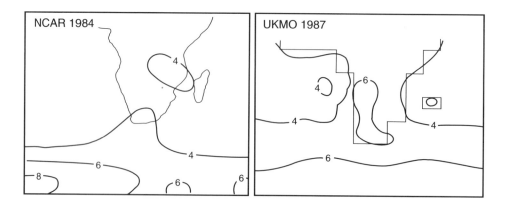

Fig. 17.8　Simulated summer (December–February) temperature increases (°C) in the next 30–50 years consequent on an instantaneous doubling of carbon dioxide; *left*: using the 1984 NCAR mixed-layer model (after Washington and Meehl, 1984); *right*: using the 1987 UKMO mixed-layer model (after Wilson and Mitchell, 1987).

don't all work equally well and major uncertainties remain associated with their use. Their results need to be validated and compared.

Some early-generation, mixed-layer climate model simulations of the future climate of southern Africa (in this case summer temperatures) are given in Figure 17.8. Increases of between 3 and 6 °C were being predicted by the mid-twenty-first century for an instantaneous doubling of CO_2. In general, in the southern African region, like elsewhere in the world, with improvements in the models, simulations of future temperature changes have been revised

downwards (see later).

In Chapter 2 the increase in certain greenhouse gases over the last few centuries, and particularly in historical times, was considered. In order to model future climates it is necessary to begin with projections of future changes in the composition of the atmosphere. This has been done by the Intergovernmental Panel on Climatic Change (IPCC). The 1992 IPCC emission scenarios include a range of possibilities. IS92a is a modest modification of 'business as usual', IS92c incorporates a drastic cut in CO_2 emissions. Others envisage shifts to lower-

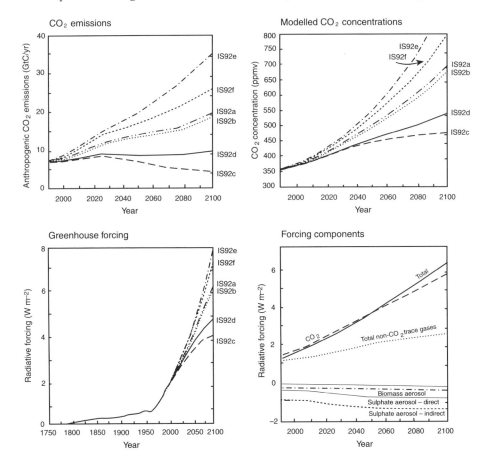

Fig. 17.9 *Upper left:* total anthropogenic CO_2 emissions under the IS92 emission scenarios; *upper right:* resulting modelled CO_2 concentrations; *lower left:* global radiative forcing from 1756 to 1990 due to changes in greenhouse-gas concentrations and tropospheric aerosol emissions and projected forcings to 2100 derived from the IS92 emission scenarios; *lower right:* radiative forcing components resulting from the IS92 emission scenarios for 1990 to 2100 (after IPCC, 1996).

carbon fuels, limiting or phasing out of a variety of trace-gas emissions, shifts to renewable energy sources, different levels of population and economic growth, and so on (IPCC, 1992). IS92a is often used in climate modelling. Projected 1990–2100 CO_2 emissions, CO_2 concentrations, the historical global radiative forcing and its growth to 2100 and the radiative forcing components resulting from the IS92a emission scenario for 1900 to 2100 are given in Figure 17.9.

Nine fully-coupled atmosphere–ocean models have been run using common initial and boundary conditions for a similar time period and greenhouse-gas emission scenario and then

compared. Southern-hemisphere-summer simulations are illustrated in Figure 17.10. In some respects the models compare well with each other and with observations; in other respects they do not. Zonally-averaged zonal winds and outgoing long-wave (terrestrial) radiation are reasonably well simulated by most models at most latitudes; mean sea-level pressure and heat flux are not as well predicted, especially at higher latitudes.

The unified UKMO/Hadley Centre model is rated as one of the most sophisticated and best models currently available for climate modelling (in climate modelling mode it is usually known as the Hadley Centre model). It was the first to

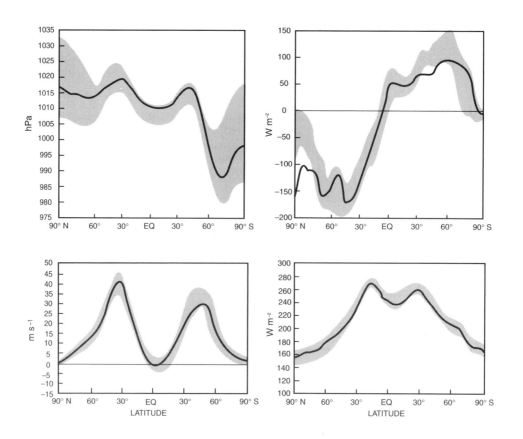

Fig. 17.10 Envelopes of zonally averaged predictions from nine coupled atmosphere–ocean GCMs compared to observations (solid line): *upper left*: mean sea-level pressure (hPa); *upper right*: surface heat flux (W m^{-2}); *lower left*: zonal wind components (m s^{-1}); and *lower right*: outgoing long-wave radiation (W m^{-2}) (after IPCC, 1996).

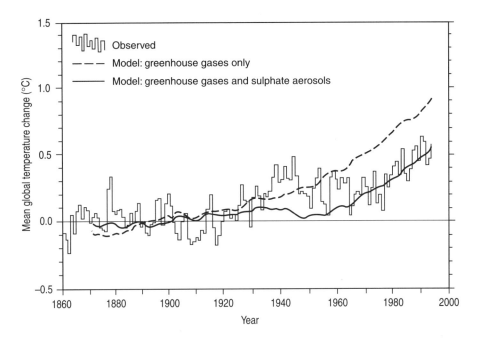

Fig. 17.11 Global warming from 1850 to 1990, together with simulations of the increase, using the Hadley Centre model incorporating greenhouse-gas forcing only and combined greenhouse-gas and sulphate-aerosol forcing (after IPCC, 1996).

incorporate the effects of aerosols in climate modelling. Using only greenhouse gases, the model overestimated global temperature changes from the mid-twentieth century onwards (Fig. 17.11). The inclusion of sulphate aerosols in the model (through changes in albedo) resulted in a more realistic simulation of conditions after 1950, but failed to predict the temporary global warming around 1940.

If the various IS92 IPCC emission scenarios are considered, then the modelled effect (in 1996) of the different emission rates on possible global warming between 1990 and 2100 may be assessed for a climate sensitivity of 2.5 °C (Fig. 17.12, *upper left*). Warming by radiative forcing will be modified by climate feedbacks which may either amplify (positive feedback) or reduce (negative feedback) the initial response. The likely equilibrium response of global surface temperature to a doubling of equivalent CO_2 concentration in the atmosphere (the climate sensitivity) has been estimated in 1990 to be in

the range of 1.5 to 4.5 °C, with a best estimate of 2.5 °C. The major feedbacks are the water-vapour feedback, the cloud radiative feedback, and those associated with ocean circulation, ice and snow albedo and land-surface/atmosphere interactions. With the same climate sensitivity, the change in extreme temperature events is likely to increase with global warming (Fig. 17.12, *upper right*). Projected global mean sea-level rise and the projected increase in sea-level rise extremes with global warming are given in Figure 17.12, *lower*. As with the temperature predictions, future sea-level changes will not occur uniformly around the world; regional variation will be considerable.

Regional warming predicted by mixed-layer models following a doubling of greenhouse gases is usually in excess of that predicted by coupled models, particularly in high latitudes. With coupled models the advection of heat in the oceans materially alters regional climatic scenarios, particularly in polar and sub-polar regions. The regional

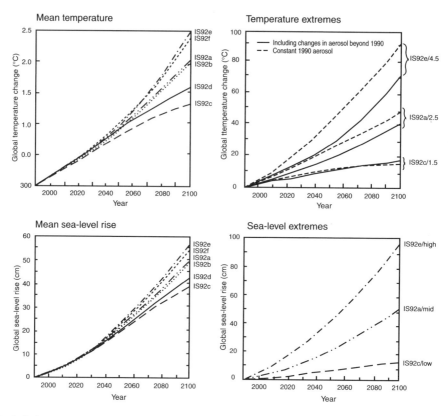

Fig. 17.12 Projected increases in: *upper left*: global temperature; *upper right*: extremes of temperature; *lower left*: projected increases in global sea level; and *lower right*: extremes of sea-level rise (after IPCC, 1996).

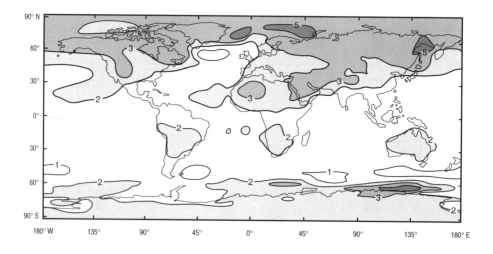

Fig. 17.13 Annual mean temperature increases (°C) simulated by the CSIRO coupled model for a 1 per cent annual increase in CO_2 until doubling (after Gordon and O'Farrell, 1996).

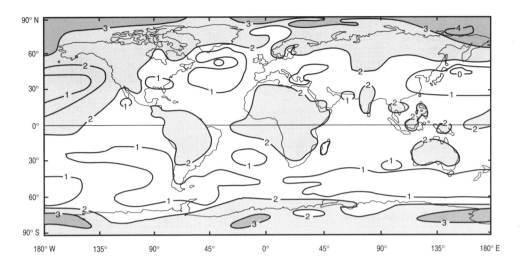

Fig. 17.14 Hadley Centre model simulations of temperature increases as a consequence of combined greenhouse-gas and sulphate-aerosol forcing with a transient increase in CO_2 until doubling in 2030–2059 (after Mitchell et al., 1995).

pattern of global warming given by an Australian coupled atmosphere–ocean model, forced by a 1 per cent annual increase in CO_2 until the time of doubling is achieved, is given in Figure 17.13.

The inclusion of sulphate aerosols in coupled modéls has an ameliorative effect on the warming process, more so in the northern hemisphere, where the sulphate-aerosol emissions are higher, than in the southern hemisphere (Fig. 17.14). Regional prediction of global change is improving constantly, but considerable uncertainty still plagues even the best simulations. The four-dimensional (latitude, longitude, altitude and time) nature of regional climatic response, coupled with the use of forcings that as yet cannot be reliably anticipated, makes regional prediction a particularly difficult task.

REGIONAL CLIMATIC-CHANGE SCENARIOS FOR SOUTHERN AFRICA

The use of the Hadley Centre model demonstrates the extent to which the inclusion of sulphate aerosols in the model act to retard the degree of global warming possibly to be expected over southern Africa. With a doubling of greenhouse gases alone, by around the middle of the twenty-first century summer temperatures (December–February) over the subcontinent may be expected to rise by up to 2 °C, and early-winter (March–May) by up to 2.5 °C (Fig. 17.15). Incorporating the aerosols lowers the increase to an anticipated maximum of 0–1.5 °C in summer and just over 1.5 °C in early winter. The retardation of greenhouse warming by sulphate aerosols is greatest over continental areas where the aerosol concentrations are highest. Simulations using the combined greenhouse-gas and sulphate-aerosol model show that the annual regional warming is the result mainly of an anticipated increase in winter minimum temperatures and not summer maximum temperatures (Fig. 17.16).

Precipitation is much more difficult to model reliably and simulations of future conditions are much more uncertain than their temperature counterparts. The combined greenhouse-gas/aerosol Hadley Centre model estimates of precipitation changes with a doubling of greenhouse gases suggests that in summer most of

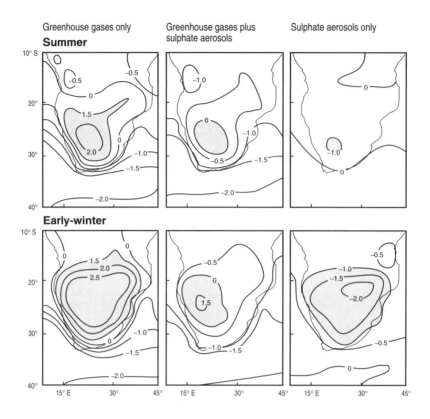

Fig. 17.15 Hadley Centre model simulations of temperature increases (°C) as a consequence of greenhouse-gas forcing alone, combined greenhouse-gas and sulphate-aerosol forcing and the aerosol effect alone for southern Africa in *upper:* summer (December–February); and *lower:* early winter (March–June) (after Joubert and Kohler, 1996).

southern Africa, except over the Angolan region, may become drier (Fig. 17.17). Winter rainfall may increase very slightly, but since this will occur at a time when almost no rainfall occurs in the summer-rainfall region, it will have no practical advantage. The winter-rainfall region of the south-western part of South Africa may become somewhat drier.

The resolution of GCMs remains coarse and downscaling to the region is a problem, since information at regional level needs to be of high resolution (10–100 km) to be useful. Downscaling may be accomplished using statistical models or by incorporating nested numerical mesoscale, regional models within GCMs. In the latter approach, the output from GCM simulations is used to provide initial and driving lateral meteorological boundary conditions for high-resolution regional climate models. In this way an increase in resolution is obtained as the regional climate model accounts for sub-GCM-grid-scale forcing. There is no feedback from the regional model to the GCM.

Both statistical downscaling and regional mesoscale modelling have been undertaken in South Africa (Fig. 17.18). Artificial neural networks have been used to provide the basis for the statistical empirical downscaling. The Australian CSIRO DARLAM limited-area regional model has been used to model regional rainfall directly. Both approaches produce more realistic detail of the spatial variation of rainfall than the driving GCM. The statistical approach offers advantages over nested modelling for southern Africa in the short term, owing to difficulties experienced with the overestimation of

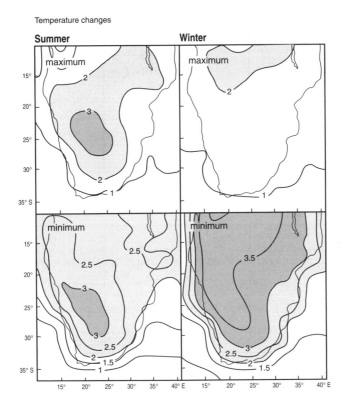

Fig. 17.16 Hadley Centre coupled model simulations of summer and winter maximum and minimum temperature increases (°C) over southern Africa as a consequence of combined greenhouse-gas and sulphate-aerosol forcing with a transient increase in CO_2 until doubling in 2030–2059 (modified after Joubert and Kohler, 1996). Shaded areas indicate the degree of warming.

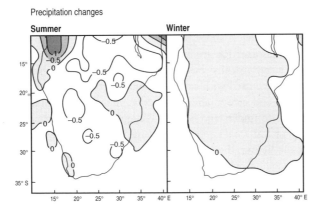

Fig. 17.17 Hadley Centre coupled model simulations of summer and winter daily precipitation changes (mm d^{-1}) over southern Africa as a consequence of combined greenhouse-gas and sulphate-aerosol forcing with a transient increase in CO_2 until doubling in 2030–2059 (after Joubert, 1997). Shading indicates drier conditions.

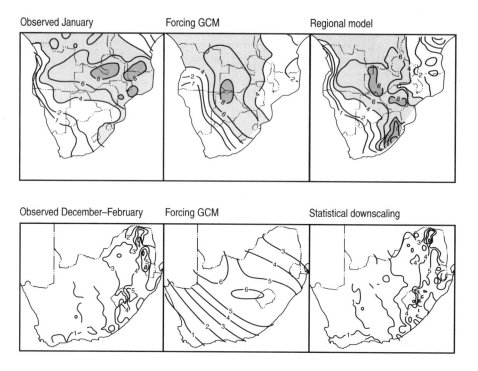

Fig. 17.18 Regional downscaling of GCM predictions; *upper.* for January rainfall (mm d⁻¹), over southern Africa using the CSIRO 9 model as the forcing GCM and the DARLAM mesoscale model as the nested regional model (after Joubert et al., 1999); *lower.* for December–February using the NCAR Genesis model as the forcing GCM and artificial neural networks for statistical downscaling (after Hewitson, 1998).

rainfall parameters by the regional model and the difficulty of convective parameterization of rainfall in regions of high relief.

If the objective of downscaling is simply to generate higher-resolution information from a coarser spatial grid, then empirically based downscaling procedures may be used with confidence given their accuracy and computational efficiency. If the objective is to understand better the processes that operate at smaller spatial scales with a view to modifying the forcing GCMs, then regional modelling, which represents the processes and feedbacks that operate at different scales, is a necessity. In the long term, there is little doubt that regional modelling will provide the best simulations of future regional conditions.

FUTURE MODELLING

In future, improved coupled atmosphere–ocean GCMs will result from advancements in existing parameterizations and by incorporating additional sub-models such as stratospheric-ozone, carbon-cycle and other biogeochemical-cycle, multiple-aerosol and land-use/land-cover change models into atmospheric GCMs. At the same time their resolution will continue to improve as computing power increases. As eddy-resolving resolution is reached in the ocean models to be coupled to the atmospheric models, and as thermohaline and other sub-models are incorporated into the ocean-component models, so significant improvements in simulating future conditions will occur. Regional nested climate modelling will continue to develop until such time as scientific and computational advances improve the parameterizations, sub-models and resolution of the driving GCM to the point where the regional model becomes redundant.

OTHER MODELLING APPLICATIONS

Climate models have been used successfully in the simulation of past conditions, notably those existing during the Pleistocene and Holocene, most notably at 125 000 BP, 18 000 BP and 9 000 BP. They have been used to assess the impact of various factors on climatic change. The effects of meridional heat transport by the oceans, global ice cover, changing solar output, increased stratospheric aerosol loading and changes in carbon dioxide content have been determined.

Climate limits the growth of crops, the availability of water, the distribution of certain diseases (like malaria), the need for energy, and many other aspects of human endeavour. Consequently, climate models provide the basis for the development of climatic scenarios that allow the impact of possible future climate and global change on human activities. The integrated assessment of global and regional change on agriculture and food provision, water and energy resources and human health is a high priority in modern life.

Sustainable development is a goal most nations are striving for. All countries are aware of the need to recognize the regional consequences of global change and the future implication of this for their people. An understanding of climatology, the climate system and its role in global change, and the prediction of climate change through climate modelling are of central importance in assuring a sustainable future.

ADDITIONAL READING

BARRY, R. G. and CHORLEY, R. J. 1998. *Atmosphere, Weather and Climate,* 7th Edition. London: Routledge.

HOUGHTON, J. T. 1997. *Global Warming.* Cambridge: Cambridge University Press.

SCORER, R. S. 1997. *Dynamics of Meteorology and Climate.* Chichester: Wiley.

HENDERSON-SELLERS, A., and MCGUFFIE, K. 1996. *A Climate Modelling Primer.* Chichester: Wiley.

HOUGHTON, J. T. 1996. *Climate Change 1995, The Science of Climate Change.* Cambridge: Cambridge University Press.

SCHNEIDER, S. H. (ed.). 1996. *Encyclopaedia of Weather and Climate*, Vols. 1 and 2. Oxford: Oxford University Press.

HARTMANN, D. L. 1994. *Global Physical Climatology.* San Diego: Academic Press.

PEIXOTO, J. P. and OORT, A. H. 1992. *Physics of Climate.* New York: American Institute of Physics.

TRENBERTH, K. E. (ed.). 1992. *Climate System Modelling.* Cambridge: Cambridge University Press.

HENDERSON-SELLERS, A. and ROBINSON, P. J. 1986. *Contemporary Climatology.* London: Longman.

HOUGHTON, J. T. (ed.). 1984. *The Global Climate.* Cambridge: Cambridge University Press.

SYMBOLS

GREEK ALPHABET

Capital letter	Small letter	Name	Capital letter	Small letter	Name
A	α	alpha	N	ν	nu
B	β	beta	Ξ	ξ	xi
Γ	γ	gamma	O	o	omicron
Δ	δ	delta	Π	π	pi
E	ε	epsilon	P	ρ	rho
Z	ζ	zeta	Σ	σ	sigma
H	η	eta	T	τ	tau
Θ	θ	theta	Υ	υ	upsilon
I	ι	iota	Φ	φ	phi
K	κ	kappa	X	χ	chi
Λ	λ	lambda	Ψ	ψ	psi
M	μ	mu	Ω	ω	omega

SYMBOLS USED IN THE TEXT

A area
     silhouette area
ΔA advection of sensible and latent heat
C circulation
     influence of condensation and evaporation on vapour content
     heat capacity (soil)
     wave speed
C_p specific heat of dry air at constant pressure
C_v specific heat of water vapour at constant pressure
D depth of an air column

E evaporation rate
E_λ irradiance per unit wavelength interval (lambda λ)
F total urban artificial heat generation
     force
     buoyancy flux
     acceleration due to friction and turbulence
G soil heat flux
     net urban heat storage
H sensible heat flux (conduction)
     half-day length
     depth of air stream
     diabatic heating
     depth of a local wind system

I	net infrared radiation		placement
$I{\uparrow}$	long-wave emission of radiation	V_T	thermal wind
$I{\downarrow}$	long-wave counter-radiation	V_i	isallobaric component of motion
K	thermal diffusivity (soil)	V_s	stack exit velocity
	bulk adiabatic transfer	W	work energy
K_h	eddy diffusivity of heat	Z	solar zenith angle
K_m	eddy viscosity		
K_s	thermal soil diffusivity	a	acceleration
K_w	eddy diffusivity of water vapour	a_λ	absorptivity
L	latent heat, of fusion (L_f), of sublima-	c	velocity of light
	tion (L_s), of vaporization (L_v)		wave speed
	wavelength of a Rossby wave	d	mean earth–sun distance
LE	latent heat flux		zero-plane displacement
N	number of isobars per unit distance		stack diameter
Q	direct-beam radiation	e	vapour pressure
	rate of air-pollution emission	e_s	saturation vapour pressure
	quantum energy	f	Coriolis parameter
R	gas constant for dry air	g	acceleration of gravity
	evaporating fraction	h	Planck's constant
R_i	Richardson number		effective stack height
R_n	net all-wave radiation		hour angle of sun
R_t	total daily radiation at the top of the		heat
	atmosphere		height of maximum flow in a local
R_w	gas constant for water vapour		wind
RH	relative humidity		height of roughness elements
S	net short-wave radiation	h_s	stack height
	energy-storage term	k	von Kármán's constant
S_0	solar constant		extinction coefficient
S_s	solar radiation at surface		friction coefficient
T	dry-bulb temperature	k_h	molecular diffusion coefficient of heat
	pressure tendency	k_s	thermal conductivity
T_d	dew-point temperature	m	mass of moist air (m_a), dry air (m_d) and
T_e	equivalent temperature		water vapour (m_w)
T_r	atmospheric transmissivity		optical thickness
T_v	virtual temperature	mw_d	molecular weight of dry air
T_w	wet-bulb temperature	mw_w	molecular weight of water vapour
U	initial velocity in Newton's equations of	n	length
	motion		distance orthogonal to streamlines
	rate of longitudinal pressure-system		cloud cover in tenths
	displacement	p	atmospheric pressure
U_m	maximum wind velocity in a local wind	q	diffuse radiation
	system		specific humidity
V	velocity	q_s	specific humidity at saturation
	terminal velocity	r	mean earth radius
	final velocity in Newton's equations of		radius of curvature
	motion		drop radius
	rate of latitudinal pressure-system dis-	s	distance

t	time	θ_w	wet-bulb potential temperature
u, v, w	wind velocities in x, y, z direction respectively	Φ	geopotential height
		Ω	angular velocity of rotation of a vortex
u_*	friction velocity		
v	volume	α	environmental lapse rate
w	precipitable water		specific volume
	vertical velocity		albedo
x	mixing ratio	ϵ	emissivity
	distance downwind		$mw_w/mw_d = 0.622$
x_s	saturated mixing ratio	ζ	relative vorticity
x_{sw}	saturated mixing ratio at wet-bulb temperature	δ	solar declination
		η	viscosity of air
x, y, z	coordinates along the three axes	λ	wave length
y	distance across wind	ν	wave frequency
z	height		kinetic viscosity
z_0	roughness length	ρ	air density
		ρ_w	water-vapour density
Γ	dry adiabatic lapse rate	σ	Stefan-Boltzmann constant
Γ'	wet adiabatic lapse rate		standard deviation
β	Bowen ratio		sigma pressure surface
	variations of the Coriolis parameter with latitude	τ	momentum flux per unit surface area
		ϕ	latitude
∂	partial differential	χ	downwind air pollution plume concentration
θ	potential temperature		
θ_e	equivalent potential temperature	ω	angular velocity of the earth's rotation

UNITS AND CONVERSIONS

PREFIXES USED TO DESCRIBE MULTIPLES OR FRACTIONS OF TEN

Prefix			Scientific notation	Decimal notation
T	tera-	one trillion (US)	10^{12}	1 000 000 000 000
G	giga-	one billion (US)	10^{9}	1 000 000 000
M	mega-	one million	10^{6}	1 000 000
k	kilo-	one thousand	10^{3}	1 000
h	hecto-	one hundred	10^{2}	100
da	deka-	ten	10	10
d	deci-	one tenth	10^{-1}	0.1
c	centi-	one hundredth	10^{-2}	0.01
m	milli-	one thousandth	10^{-3}	0.001
μ	micro-	one millionth	10^{-6}	0.000 001
n	nano-	one billionth (US)	10^{-9}	0.000 000 001
p	pico-	one trillionth (US)	10^{-12}	0.000 000 000 001

SOME IMPORTANT CONSTANTS

Quantity	Magnitude and units
Angular velocity of the earth's rotation	7.29×10^{-5} rad s^{-1}
Density of air at 0 °C and 1 000 hPa	1.275 kg m^{-3}
Dry adiabatic lapse rate	9.8 °C km^{-1}
Gas constant for dry air	287 J kg^{-1} °C^{-1}
Gas constant for water vapour	461 J kg^{-1} °C^{-1}
Gravitational acceleration	9.80 665 m s^{-2}
Mean earth radius	6.37×10^{6} m
Mean solar distance	1.50×10^{11} m
Planck's constant	6.62×10^{-34} J s^{-1}

Quantity	Magnitude and units
Solar constant	$1\ 370 \pm 20$ W m^{-2}
Specific heat of air at constant pressure	$1\ 010$ J kg^{-1} °C^{-1}
Specific heat of water vapour	$1\ 880$ J kg^{-1} °C^{-1}
Specific heat of carbon dioxide	850 J kg^{-1} °C^{-1}
Standard atmospheric pressure	$1\ 013.3$ hPa
Stefan-Boltzmann constant	5.67×10^{-8} W m^{-2} K^{-4}
Velocity of light	3×10^{8} m s^{-1}
Von Kármán's constant	0.4

SYSTEME INTERNATIONAL (SI) UNITS

SI base units

Base unit	Dimensions	SI units
Length	L	m (metre)
Mass	M	kg (kilogram)
Time	T	s (second)
Temperature	θ	K (Kelvin)[a]

[a] The Kelvin scale gives absolute temperature with absolute zero where theoretically molecular motion ceases at -273 °C and a body contains no heat. One degree Kelvin equals one degree Celsius and, since zero degrees Celsius is the freezing point of water, the two scales are linked by K = °C + 273.

Derived units

Velocity: a vector quantity; hence, for its complete specification velocity should include a direction as well as a speed.

Force: the newton (N) is the force required to accelerate a body with a mass of 1 kg at 1 metre per second per second.

Pressure: the pascal (Pa) is the pressure exerted by a force of 1 N evenly distributed over an area of 1 square metre.

Work, energy: the joule (J) is the energy required to displace a force of 1 N through a distance of 1 metre.

Power: the watt (W) is the power required to equal the rate of working of 1 joule per second.

SI Units and some equivalents

Measure	Name of unit (symbol)	Value	Approximate conversions	
Length	metre (m)	base unit	1 m	= 3.281 feet
	micrometre (μm)	10^{-6} m	1 mm	= 10^{4} Ångström units
	millimetre (mm)	10^{-3} m		
	kilometre (km)	$1\ 000$ m	1 km	= 0.621 mile
				= 0.540 nautical mile
Mass	kilogram (kg)	base unit	1 kg	= 2.205 pounds
	milligram (mg)	10^{-6} kg	1 g	= 0.0353 ounce
	gram (g)	10^{-3} kg	*1 t	= 1.102 US tons
	tonne (t)	10^{3} kg		= 0.984 ton

Time	second (s)	base unit		
	minute (min)	60 s		
	hour (h)	3 600 s		
	day (d)	86 400 s		
	year (a)			
Area	square metre (m^2)	SI unit	*1 cm^2	= 0.155 square inch
	square centimetre (cm^2)	10^{-4} m^2	1 m^2	= 10.76 square feet
	hectare (ha)	10^4 m^2	*1 ha	= 2.471 acres
	square kilometre (km^2)	10^6 m^2	*1 km^2	= 0.386 square mile
Volume	cubic metre (m^3)	SI unit	*1 cm^3	= 0.061 cubic inch
	cubic centimetre (cm^3)	10^{-6} m	1 m^3	= 35.31 cubic feet
Volume (fluids only)	litre (l)	10^{-3} m^3	*1 ml	= 0.035 fluid ounce
	millilitre (ml)	10^{-3} l	*1 l	= 1.760 pints
	megalitre (Ml)	10^6 m^3		= 0.220 gallon (UK)
				= 0.264 gallon (US)
Velocity	metre per second (m s^{-1})	SI unit	1 m s^{-1}	= 3.281 feet per second
	kilometre per hour (km h^{-1})	0.278 m s^{-1}		= 2.237 miles per hour
				= 1.944 knots
				= 3.600 kilometres per hour
Energy, Work	joule (J) (i.e., 1 kg m^2 s^{-2})	SI unit	1 J	= 0.239 calorie
	kilojoule (kJ)	10^3 J	1 cal	= 10^7 ergs
	megajoule (MJ)	10^6 J		= 4.186 joules
Power	watt (W) (i.e., J s^{-1} or 1 kg m^{-2} s^{-3})	SI unit	1 mW	= 0.014 calorie per minute
	milliwatt (mW)	10^{-3} W	1 W	= 0.239 calorie per second
	kilowatt (kW)	10^3 W	*1 kW	= 1.341 horsepower
	megawatt (MW)	10^6 W		
Energy, Flux	watt per square metre (W m^{-2})	SI unit	1 mW cm^{-2}	= 0.014 calorie per square centimetre per minute
Density	milliwatt per square centimetre (mW cm^{-2})	10 W m^{-2}	1 W m^{-2}	= 0.001 calorie per square centimetre per minute
			1 ly	= 1 calorie per square centimetre per minute
				= 697.3 watts per square metre

Density	kilogram per cubic metre (kg m^{-3})	SI unit	*1 g cm^{-3}	= 0.036 pound per cubic inch
	gram per cubic metre (g m^{-3})	10^{-3} kg m		
	gram per cubic centimetre (g cm^{-3})	10^{3} kg m		
	tonne per cubic metre (t m^{-3})	10^{6} kg m		
Force	newton (N) (i.e., 1 kg m s^{-2})	SI unit	1 N	= 10 dynes
	kilonewton (kN)	10^{3} N		
	meganewton (MN)	10^{6} N	*1 kN	= 0.100 ton force
Pressure	pascal (Pa) (i.e., 1 kg m^{-1} s^{-2})	SI unit (N m^{-2})	1 Pa	= 10 dynes per square centimetre
	hectopascal (hPa) (millibar, mb)	10^{2} Pa	1 mb	= 100 Pa = 1 hPa = 0.750 millimetre mercury
			1 std atmosphere	= 1 013.25 mb = 1 013.25 hPa = 760 millimetre mercury
Tempera-ture	Kelvin (K) degree Celsius (°C)	base unit K − 273	°C °F	= 5/9 (°F − 32) = (9 × °C) /5 + 32
Angular	radian (rad)	SI unit	1 radian	= 57°18′
	milliradian (mrad)	10^{-3} rad	1 radian of	
	degree (...°)	π/180 rad	latitude	= 6 129 km on Earth
	minute (...′)	1/60°		
	second (...″)	1/60′	1 rad s^{-1}	= 2.063 × 10^{5} degrees per hour

* Not base SI units but derived from them.

GLOSSARY

absolute acceleration acceleration relative to a fixed reference.

absolute instability the stability category that exists when the environmental (measured) lapse rate is greater than the dry adiabatic lapse rate.

absolute stability the stability category that exists when the environmental lapse rate is less than the wet (saturated) adiabatic lapse rate.

absolute vorticity vorticity measured relative to a fixed reference.

absolute zero the zero point on the Kelvin temperature scale when all molecular motion ceases (0 K = −273.16 °C).

absorption the process by which incident radiation is absorbed into a body and retained without reflection or transmission.

absorption band wavelengths over which the absorption of radiation by a particular substance, usually a gas, is large.

absorptivity the fraction of incident electromagnetic radiation absorbed by a surface.

acceleration the rate of change of velocity with time.

accretion the growth of water drops or ice particles by the collection of smaller droplets or particles.

adiabatic without change of heat.

advection horizontal transport (of mass, heat, etc.).

advection fog fog formed by the advection of warm air over a colder surface.

aerodynamic roughness length the height at which wind speed falls to zero with downward extrapolation of a log wind profile.

aerosol a suspension of very small particles in air.

air mass a synoptic-scale volume of air with uniform properties.

air parcel a small parcel of air undergoing little mixing with its environment.

Aitken nuclei aerosols less than 0.2 μm in diameter.

albedo the fraction of incident radiation reflected by a surface.

alto a cloud-type prefix referring to middle-level cloud (e.g., altocumulus).

altostratus middle-level sheet cloud.

ambient air the air of the surrounding environment.

anabatic flow warm airflow up sloping terrain.

analogue model a model based on a known historical situation that has similar features to those anticipated in the future.

anemometer an instrument for measuring wind speed or run of wind.

angular momentum the product of the tangential momentum of a body and its radial distance from the axis of rotation.

angular velocity the rate of the turning of air about an axis, expressed in angle turned per unit time (usually radians per second).

anticyclone a closed high-pressure system around which winds blow in an anticlockwise direction in the southern hemisphere.

anvil the extended ice cloud capping a cumulonimbus cloud.

aphelion the time at which the sun and earth are at their greatest distance apart.

atmosphere the envelope of air surrounding the earth and bound to it by the earth's gravitation.

atmospheric boundary layer or **planetary boundary layer** the layer of air most directly influenced by the underlying surface, including the laminar, surface and Ekman boundary layers.

atmospheric pressure the pressure created by the mass of air above a point or level; the total force per unit area is the pressure.

atmospheric waves horizontal or vertical undulations of airflow on scales from the metric to the hemispheric.

atmospheric window the relative gap between 8 and 11 μm in the infrared absorption spectrum.

attenuation any process in which a diminution of energy occurs with increasing distance from the energy source.

available potential energy potential energy that could be converted to kinetic energy by a redistribution of air parcels.

azimuth a direction specified by the angle taken anticlockwise from zero at true north.

backing a wind-direction change involving a decrease in azimuth.

baroclinic a state in which isopycnic (or isothermal) and isobaric surfaces intersect.

baroclinic instability dynamic instability associated with baroclinic zones.

barograph a barometer with a graphical output.

barometer an instrument for measuring atmospheric pressure.

barotropic a state in which isopycnic (or isothermal) and isobaric surfaces are parallel.

Bergeron-Findeison theory (or **process**) the preferential growth of ice crystals in clouds consisting largely of ice and supercooled water.

black body an ideal body that emits all the incident radiation it has absorbed.

blocking high a nearly stationary anticyclone in the mid-latitudes that inhibits the west–east progressions of cyclones.

boundary layer the layer directly influenced by a surface.

Bowen ratio the vertical flux of sensible heat expressed as a fraction of the latent heat flux.

buoyancy the upward force on a parcel of fluid as a result of the density difference between it and the surrounding fluid.

buster a sudden (and often violent) wind shift from a north-westerly to south-westerly direction to the rear of a coastal low.

Buys-Ballot's law if observers stand with their backs to the wind, then lower pressure will be on their right in the southern hemisphere and left in the northern hemisphere.

cellular convection convection organized in quasi-regular cells, as in stratocumulus, or in the ordered clustering of cumulus clouds.

Celsius scale a temperature scale with the freezing and boiling points of pure water at standard atmospheric pressure defined as 0 and 100 degrees (°C) respectively.

centrifugal directed away from the centre of a circle (as with a centrifugal force).

centripetal directed toward the centre of a circle (as in centripetal acceleration).

chromosphere the layer, approximately 2 500 km deep, immediately above the surface of the sun (the photosphere), in which temperature increases to values of 106 K at its outer limit.

cirrus a fibrous cloud composed of ice particles.

climate the integrated effect of weather typical of a region or site.

cloud a dense concentration of water droplets and/or ice crystals, each of which has a diameter of order 10 μm.

cloud seeding the encouragement of precipitation from clouds by accelerating droplet formation and enhancing the efficiency of precipitation-producing mechanisms.

cold front see **front**.

collision and **coalescence** enhanced precipitation development by the growth of larger particles or droplets falling through populations of smaller ones.

condensation the growth of water or ice by diffusion from contiguous vapour (in the case of ice growth, the process is often termed sublimation).

condensation nuclei small particles that act as nuclei for the formation of individual cloud crystals or droplets.

conditional instability the range of temperature profiles between which a wet air parcel is unstable and a dry parcel is stable.

conduction heat transfer by molecular agitation and impact.

conservative a term applied to any property that remains constant when acted upon by some outside process.

contact nucleation the process by which a cloud droplet freezes when it comes into contact with a freezing nucleus.

continuity equation the conservation of the mass of a body of moving air in which the net mass flux into a fixed volume is related to the rate of accumulation of air mass therein.

contour a geopotential-height contour of an isobaric surface.

convection small-scale vertical exchange of air parcels driven either by buoyancy (thermal or free convection) or wind shear (mechanical or forced convection). See **slope convection** for application to the large scale.

convective instability a state in which the uplift of layers of air is prompted by a favourable thermal stratification and moisture distribution in the atmosphere.

convergence negative divergence.

Coriolis force the apparent force deflecting any object moving over the surface of a rotating body (to the right/left in the northern/southern hemisphere) without altering its speed.

Coriolis parameter the factor $2\omega\sin\phi$, where ω is the earth's angular velocity and ϕ is the angle of latitude.

covariance the average value of the product of deviations of matched pairs of values (e.g., of vertical wind speeds and temperatures) from their respective means.

cryosphere the earth's snow and ocean- and land-ice masses.

cumulus the family of cauliflower-shaped clouds ranging from small cumulus humilis to large cumulonimbus. (On its own the term usually denotes cumulus humilis or only slightly larger clouds.)

cumulus congestus large rapidly growing cumulus clouds.

cumulonimbus a massive precipitating cumulus cloud, usually with an anvil.

cut-off low a deep, cold low that has become displaced equatorward from the normal path of west–east progressing cyclones in the westerly airstream.

cyclogenesis cyclone formation in extra-tropical cyclones.

cyclone a synoptic-scale low-pressure system with winds rotating clockwise in the southern hemisphere (also the local name for a severe tropical storm in the Indian Ocean).

cyclone wave the wave-like deformation of flow in the middle and upper troposphere associated with an extra-tropical cyclone.

cyclonic shear horizontal flow sheared in such a way as to cause an embedded parcel to rotate cyclonically as it moves across the flow in the direction of increasing velocity.

cyclostrophic flow horizontal circular air motion in which the centrifugal force balances the pressure-gradient force.

declination the angular distance measured by drawing an angle from a point on the equator at noon, to the centre of the earth and to the sun.

density mass concentration expressed as mass per unit volume.

dendrochronology the analysis of the growth patterns of tree rings.

deposition nuclei particles in the air that serve as centres upon which ice crystals may form directly from the vapour phase.

depression an extra-tropical cyclone.

detrainment loss of mass from a rising thermal to surrounding air (the reverse of entrainment).

dew water condensed onto a solid surface that has cooled to the dew-point temperature of the contiguous air.

dew point or **dew-point temperature** the temperature at which a parcel of air becomes saturated during isobaric cooling with conservation of vapour content.

diabatic heating all forms of non-adiabatic heating in the atmosphere.

diffuse radiation scattered or reflected short-wave radiation reaching the earth's surface.

diffusion the exchange of fluid elements between regions in space by apparently random molecular or turbulent motion.

diffusivity the rate of diffusion of a property.

direct radiation unattenuated short-wave radiation.

diurnal with a daily cycle.

divergence (of velocity) the resultant rate of the stretching of a fluid as given by the sum of the longitudinal gradients of the three components of flow velocity. It refers in particular to the acceleration of air in the local horizontal.

downdraught the rapidly sinking column of air in a thundercloud.

downwash the downward movement of air behind an obstacle caused by the reduction of pressure to the lee.

drizzle precipitation of water droplets with radii of about 100 μm.

dry adiabat a line describing a dry adiabatic process on a thermodynamic diagram.

dry adiabatic an adiabatic movement or process of an air parcel in which no heat enters or leaves the parcel (also called isentropic, because entropy is conserved).

dry adiabatic lapse rate the temperature lapse rate equal to the value found in dry adiabatic ascent or descent of an air parcel (1 °C per 100 m).

dry-bulb temperature the temperature registered by a dry-bulb thermometer.

easterly wave a synoptic-scale wave disturbance in the low-level tropospheric easterlies of the tropics.

eccentricity a measure of the deviation of an ellipse from a circle expressed as the ratio of the distance between the foci to the length of the major axis of an ellipse (the eccentricity of the earth's orbit is 0.018 at present).

ecliptic the great circle the sun appears to describe on the celestial sphere (23.5° to the equator at present).

eddy an individual transient element of turbulence and usually thought of as a translating and rotating air parcel.

eddy diffusion the turbulent diffusion of properties in which eddies are considered to play a role analogous to that of molecules, but on a much larger scale.

eddy viscosity the apparent viscosity arising from turbulence.

effective temperature the temperature of a radiating black body.

electromagnetic radiation travelling electric and magnetic waves.

El Niño the reversal in direction of the Walker Circulation, which is associated with the replacement of the cool upwelling Peruvian coastal current by warm equatorial water.

Ekman layer the surface layer of the atmosphere or ocean in which frictional effects are discernible.

emissivity the actual emittance of radiation expressed as a fraction of the black-body value.

emittance the flux of electromagnetic radiant energy emitted by unit area of a radiating surface.

energy the capacity to do work (kinetic, gravitational potential, radiant and heat energies are important in meteorology).

ENSO El Niño-Southern Oscillation.

enthalpy sensible heat not associated with a change in state.

entrainment incorporation of surrounding air by a rising thermal.

equation of motion equation of relative acceleration with net force acting on an air parcel of unit mass.

equation of state the equation relating pressure, density and temperature of a parcel of an ideal gas or mixture of ideal gases.

equinox times of year when the sun is vertically overhead at the equator (21 March and 22 September) when daylight and night are equally long.

equivalent barotropic a situation in which temperature gradients are such that isotherms are parallel to the isobars.

equivalent potential temperature equivalent temperature reduced dry adiabatically to a standard level of 1 000 hPa.

equivalent temperature temperature attained if, at constant pressure, all the water vapour in the air is condensed and the latent heat so released is added to the air.

evaporation reverse of condensation.

evapotranspiration the combined process of evaporation and transpiration.

exponential growth or **decay** profile of growth (or decay) of a quantity whose gradient is directly proportional to the quantity itself.

exosphere outermost portion of the atmosphere beginning at a height of about 500 km.

extra-tropical cyclone a cyclone outside the tropics, distinguished from a tropical cyclone by greater scale, the presence of one or more fronts and the absence of great central intensity.

fallspeed see **terminal velocity**.

feedback mechanism process by which change introduced into a system is transmitted either to enhance the change (positive feedback) or to reduce the change (negative feedback).

fetch distance upwind from point of observation to a significant location.

flux the rate of flow of a quantity per unit time.

flux density the flux per unit area.

fog a dense cloud in contact with a land or water surface.

föhn a strong, dry lee wind produced by prior enforced ascent of air over mountains (usually refers to the European Alps).

free atmosphere the atmosphere above the atmospheric boundary layer.

freezing fog a fog of supercooled water droplets, freezing on impact with solid objects.

freezing nuclei solid aerosols that act as nuclei for the freezing of supercooled cloud droplets.

friction velocity a basic wind-speed parameter.

front a synoptic-scale swath of cloud and precipitation associated with a significant horizontal temperature gradient (a front is warm when warm air replaces cold on the passage of the front; with a cold front cold air replaces warm).

frontal zone a region of continually or seasonally preferred cyclogenesis.

frontogenesis initial formation of a frontal zone.

frost the deposition of ice on a land surface by diffusion and sublimation.

fumigation the process by which atmospheric pollutants are carried downward to the surface by turbulence and convection.

gale a wind whose ten-minute average speed at a height of 10 m equals at least 37 knots.

gas constant the constant in the equation of state for an ideal gas (in the universal form a single universal gas constant applies to all gases; in meteorology a different specific gas constant applies to each gas of uniform mixture).

Gaussian or **normal distribution** a bell-shaped, symmetrical distribution with its mean, mode and median at the point of symmetry.

General Circulation Model (GCM) a model in which the three-dimensional general circulation of the earth's atmosphere is modelled.

geopotential height the height of a surface in the atmosphere that is proportional to the potential energy of unit mass at this height (for most purposes ordinary geometric height is equivalent to geopotential height).

geostrophic wind a horizontal wind produced by the balance of the Coriolis force and the horizontal pressure-gradient force.

glaciation the process whereby a supercooled water cloud is converted into an ice cloud.

gradient the spatial rate of change of a quantity in the direction of maximum rate of increase (as in temperature gradient).

gradient wind a horizontal wind produced by the balance of the Coriolis, centrifugal and horizontal pressure-gradient forces with cyclonic or anticyclonic curvature of flow.

graupel a snow pellet formed by ice-crystal growth by riming.

gravitational acceleration the downward acceleration of a body falling freely in a vacuum (equivalently it is the gravitational force per unit mass of the body).

gravity waves waves in which gravity is the predominant restoring force.

greenhouse effect the increase of tropospheric temperature as a result of the atmosphere's transparency to solar radiation and partial opaqueness to terrestrial radiation.

gust a short positive departure from the average wind speed.

Hadley circulation the meridional circulation of air in low latitudes, consisting of two opposing cells, each with air rising in the Inter-Tropical Convergence Zone and sinking in the adjacent subtropical zone.

hail one-millimetre or larger precipitation particles of ice, formed by the layered accretion of ice crystals and rapidly freezing supercooled water droplets.

haze a suspension of small, non-aqueous, solid particles in the atmosphere giving it a dusky, opalescent or milky appearance.

heat the total molecular energy of a body which is stored in the forms of sensible heat and/or latent heat.

heat capacity the amount of heat required to raise the temperature of a body by one degree Kelvin. In air its value differs depending on whether heating occurs at constant volume or constant pressure, the latter being the larger.

heterosphere the region of the atmosphere above 80–100 km.

heterogeneous nucleation the process by which cloud droplets form on atmospheric aerosols.

highs regions of raised atmospheric pressure (also known as anticyclones).

homosphere the region of the atmosphere below approximately 80 km.

homogeneous (spontaneous) nucleation the process by which pure water droplets form by condensation in a supersaturated vapour without the presence of aerosols.

humidity the vapour content of air (see **relative** and **specific humidity**).

humidity mixing ratio the mass of vapour as a fraction of the mass of dry air with which it is mixed in a moist air parcel; numerically indistinguishable from specific humidity in all but the most humid air.

hurricane see **tropical cyclone**.

hydrologic cycle the network of pathways of water throughout oceans, land surface and atmosphere.

hydrosphere the layer of water that nearly envelopes the earth in the form of oceans and inland seas.

hydrostatic equation the formal expression of pure hydrostatic equilibrium.

hydrostatic equilibrium a balance between the downward gravitational force acting on an air parcel and the upward pressure-gradient force arising from the general decrease of pressure with height.

hygrograph a recording hygrometer.

hygrometer an instrument measuring humidity, often by measuring relative humidity.

hygroscopic substances attracting and absorbing ambient water vapour.

ice age a period of time when the earth's cryosphere greatly exceeds its present area.

ideal gas a gas that behaves as if its molecules were infinitely small, interacting only by perfectly elastic collision at the instant of collision, and therefore obeying the equation of state for an ideal gas.

index cycle the term used to describe alternation between periods of zonal and meridional flow (the index is the pressure difference between two latitude zones).

infrared radiation electromagnetic radiation lying beyond the red end of the visible spectrum, but not far beyond.

insolation solar irradiance of a surface, usually

totalled over a finite period such as a day.

interglacial the period between two ice ages.

internal energy the sensible heat capacity of an ideal gas as assessed from its specific heat at constant volume.

Inter-Tropical Convergence Zone (ITCZ) the zone of persistent convergence of airflow in the lower troposphere in low latitudes.

inversion an increase of atmospheric temperature with height.

ionosphere the region of the upper atmosphere (usually reckoned to start 50 km above the earth's surface) that is in a state of significant ionization.

irradiance the value obtained by dividing the radiant flux by the area through which it passes.

isentropic with constant potential temperature.

isallobar an isopleth of pressure tendency.

isobar an isopleth of atmospheric pressure.

isopleth a line of equal value on a map.

isopycnal or **isopycnic** an isopleth of air density.

isotach an isopleth of wind speed.

isotherm an isopleth of air temperature.

isotropic independent of direction (microscale turbulence is nearly isotropic; larger scales are anisotropic because they tend to be larger horizontally than vertically).

jet stream a relatively narrow and shallow stream of fast-flowing air, usually in the high troposphere.

joule the unit of energy.

katabatic winds cool airflow down sloping terrain.

Kelvin the unit or scale of absolute temperature.

Kelvin wave vertical waves that result from shear instability between two masses of different densities.

kinematic properties of flow that can be deduced without reference to Newton's laws.

kinetic energy energy of motion.

knot a wind speed of one nautical mile per hour or 0.515 m s^{-1}.

laminar flow smooth, viscosity-dominated flow.

land breeze the offshore component of a local diurnal wind cycle that occurs on coastlines due to differences in land and sea temperature.

lapse rate see **temperature lapse rate**.

latent heat the heat given out when gases and liquids liquefy or solidify, and which is absorbed when solids or liquids melt or evaporate.

lee trough a synoptic-scale low-pressure system formed or maintained in the lee of mountains or plateaux.

lifting condensation level the level at which air would become saturated if lifted dry adiabatically conserving its vapour content.

lightning electric discharges produced in and around thunderclouds.

log wind profile the linear profile of wind speed in the surface boundary layer when plotted against the logarithm of height.

lysimeter an instrument for measuring the amount of water lost by evapotranspiration.

magnetosphere the portion of the high atmosphere where particle motion is controlled by the earth's magnetic field.

mass the quantity of material in a body as assessed by its reluctance to accelerate when acted on by a force when free of all other constraints.

mechanical convection vertical exchange of air parcels consequent upon turbulence generated by vertical wind shear.

melting band the horizontal band of enhanced radar echo produced by snow melting at the 0 °C level in precipitating clouds.

meridional along a line of longitude.

mesopause the top of the mesosphere.

mesoscale spatial scales intermediate between small and synoptic scales of weather systems.

mesosphere the region of the upper atmosphere lying between the stratosphere and the thermosphere.

microphysics of clouds physical processes active on the scale of individual cloud and precipitation droplets and particles.

Mie scattering scattering of electromagnetic radiation by spherical particles with much of the radiation scattered in a forward direction.

Milankovitch periodicities periodic changes in the earth's orbital parameters (and which are believed to contribute to the occurrence of the ice ages).

millibar a now-redundant meteorological unit of pressure (mb); equals 100 pascals (1 hectopascal).

mixing layer the layer of air (usually sub-inversion) within which pollutants are mixed by turbulence.

momentum the product of the mass and velocity of a body.

monsoon a system showing a pronounced seasonal variation of airflow in the lower troposphere in tropical and subtropical regions.

mountain–plain wind a cool regional, mainly nocturnal wind blowing from a mountain region or plateau towards contiguous plains.

mountain wind a cool, mainly nocturnal local wind blowing down a valley.

neutral stability the state of a system in which the environmental lapse rate equals the dry adiabatic lapse rate.

nimbostratus an extensive layer of cloud precipitating rain and/or snow.

noctilucent clouds tenuous, at times brilliant clouds occurring at great heights in the atmosphere and visible against a dark night sky at high latitudes.

noise random fluctuations in a parameter caused by effects other than the one being studied.

normalize to divide a quantity by a more fundamental quantity of the same dimensions to produce a non-dimensional ratio.

Normand's theorem states that on a tephigram the dry adiabat drawn through the dry-bulb temperature, the wet adiabat through the wet-bulb temperature and the dew-point line through the dew-point temperature intersect at condensation level.

occluded front the front formed by the merging of a cold and a warm front.

optical thickness a measure of the attenuation of solar radiation by the atmosphere due to scattering and absorption (the greater the optical thickness, the greater the attenuation).

order of magnitude the typical magnitude of a quantity to the nearest integral power of 10.

orographic occurring as a result of relief of the ground surface.

oxidant any oxidizing agent, a substance that acquires electrons in a chemical reaction (ozone, O_3, and atomic oxygen, O, are examples of very effective oxidants).

ozone a form of oxygen containing three atoms per molecule and which absorbs ultraviolet radiation. The presence of this gas in the ozonosphere screens the earth from the harmful effects of ultraviolet radiation.

parameterization a representation of physical effects in terms of oversimplified parameters.

partial pressure the pressure exerted individually by a particular component of a gaseous mixture.

pascal the SI unit of pressure (Pa).

period the interval between consecutive similar stages in a repeating cycle of events.

perturbation any departure introduced into an assumed steady state of a system.

plain–mountain wind a warm regional, mainly daytime wind blowing from a plain towards a mountain range or plateau.

planetary boundary layer the atmospheric boundary layer or Ekman layer (i.e., the layer of the atmosphere from the surface to the level where the frictional influence is absent).

photochemistry the study of chemical reactions that take place when substances are exposed to ultraviolet electromagnetic energy, especially ultraviolet and short-wavelength visible radiation.

photodissociation molecular dissociation caused by the absorption of photons of solar radiation.

photoionization ionization caused by the absorption of photons of solar radiation.

photon the elementary quantity of radiant energy.

photosphere a thin, tenuous layer forming the 'surface' of the sun with an effective temperature of about 6 000 K.

photosynthesis the synthesis of water and carbon dioxide to form oxygen and organic compounds in the presence of sunlight.

Planck curve the curve describing the variation, with wavelength, of the amount of energy being radiated by a black body at a fixed temperature.

Poisson's equation the equation relating initial and final absolute temperatures and pressures of an ideal gas undergoing dry adiabatic compression or decompression.

polar front the middle-latitude frontal zone separating air flowing from tropical and polar source regions (a favoured site for cyclogenesis).

potential energy the energy that a body possesses as a result of its position in a field of gravity.

potential temperature the temperature a parcel of dry air would have if reduced adiabatically to a standard pressure of 1 000 hPa.

power energy input or output per unit time.

precipitable water the depth of water that would result from condensation of all the water vapour in a column of air extending from the surface to the top of the atmosphere.

precipitation ice particles or water droplets large enough (about 100 μm or larger) to fall at least 100 m below cloud base before evaporating (see **drizzle**, **hail**, **rain** and **snow**).

pressure the apparently continuous and isotropic force exerted on unit area of any real or imaginary surface because of the bombardment by molecules of contiguous fluid.

pressure-gradient force the net force on an air parcel arising from its location in a pressure gradient.

pressure tendency the rate of change of pressure at a fixed location.

psychrometer an instrument measuring vapour content by means of wet- and dry-bulb thermometry.

radian the angle subtended at its centre by an arc of a circle equal to its radius.

radiant flux the flow of electromagnetic radiant energy per unit time.

radiant flux density the radiant flux through unit area.

radiation see **electromagnetic radiation**.

radiation fog fog produced by net radiative cooling of the underlying surface.

radiometer an instrument to measure fluxes of electromagnetic radiation.

radiosonde a balloon-borne thermometer, barometer, hygrometer and radar reflector.

rain precipitation in the form of millimetre-sized water droplets (as distinct from drizzle).

rain day a 24-hour period during which 0.2 mm or more of precipitation falls.

rainout the deposition of pollution particles in precipitation.

rawinsonde a radiosonde that also measures wind speed and direction.

Rayleigh scattering scattering of electromagnetic radiation by spherical particles (such as air molecules) of radius less than one-tenth the wavelength of the incident radiation.

reflectivity the proportion of incident radiation reflected by a surface, expressed as a fraction or percentage.

relative humidity the vapour content of the air (measured as vapour density or pressure) as a percentage of the vapour content needed to saturate air at the same temperature.

relative vorticity vorticity about the local vertical relative to the local tangential plane of the earth.

residence time the average time spent by a particle in one particular component of a system in dynamic equilibrium.

Richardson number the dimensionless ratio of buoyancy forces and velocity shears typical of a particular flow regime.

ridge an elongated local region of high pressure.

riming the process by which ice particles in a mixed cloud increase in mass by colliding with supercooled droplets which then freeze onto them.

Rossby wave the breakdown of zonal flow into

large-scale wave motion when the temperature gradient across a rotating fluid reaches a critical value.

run of wind the length of airflow registered by an anemometer in a certain period (often an hour or a day).

satellite (geostationary) a satellite whose high-altitude orbit (about 35 000 km) is in the equatorial plane and whose orbital velocity matches that of the earth so that its position remains constant with respect to the earth.

satellite (polar orbiting) a satellite whose orbit is approximately sun-synchronous, low altitude (about 1 000 km), normal to the equator and passes close to the poles on each orbit.

saturated adiabat a line on a tephigram representing a saturated adiabatic process.

saturated adiabatic process an adiabatic process in which air is kept in a saturated state.

saturation the state of water vapour that is in dynamic equilibrium with a plane surface of pure water or ice through balanced fluxes of evaporating and condensing water molecules.

scale analysis the determination of the relative significance of magnitudes of individual terms in dynamic and thermodynamic equations.

scale height the vertical height interval in which a specified atmospheric property (such as pressure) decreases by a specified proportion (such as 1/10).

scattering the process of chaotic deflection of electromagnetic radiation by impact on small particles (usually refers to solar photons and occurs as Rayleigh or Mie scattering).

scavenging the precipitation of airborne particulates by rain or snow.

sea breeze the onshore component of a local diurnal wind cycle that occurs on coastlines due to differences in land and sea temperature.

seeding of clouds see **cloud seeding**.

sensible heat the heat energy able to be sensed (e.g., with a thermometer) and involving no change in state, in contrast to latent heat.

severe local storm a storm associated with the most active cumulonimbus clouds.

shear the gradient of the direction and/or speed across fluid flow.

shear instability the development of wave-like motion in a stably stratified atmosphere when the vertical wind shear exceeds a critical value.

shear stress the tangential force per unit area of sheared fluid.

shelter belt a zone of protection downwind from a line of increased surface roughness (e.g., a belt of trees).

slope convection synoptic-scale ascent and descent associated with extra-tropical cyclones.

smog a mixture of smoke and fog.

snow precipitation in the form of well-developed dendritic ice crystals usually in aggregated flakes.

solar altitude the vertical direction of the sun above the horizon expressed in degrees.

solar azimuth the horizontal direction of the sun relative to true north expressed in degrees.

solar constant the amount of energy passing in unit time through a unit surface area perpendicular to the sun's rays at the outer edge of the atmosphere at the mean distance between the earth and the sun (currently believed to be 1 370 W m^{-2}).

solarimeter an instrument to measure solar irradiance.

solar radiation electromagnetic radiation from the sun, especially in the wavelength range 0.3 to 3 μm.

solar wind a stream of interplanetary gas outwards from the sun towards the earth, near which it interacts with the earth's magnetic field.

solar zenith angle the vertical direction of the sun relative to the zenith expressed in degrees (the reciprocal of solar altitude).

solstice the times of year when the sun is vertically overhead at the Tropic of Capricorn (21 December) or the Tropic of Cancer

(21 June).

Southern Oscillation a fluctuation in inter-tropical pressure, wind, sea-surface tempera-ture and rainfall and an exchange of air between the south-east Pacific subtropical high and the Indonesian equatorial low.

specific heat the heat needed to raise the tem-perature of unit mass of a particular substance by one degree Kelvin.

specific humidity the mass of water vapour as a proportion of the total mass of moist air of which it forms a part.

spectral analysis a type of Fourier (harmonic) analysis which identifies cycles in atmospher-ic features.

spectrum the distribution of variance or power with wavelength or frequency.

squall a sudden onset of strong, gusty winds.

squall line or **line squall** a linear organization of vigorous cumulonimbus.

stability the condition of a body or system that responds to a specified disturbance by oppos-ition and suppression. Often used in meteor-ology to refer to convective stability in partic-ular.

steam fog fog formed by evaporation from a relatively warm wet surface into cooler air.

storm a general term applied to any type of weather system associated with strong surface winds.

stratopause the boundary between the strato-sphere and mesosphere.

stratosphere the atmospheric layer between the troposphere and mesosphere, character-ized by an increase of temperature with height.

streamline a line whose tangent is parallel to fluid flow.

sublimation the transition of a substance directly from the solid to the vapour phase or vice versa.

subsidence inversion a temperature inversion formed by subsidence in an anticyclone.

subtropical high pressure areas of raised sur-face pressure between latitudes 20° and 40°.

sunspot a relatively dark area on the sun's sur-face.

superadiabatic lapse rate a lapse rate which exceeds the dry adiabatic lapse rate.

supersaturation vapour pressure or density greater than is required for saturation at the prevailing temperature.

synoptic observations simultaneous observa-tions taken at recognized weather stations.

synoptic scale the minimum horizontal spatial scale of weather systems that may be defined in a synoptic observation network.

teleconnections linkages over great distances of atmospheric and oceanic variables, usually taken to be similar variations between regions caused by a common mechanism.

temperature the concentration of sensible heat in a body and measured on the Celsius and Kelvin scales.

temperature lapse rate the rate of fall of tem-perature with increasing height (see **dry** and **saturated adiabatic**).

tephigram a thermodynamic diagram with temperature and potential temperature as perpendicular axes.

terminal velocity or **fallspeed** the speed at which a mass of a body is balanced by its drag as it falls through a fluid.

terrestrial radiation electromagnetic radiation emitted by materials on the surface or in the atmosphere.

thermal a buoyant eddy, usually about 100 m in diameter or larger.

thermal conductivity the ability of a substance to conduct heat by molecular motion.

thermal convection buoyant convection driven by temperature differences between air parcels and the ambient air.

thermal diffusivity the ratio of the thermal conductivity to the heat capacity and deter-minant of the rate of heating of a substance.

thermal wind the vertical shear of geostrophic wind as a consequence of the occurrence of horizontal temperature gradients.

thermocline a vertical temperature gradient in a water body, which is appreciably greater than the gradients above or below it.

thermodynamic diagram a diagram whose

axes are thermodynamic parameters chosen so that areas on the diagram represent energy.

thermodynamics the science of the relationships between the different forms of energy; particularly the ways in which energy can be converted from one form to another and work be done.

thermograph a recording thermometer.

thermohaline circulation circulation in water caused by changes in density brought about by the combined effects of variations in temperature and salinity.

thermopause the top of the thermosphere.

thermosphere the atmospheric shell extending from the top of the mesosphere to outer space.

thickness the vertical depth of a slab of air bounded by chosen isobaric surfaces.

thunder the audible noise accompanying lightning.

tornado an intense, cloud-cored vortex extending from the base of a severe local storm to the surface.

torque the turning moment of a force.

total potential energy the sum of the internal and gravitational potential energies of an atmospheric column.

trade winds the low-level tropical easterly winds blowing obliquely from a subtropical high towards the Inter-Tropical Convergence Zone and which are equatorward of the Hadley cells.

trajectory the path traced by a parcel of moving air.

transmissivity the degree to which electromagnetic radiation is able to pass through the atmosphere.

tropical cyclone a cyclonic disturbance in tropical regions; when surface winds exceed 33 m s^{-1} it is called a hurricane in the Atlantic, a typhoon in the Pacific and a cyclone in the Indian Ocean.

tropopause the stable discontinuity marking the upper limit of the troposphere.

troposphere the layer of air continually stirred by weather systems and bounded by the surface and the tropopause.

trough an elongated local region of low atmospheric pressure.

turbidity a condition that reduces the transparency of the atmosphere to solar radiation.

turbosphere the atmospheric layer in which convection and turbulence are the dominant mixing processes (from the surface to about 90 km above sea level).

turbulence a state of fluid flow in which instantaneous velocities exhibit irregular and apparently random fluctuations capable of transporting atmospheric properties (e.g., heat, water vapour, momentum, etc.) at rates far in excess of molecular processes.

typhoon see **tropical cyclone**.

ultraviolet radiation electromagnetic radiation between the violet end of the visible spectrum and X-radiation.

unstable the condition of a system that amplifies a particular type of imposed disturbance, the type of instability depending on the type of disturbance (e.g., convective, dynamic and baroclinic).

upper-air observations synoptic observations made by radiosonde or equivalent techniques.

urban heat island the localized warming exerted by an urban area.

vacillation an irregular fluctuation in airflow varying between minimum and maximum sinuosity.

valley wind a warm daytime wind blowing up a valley.

vapour the gaseous form of a substance (also used as an abbreviation for water vapour).

vapour pressure the partial pressure of water vapour.

vector a vector quantity having both magnitude and direction.

veering a wind direction change involving an increase in azimuth.

velocity the rate of change with time of the position of a body.

Venturi effect the increased velocity of a fluid flowing through a constriction.

virtual temperature temperature at which a

fictitious volume of dry air would have the same density as the same volume of moist air.

viscosity the friction in gases and liquids arising from molecular exchange and impact (often called molecular viscosity to distinguish it from eddy viscosity).

vorticity the measure of fluid rotation (in synoptic meteorology the term usually means the relative vorticity about the local vertical).

wake capture the process by which small cloud droplets are pulled into the wake of larger droplets with which they collide and coalesce.

Walker Circulation a thermally driven longitudinal cellular circulation extending across the Pacific Ocean from Indonesia to close to the Peruvian coast and forming a component of the Southern Oscillation.

warm front see **front**.

warm sector the relatively warm area between a warm front and the following cold front.

washout pollution particles removed from the atmosphere by being incorporated into falling raindrops by collision and coalescence.

watt one joule per second – the SI unit of power.

wave cyclone an extra-tropical cyclone (applied especially to the immature phase).

wavelength the shortest distance between adjacent wave crests (or any other similar parts) in a train of waves.

waves regular distortions of the flow field that repeat themselves in space and time.

wet-bulb depression the difference between dry-bulb and wet-bulb temperatures.

wet-bulb potential temperature the wet-bulb temperature after an air parcel is reduced adiabatically to 1 000 hPa.

wet-bulb temperature the temperature registered by a wet-bulb thermometer.

wind horizontal air motion relative to the earth's surface.

wind chill the cooling of a body caused by the wind removing sensible and latent heat.

wind shear the variation of wind speed (and/or direction) with height.

wind stress the horizontal stress exerted on a surface because of adjacent wind shear.

work a form of energy arising from the motion of a system against a force and existing only in the process of energy conversion.

zenith angle the angular distance between the sun and the local vertical (zenith).

zonal along a line of latitude.

zonal index a pressure index of the sinuosity of large-scale tropospheric flow.

REFERENCES

ACOCKS, J. P. H. 1953. Veld types of South Africa. *Memoirs of the Botanical Survey of South Africa*, 28. 192 pp.

ALLAN, R. A., LINDESAY, J. A. and PARKER, D. 1996. *El Niño, Southern Oscillation and Climatic Variability*. CSIRO, Australia. 405 pp.

ALLEY, R. B., MAYEWSKI, P. A., SOWERS, T., STUIVER, M., TAYLOR, K. C. and CLARK, P. U. 1997. Holocene climatic instability: A prominent, widespread event 8 200 yr ago. *Geology*, 25: 483–486.

ANGELL, J. K., PACK, D. H., DICKSON, C. R. and HOEKER, W. H. 1971. Urban influence on night-time airflow estimated from tetroon flights. *Journal of Applied Meteorology*, 10: 194–204.

ATKINSON, B. W. 1968. A preliminary examination of the possible effect of London's urban area on the distribution of thunder rainfall, 1951–60. *Transactions of the Institute of British Geographers*, 44: 97–118.

———. 1981. *Meso-scale Atmospheric Circulations*. London: Academic Press. 495 pp.

BANG, N. D. 1971. The southern Benguela Current region in February 1966. Part II: Bythythermography and air-sea interactions. *Deep Sea Research*, 18: 209–224.

BARRY, R. G. and CHORLEY, R. J. 1998. *Atmosphere, Weather and Climate*, 7th Edition, Routledge, London. 409 pp.

BERGERON, T. 1954. The problem of tropical hurricanes. *Quarterly Journal of the Royal Meteorological Society*, 80: 131–164.

BIERLY, E. W. and HEWSON, E. W. 1962. Some restrictive meteorological conditions to be considered in the design of stacks. *Journal of Applied Meteorology*, 1: 383–390.

BLAKE, N. J., BLAKE, D. R., SIVE, B. C., CHEN, T. Y., ROWLAND, F. S., COLLINS, J. E., SACHSE, G. W. and ANDERSON, B. E. 1996. Biomass burning emissions and vertical distribution of atmospheric methyl halides and other reduced carbon gases during the TRACE-A experiment. *Journal of Geophysical Research*, 101: 24 151–24 164.

BOND, G., BROECKER, W. S., JOHNSEN, S., JOUZEL, J., LABEYRIE, L. D., MCMANUS, J. and TAYLOR, K. 1993. Correlations between climate records from North Atlantic sediments and Greenland ice. *Nature*, 365: 143–147.

BROECKER, W. S. and DENTON, G. H. 1990. What drives glacial cycles? *Scientific American*, 262: 43–50.

BROWNING, K. A. 1964. Airflow and precipitation trajectories within severe local storms which travel to the right of winds. *Journal of the Atmospheric Sciences*, 21: 634–639.

———. 1985. Conceptual models of precipitation systems. *Meteorological Magazine*, 114: 293–319.

BROWNING, K. A. and FOOTE, G. B. 1976. Airflow and local growth in supercell storms and some implications for hail suppression.

Quarterly Journal of the Royal Meteorological Society, 102: 499–534.

BROWNING, K. A. and LUDLAM, F. H. 1962. Airflow in convective storms. *Quarterly Journal of the Royal Meteorological Society*, 88: 117–135.

BYERS, H. R. 1974. *General Meteorology*, 4th Edition. New York: McGraw-Hill. 461 pp.

CARLSON, T. N. 1980. Airflow through midlatitude cyclones and the comma cloud pattern. *Monthly Weather Review*, 108: 1 498–1 509.

CARTE, A. E. 1979. Sustained storms on the Transvaal Highveld. *South African Geographical Journal*, 61: 39–56.

———. 1981. Morphology of persistent storms in the Transvaal on 16/17 October 1978. *Beitrage zur Physik der Atmosphare*, 54: 86–100.

CHANDLER, T. J. 1965. *The Climate of London*. London: Hutchinson. 292 pp.

CHANGNON, S. A. 1972. Urban effects on thunderstorm and hailstorm frequencies. Preprints, American Meteorological Society Conference on the Urban Environment, Philadelphia, 31 October–2 November 1972: 177–185.

CHISHOLM, A. J. and RENICK, J. H. 1972. The kinematics of multicell and supercell Alberta hailstorms. Alberta Hail Studies, Research Council of Alberta Hail Studies, Report 72–2: 24–31.

CLARKE, J. F. 1969. Nocturnal boundary layer over Cincinnati, Ohio. *Monthly Weather Review*, 97: 582–589.

COSIJN, C. and TYSON, P. D. 1996. Stable discontinuities in the atmosphere over South Africa. *South African Journal of Science*, 92: 381–386.

CRITCHFIELD, H. J. 1974. *General Climatology*, 2nd Edition. Englewood Cliffs: Prentice Hall. 420 pp.

CROWELL, J. C. and FRAKES, L. A. 1975. Late Palaeozoic glaciation. In Campell, K. S. W. (ed.). *Gondwana Geology*. Canberra: Australian National University. 313–331.

DE ANGELIS, M., JOUZEL, J., LORIUS, C.,

MERLIVAT, L., PETIT, J. R. and RAYNAUD, D. 1984. Ice-age data for climate modelling from an Antarctic (Dome C) ice core. In Berger, A. L. and Nicolis, C. (eds.). *New Perspectives in Climate Modelling*. Amsterdam: Elsevier. 23–45.

DE WIT, M., JEFFERY, M. and NICOLAYSEN, L. 1986. *Geological map of Gondwana*. American Association of Petroleum Geologists and University of the Witwatersrand, 1988.

DEFANT, A. 1961. *Physical Oceanography*, Vol. 1. London: Pergamon. 598 pp.

DESOUZA, R. L. 1972. A study of atmospheric flow over a tropical island. Unpublished MSc thesis, Department of Meteorology, Florida State University. 203 pp.

DEVEREAUX, I. 1967. Oxygen isotope palaeotemperature measurements on New Zealand Tertiary fossils. *New Zealand Journal of Science*, 10: 988–1 011.

DIAB, R. D. 1975. Stability and mixing layer characteristics over Southern Africa. Unpublished MA thesis, University of Natal, Durban. 203 pp.

DUNN, P. 1985. An investigation into tropical cyclones in the south-west Indian Ocean. *Flood Studies:* Technical Note No. 1, Department of Water Affairs, Pretoria. 33 pp.

DYER, A.J. 1975. An international initiative in observing the global atmosphere. *Search*, 6: 29–33.

EAGLEMAN, J. R. 1980. *Meteorology, the Atmosphere in Motion*. New York: D. Van Nostrand Company. 384 pp.

EAST, W. R. and MARSHALL, J. S. 1954. Turbulence in clouds as a factor in precipitation. *Quarterly Journal of the Royal Meteorological Society*, 80: 26–47.

ERNST, J. A. 1976. SMS-1 night-time infra-red imagery of low-level mountain waves. *Monthly Weather Review*, 104: 207–209.

FLEAGLE, R. and BUSINGER, J. 1963. *An Introduction to Atmospheric Physics*. New York: Academic Press. 346 pp.

FLOHN, H. 1960. Indian Meteorological Department publication cited in Barry and

Chorley, 1998.

———. 1969. Local wind systems. In Landsberg, H. E. (ed.). *World Survey of Climatology*, Vol. 2. Amsterdam: Elsevier. 139–171.

FUGGLE, R. F. and OKE, T. R. 1970. Infra-red flux divergence and the urban heat island. *Urban Climates*, W. M. O. Technical Note 108: 70–78.

FUJITA, T. 1955. Results of detailed synoptic studies of squall lines. *Tellus*, 7: 405–436.

GARSTANG, M. 1965. Diurnal and semi-diurnal variations in sensible and latent heat exchange, cloudiness and precipitation over the western tropical Atlantic. Proceedings of the U. S. Army Conference on Tropical Meteorology, Miami, Florida. 123–136.

GARSTANG, M., EMMITT, G. D. and HOUSTON, S. 1985. *Mesoscale studies and developments*. Programme for Atmospheric Water Supply, Annual Report 1984/85, Water Research Commission Pretoria. i-iv and 54–72.

GARSTANG, M., TYSON, P. D., SWAP, R., EDWARDS, M., KALLBERG, P. and LINDESAY, J. A. 1996. Horizontal and vertical transport of air over southern Africa. *Journal of Geophysical Research*, 101: 23 721–23 736.

GEDZELMAN, S. D. 1980. *The Science and Wonders of the Atmosphere*. New York: John Wiley and Sons. 535 pp.

GHIL, M. and MCWILLIAMS. 1994. Workshop tackles oceanic thermohaline circulation. *EOS, Transactions, American Geophysical Union*, 75: 42 493–42 498.

GORDON, H. B. and O'FARRELL, S. P. 1996. Transient climate change in the CSIRO coupled model with dynamic sea ice. *Monthly Weather Review*, 125: 875-907.

HARRISON, M. S. J. 1984. The annual rainfall cycle over the central interior of South Africa. *South African Geographical Journal*, 66: 47–64.

———. 1986. A synoptic climatology of South African rainfall variations. Unpublished PhD thesis, University of the Witwatersrand. 341 pp.

HASS, W. A. ET AL. 1967. An analysis of low-level constant volume balloon (tetroon) flights over New York City. *Quarterly Journal of the Royal Meteorological Society*, 93: 483–493.

HASTENRATH, S. 1985. *Climate and Circulation of the Tropics*. Dordrecht: D. Reidel. 455 pp.

HAYWARD, L. Q. and VAN DEN BERG, H. J. C. 1970. The Eastern Cape floods of 24–28 August 1970. *South African Weather Bureau Newsletter*, 257: 129–141.

HELD, G. and CARTE, A. E. 1979. Hailstorms in the Transvaal during January 1975. *South African Geographical Journal*, 61: 128–142.

HENDERSON-SELLERS, A. and ROBINSON, P. J. 1986. *Contemporary Climatology*. Longman.

HERMAN, J. R., BHARTIA, P. B., TORRES, O., HSU, C., SEFTOR, C. and CELARIER, E. 1997. Global distribution of UV-absorbing aerosols from Nimbus-7/TOMS data. *Journal of Geophysical Research*, 102: 16 911–16 922.

HESS, S. L. 1959. *Introduction to Theoretical Meteorology*. New York: Henry Holt and Company. 362 pp.

HEWITSON, B. C. 1998. Deriving regional climate scenarios from general circulation models. Water Research Commission, Report 751/1/99. 40 pp.

HEWSON, E. W. and LONGLEY, R. W. 1944. *Meteorology Theoretical and Applied*. New York: John Wiley and Sons. 468 pp.

IMBRIE, J. ET AL. 1984. The orbital theory of Pleistocene climate: support from a revised chronology of the marine $\delta^{18}O$ record. In Berger, A. L., Imbrie, J., Hays, J., Kukla, G. and Saltzman, B. (eds.). *Milankovitch and Climate*, Part 1. Dordrecht: D. Reidel. 269–305.

IPCC. 1990. *Climate Change, The IPCC Scientific Assessment*. Editors: Houghton, J. T., Jenkins, G. J. and Ephraums, J. J. Cambridge: Cambridge University Press. 364 pp.

———. 1992. *Climate Change 1992, The Supplementary Report to the IPCC Scientific Assessment*. Editors: Houghton, J. T., Callender, B. A. and Varney, S. K. Cambridge: Cambridge University Press. 200 pp.

————. 1996. *Climate Change, 1995, The Science of Climate Change.* Editors: Houghton, J. T., Miera Filho, L. G., Callander, B. A., Harris, N., Kattenburg, A. and Maskell, K. Cambridge: Cambridge University Press. 572 pp.

JACKSON, S. P. 1961. *Climatological Atlas of Africa,* CCTA/CSA, Nairobi, 55 plates.

JIANG, Y. B., YUNG, Y. L. and ZUREK, R. W. 1996. Decadal evolution of the Antarctic ozone hole. *Journal of Geophysical Research,* 101 (D4): 8 985–9 000.

JOUBERT, A. M. 1997. Modelling present and future climates over southern Africa. Unpublished PhD thesis, University of the Witwatersrand. 229 pp.

JOUBERT, A. M. and KOHLER, M. O. 1996. Projected temperature increases over southern Africa due to increasing levels of greenhouse gases and sulphate aerosols. *South African Journal of Science,* 92: 524–526.

JOUBERT, A. M., KATZFEY, J. J., MCGREGOR, J. L. and NGUYEN, K. C. 1999. Simulating mid-summer climate over southern Africa using a nested regional climate model. *Journal of Geophysical Research,* in press.

JOUZEL, J. 1998. Palaeoclimate record of climate: polar ice cores. *CLIVAR Exchanges,* 3: 6–8.

JURY, M., ROUAULT, M., WEEKS, S., and SCHORMAN, M. 1997. Atmospheric boundary-layer fluxes and structure across a land-sea transition zone in south-eastern Africa. *Boundary Layer Meteorology,* 83: 311–330.

JURY, M. J., MCQUEEN, C. A. and LEVEY, K. M. 1994. SOI and QBO signals in the African region. *Theoretical and Applied Climatology,* 50: 103–115.

LAMB, H. H. 1969. Climatic fluctuations. In Landsberg, H. E. (ed.). *World Survey of Climatology,* Vol. 2. Amsterdam: Elsevier. 173–249.

LANGENBERG, H. M. 1974. Ventilation potential of the atmosphere over Colenso, Natal – an investigation of air movement and thermal stability. Report to the Director, Natal Town and Regional Planning Commission, CSIR, APRG/74/20.

LEAHEY, D. M. 1969. An urban heat island model. Report TR-69–11, Department of Meteorology and Oceanography, New York University. 70 pp.

LENGOASA, J. R. 1987. Atmospheric circulation and temperature fields over South Africa. Unpublished MSc thesis, University of the Witwatersrand, Johannesburg. 70 pp.

LINDESAY, J. A. 1987. Relationships between the Southern Oscillation and atmospheric circulation changes over southern Africa, 1957 to 1982. Unpublished PhD thesis, University of the Witwatersrand. 284 pp.

LINDESAY, J. A. and ALLAN, R. A. 1992. Modulation of summer rainfall in southern Africa and Australia with teleconnection patterns across the Indian Ocean. Paper presented at the 27th International Congress of the International Geographical Union, Washington, 10–14 August 1992.

LOCKWOOD, J. C. 1979. *Causes of Climate.* London: Edward Arnold. 260 pp.

LONDON, J. and SASAMORI, T. 1971. Radiative energy budget of the atmosphere. *Space Research,* 11: 639–649.

LONGLEY, R. W. 1976. Weather and weather maps of South Africa. South African Weather Bureau Technical Paper, No. 3. 76 pp.

LUDLAM, F. H. 1980. *Clouds and Storms.* University Park: Pennsylvania State University Press. 405 pp.

MALAN, D. J. 1963. *Physics of Lightning.* London: English Universities Press. 176 pp.

MASON, B. J. 1962. *Clouds, Rain and Rainmaking.* Cambridge: Cambridge University Press. 145 pp.

MASON, S. J. 1992. Sea surface temperatures and South African rainfall variability. Unpublished PhD thesis, University of the Witwatersrand, Johannesburg. 235 pp.

MASON, S. J and JURY, M. J. 1997. Climate variability and change over southern Africa: a reflection on underlying processes. *Progress in Physical Geography,* 21: 23–50.

MATHER, G. K., TERBLANCHE, D. E. and

STEFFENS, F. E. 1997. The National Precipitation Research Programme, Final Report 1993–1996. Water Research Commission Report No. 726/1/97, Pretoria. 147.

MCGEE, O. S. 1986. The distribution of water vapour in the atmosphere over South Africa. *South African Geographical Journal*, 68: 117–131.

MCGEE, O. S. and HASTENRATH, S. L. 1966. Harmonic analysis of the rainfall over South Africa. *Notos*, 15: 79–90.

MCILVEEN, J. F. R. 1986. *Basic Meteorology*. Wokingham: Van Nostrand Reinhold. 457 pp.

MCINTOSH, D. H. and THOM, A. S. 1969. *Essentials of Meteorology*. London: Wykeham Publications. 240 pp.

MEESE, D. A., GOW, A. J., GROOTES, P., MAYEWS-KI, P. A., RAM, M., STUIVER, M., TAYLOR, K. C., WADDINGTON, E. D. and ZIELINSKI, G. A. 1994. The accumulation record from the GISP2 core as an indicator of climate change throughout the Holocene. *Science*, 266: 1 680–1 682.

MILLER, A. and THOMPSON, J. C. 1970. *Elements of Meteorology*, 3rd Edition. Columbus: Charles E. Merrill Publishing Company. 383 pp.

MITCHELL, J. F. B., DAVIS, R. A., INGRAM, W. J. and SENIOR, C. A. 1995. On surface temperature, greenhouse gases and aerosols. *Journal of Climate*, 10: 2 364–2 386.

MUNN, R. E. 1966. *Descriptive Micrometeorology*. New York: Academic Press. 245 pp.

NEWELL, R. E. 1964. The circulation of the upper atmosphere. *Scientific American*, 210: 62–74.

NEWELL, R. E., KIDSON, J. W., VINCENT, D. G. and BOER, G. J. 1972. *The General Circulation of the Tropical Atmosphere and Interactions with Extratropical Latitudes*, Vol. I. Cambridge, Massachusetts: MIT Press. 258 pp.

NEWELL, R. E., VINCENT, D. G., DOPPLICK, T. G., FERRUZZA, D. and KIDSON, J. W. 1969. The energy balance of the global atmosphere. In Corby, G. A. (ed.). *The Global Circulation of the Atmosphere*. London: Royal Meteorological Society. 257.

NICHOLSON, S. E. 1986. The nature of rainfall variability in Africa south of the equator. *Journal of Climatology*, 6: 515–530.

NICHOLSON, S. E. and ENTEKHABI, D. 1986. The quasi-periodic behavior of rainfall variability in Africa and its relationship to the Southern Oscillation. *Archiv fur Meteorologie, Geophysik und Bioklimatologie*, Ser A., 34: 311–348.

NIILER, P. P. 1992. The ocean circulation. In Trenberth, K. (ed.). *Climate System Modelling*. Cambridge: Cambridge University Press. 117–148.

NORTON, I. O. and SCLATER, J. C. 1979. A model for the evolution of the Indian Ocean and breakup of Gondwanaland. *Journal of Geophysical Research*, 84: 6 803–6 830.

OBERHUBER, J. M., ROECKNER, E., CHRISTOPH, M., ESCH, M. and LATIF, M. 1998. Predicting the '97 El Niño with a global climate model. *Geophysical Research Letters*, 25: 2 273–2 276.

OKE, T. R. 1978. *Boundary Layer Climates*. London: Methuen and Co. Ltd. 372 pp.

OLIVIER, J. 1990. Some temporal aspects of hail in the Transvaal. *South African Geographer*, 16: 39–53.

OORT, A. H. 1983. Global atmospheric circulation statistics, 1958–1973. NOAA Professional Paper 14, U. S. Government Printing Office, Washington, D. C. 180 pp.

PALMEN, E. and NEWTON, C. W. 1969. *Atmospheric Circulation Systems: their Structure and Physical Interpretation*. London: Academic Press. 603 pp.

PARTRIDGE, T. C. 1997. Cainozoic environmental change over Southern Africa, with special emphasis on the last 20 000 years. *Progress in Physical Geography*, 21: 3–22.

PARTRIDGE, T. C., DEMENOCAL, P. B., LORENTZ, S. A., PAIKER, M. J. and VOGEL, J. C. 1997. Orbital forcing of climate over South Africa: a 2 000-year rainfall record from the Pretoria Saltpan. *Quaternary Science Reviews*, 16: 1 125–1 133.

PEDGLEY, D. E. 1962. A meso-synoptic analysis of the thunderstorms on 28 August 1958. *Geophysical Memoirs*, U. K. Meteorological Office, 14 (1). 30 pp.

PETTERSSEN, S. 1969. *Introduction to Meteorology*, 3rd Edition. New York: McGraw-Hill. 333 pp.

POLLARD, R. T. and SMYTHE-WRIGHT, D. 1996. Understanding Ocean Circulation. *UK WOCE – The First Six Years*. Natural Environment Research Council, UK. 30.

POOLMAN, E. 1986. Voorspelling van weerselemente met behulp van weermodelroosterdata. Paper presented at the Fourth Annual Conference of the South African Society for Atmospheric Sciences, Pretoria, 13–14 October 1987.

PRELL, W. L., HUTSON, W. H. and WILLIAMS, D. F. 1979. The Subtropical Convergence and late Quaternary circulation in the Southern Indian Ocean. *Marine Micropalaeontology*, 4: 225–234.

PRIESTLEY, C. H. B. 1959. *Turbulent Transfer in the Lower Atmosphere*. Chicago: University of Chicago Press. 130 pp.

RAYNER, P. J. and LAW, R. M. 1995. A comparison of modelled responses to prescribed CO_2 sources, CSIRO Australia, Division of Atmospheric Research, Technical Paper No 36. 82 pp.

REYNAUD, D., ET AL. 1993. The ice core record of greenhouse gases. *Science*, 259: 926–934.

SCHULZE, B. R. 1972. South Africa. In Landsberg, H. E. (ed.). *World Survey of Climatology*, Vol. 10. Amsterdam: Elsevier. 501-586.

————. 1984. Climate of South Africa: Part 8: General Survey, WB28. South African Weather Bureau, 5th Edition. 330 pp.

SCHULZE, G. C. 1987. Material prepared and provided by the South African Weather Bureau.

SCHULZE, R. E. and MCGEE, O. S. 1976. Climatic indices and classification in relation to the biogeography of southern Africa. In Werger, M. J. A. (ed.). *Biogeography and Ecology of Southern Africa*. The Hague: W. Junk. 19–54.

SCOTT, L. and THACKERAY, J. F. 1987. Multivariate analysis of late Pleistocene and Holocene pollen spectra from Wonderkrater, Transvaal, South Africa. *South African Journal of Science*, 83: 93–98.

SELLERS, W. D. 1965. *Physical Climatology*. Chicago: University of Chicago Press. 272 pp.

SHACKELTON, N. J. 1977. The oxygen isotope stratigraphy record of the late Pleistocene. *Philosophical Transactions of the Royal Society*, 280: 169–182.

SHACKLETON, N. J. and KENNETT, J. P. 1975. Palaeotemperature history of the Cenozoic and the initiation of Antarctic glaciation: oxygen and carbon isotope analyses in DSDP sites 277, 279 and 281. In Initial Reports of the Deep Sea Drilling Project, Volume 29, Washington D. C. United States Government Printing Office. 743–755.

SHACKLETON, N. J. and OPDYKE, N. D. 1973. Oxygen isotope and palaeomagnetic stratigraphy of Equatorial Pacific Core V 28-239: oxygen isotope temperatures and ice volumes on a 10Y year and 10C year scale. *Quaternary Research*, 3: 39–55.

SHUKLA, J. 1998. Seasonal predictions: ENSO and TOGA. Proceedings of the Conference on the World Climate Research Programme: Achievements, Benefits and Challenges, Geneva. 37–48.

SIMMONS, A. J. and BENGTSSON, L. 1984. Atmospheric general circulation models: their design and use for climatic studies. In Houghton, J. T. (ed.). *The Global Climate*. Cambridge: Cambridge University Press. 37–62.

SMAGORINSKY, J. 1974. Global atmospheric modelling and the numerical accumulation of climate. In Hess, W. N. (ed.). *Weather and Climate Modification*. New York: John Wiley and Sons. 842 pp.

SPARROW, M., HEYWOOD, K., BROWN, J. and STEVENS, D. 1996. Current structure of the South Indian Ocean. In Pollard, R. T. and Smythe-Wright, D. (eds.). *Understanding Ocean Circulation, UK WOCE – The First Six*

Years. Natural Environment Research Council, UK. 24.

STEYN, P. C. L. 1984. The relationship between the 300 hPa circulation pattern and rainfall classification in the Bethlehem region. Bethlehem Precipitation Research Project Report, No. 24, South African Weather Bureau, Pretoria. 19 pp.

STRAHLER, A. N. 1963. *Physical Geography*. New York: John Wiley and Sons. 733 pp.

STRETEN, N. A. 1973. Some characteristics of satellite-observed bands of persistent cloudiness over the Southern Hemisphere. *Monthly Weather Review*, 101: 486–495.

TALJAARD, J. J. 1958. South African air-masses: their properties, movement and associated weather. Unpublished PhD thesis, University of the Witwatersrand. 221 pp.

———. 1967. Development, distribution and movement of cyclones and anticyclones in the Southern Hemisphere during the IGY. *Journal of Applied Meteorology*, 6: 973–987.

———. 1972. Synoptic meteorology of the Southern Hemisphere. In Newton, C. W. (ed.). *Meteorology of the Southern Hemisphere*. Meteorological Monographs 35: 139–213.

———. 1981. The anomalous climate and weather systems of January to March 1974. South African Weather Bureau Technical Paper, No. 9. 92 pp.

———. 1981. Upper-air circulation, temperature and humidity over southern Africa. *South African Weather Bureau Technical Paper*, No. 10. 94 pp.

———. 1982. Cut-off lows and heavy rain over the Republic. *South African Weather Bureau Newsletter*, No. 403: 155–156.

TALJAARD, J. J., SCHMITT, W. and VAN LOON, H. 1961. Frontal analyses with application to the Southern Hemisphere. *Notos*, 10: 25–58.

TALJAARD, J. J., VAN LOON, H., CRUTCHER, H. L. and JENNE, R. L. 1969. Climate of the upper air: Part I – Southern Hemisphere. Vol. I Temperatures, dew points, and heights at selected pressure levels. NAVAIR 50-1C-55, Chief Naval Operations, Washington, D. C. 135 pp.

TALMA, A. S. and VOGEL, J. C. 1992. Late Quaternary paleotemperatures derived from a speleothem from Cango Caves, Cape Province, South Africa. *Quaternary Research*, 37: 203–213.

TAPPER, N. J., TYSON, P. D., OWENS, I. F. and HASTIE, W. J. 1981. Modelling the winter urban heat island over Christchurch, New Zealand. *Journal of Applied Meteorology*, 20: 365–376.

TAYLOR and KENT. 1996. Estimates of air-sea fluxes from climatological data. In Pollard, R. T. and Smythe-Wright, D. (eds.). *Understanding Ocean Circulation, UK WOCE – The First Six Years*. Natural Environment Research Council, UK. 5–6.

THACKERAY, J. F. 1987. Late Quaternary environmental changes inferred from small mammalian fauna, southern Africa. *Climatic Change*, 10: 1–21.

———. 1990. Temperature indices from late Quaternary sequences in South Africa: comparisons with the Vostok core. *South African Geographical Journal*, 72: 47–49.

———. 1992. Chronology of late Pleistocene deposits associated with Homo sapiens at Klasies River Mouth, South Africa. *Palaeoecology of Africa*, 23: 177–191.

TOSEN, G. R. 1997. The winter nocturnal low-level jet over the Mpumalanga Highveld. Unpublished MSc dissertation, University of the Witwatersrand, 118 pp.

TRIEGAARDT, D. O. and KITS, A. 1963. Die drukveld by verskillende vlakke oor suidelike Afrika en aangrensende oseane tydens vyfdaagse reen- en droë periodes in suid-Transvaal en noord-Vrystaat gedurende die 1960–1961 somer. *South African Weather Bureau Newsletter*, No. 168. 37–43.

TROUP, A. J. and STRETEN, N. A. 1972. Satellite-observed Southern Hemisphere cloud vortices in relation to conventional observations. *Journal of Applied Meteorology*, 11: 909–917.

UNCOD. 1977. Report of the United Nations Conference on Desertification. Nairobi, Kenya. Conference 74/31. 4–6.

VAN LOON, H. 1972. Pressure in the Southern Hemisphere. In Newton, C. W. (ed.).

Meteorology of the Southern Hemisphere. Meteorological Monographs 35: 59–86.

VAN LOON, H. and JENNE, R. L. 1972. The zonal harmonic standing waves in the Southern Hemisphere. *Journal of Geophysical Research,* 77: 992–1 003.

VAN LOON, H. and ROGERS, J. C. 1981. Aspects of the half-yearly oscillation on the southern hemisphere. Research Centre for Atmospheric Physics and Climatology, Publication No. 5, Academy of Athens. 45 pp.

———. 1984. Interannual variations in the half-yearly cycle of pressure gradients and zonal wind at sea level on the Southern Hemisphere. *Tellus,* 36A: 76–86.

VAN LOON, H., TALJAARD, J. J., JENNE, R. L. and CRUTCHER, H. L. 1971. Climate of the upper air: Southern Hemisphere Vol. II Zonal geostrophic winds. NCAR TN/STR-57 and NAVAIR 50-1C-56, NCAR, Boulder, Colorado. 43 pp.

VOGEL, C. H. 1987. Documentary evidence: a means of determining the climate of the Cape Colony. Unpublished MA thesis, University of the Witwatersrand. 190 pp.

VON GOGH, R. G. 1978. Elements of the wintertime temperature and wind structure over Pretoria. Environmental Studies Occasional Paper No. 20, Department of Geography and Environmental Studies, University of the Witwatersrand, Johannesburg. 45 pp.

VOWINCKEL, E. 1955. Southern Hemisphere weather map analysis: five-year mean pressures. *Notos,* 4: 17–50.

———. 1956. Ein Beitrag zur Witterungsklimatologie des suedlichen Mozambiquekanals. Miscelanea Geofisica Publicada Pelo Servico Meteorologico de Angola em Comemoracao do X Aniversario do Servico Meteorologico Nacional, Luanda. 63–86.

WALKER, N. D. 1990. Links between South African summer rainfall and temperature variability of the Agulhas and Benguela current systems. *Journal of Geophysical Research,* 95 (C3): 3 297–3 319.

WALLACE, J. M. and HOBBS, P. V. 1977. *Atmospheric Science, an Introductory Survey.*

New York: Academic Press. 467 pp.

WASHINGTON, W. M. and MEEHL, G. A. 1984. Seasonal cycle experiments on the climate sensitivity due to a doubling of CO_2 with an atmospheric general circulation model coupled to a simple mixed-layer ocean model. *Journal of Geophysical Research,* 89: 9 475–9 503.

WEATHER BUREAU, SOUTH AFRICA. 1954. *Climate of South Africa, Part 1, Climate Statistics, WB19.* South African Weather Bureau, Pretoria. 160 pp.

WILSON, C. A. and MITCHELL, J. F. B. 1987. A doubled CO_2 sensitivity experiment with a global climate model including a simple ocean. *Journal of Geophysical Research,* 92 (D12): 13 315–13 343.

WILTON, J. 1971. A study of surface temperature and humidity patterns over Pietermaritzburg, South Africa. Paper presented to the Third South African Students' Geographical Conference, University of Natal, Pietermaritzburg. 12 pp.

WMO. 1990. The role of the World Meteorological Organization in the International Decade for Natural Disaster Reduction. WMO, No. 745, World Meteorological Organization, Geneva. 32 pp.

———. 1997. WMO Statement on the Status of Climate in 1997, WMO-No. 877, Geneva. 12 pp.

WOODS, J. D. 1984. The upper ocean and air-sea interaction in global climate. In Houghton, J. T. (ed.). *The Global Climate.* Cambridge: Cambridge University Press. 141–187.

WUNSCH, C. 1994. Tracer inverse problems. In Anderson, D. L. T. and Willebrand, J. (eds.). *Ocean Circulation Models: Combining Dynamics and Data.* Hingham, MA: Kluwer Academic. 1–7.

WYRTKI, 1982. The Southern Oscillation, ocean-atmosphere interaction and El Niño. *Marine Technology Society Journal,* 16: 3–10.

ZHANG, Y., WALLACE, J. M. and BATTISTI, D. S. 1997. ENSO-like interdecadal variability: 1900–1993. *Journal of Climate,* 10: 1 004–1 020.

Acknowledgements

The author and publisher would like to thank those who provided illustrative material and are grateful to the following for permission to reproduce copyright material. (In addition to this list, citations are made in figure captions and full bibliographical details are given in References on page 377.)

Figures 7.11 and 7.18: Figure 4.13 from *An introduction to atmospheric physics* by Robert G. Fleagle and Jorst A. Businger, copyright © 1963 by Academic Press, reproduced by permission of the publisher; **Figure 16.6**: from 'Late Quaternary paleotemperatures derived from a speleothem from Cango caves, Cape Province, South Africa' by Talma and Vogel in *Quaternary Research*, Volume 37, 203–213, copyright © 1992 by Academic Press, reproduced by permission of the publisher; **Figure 5.3**: Figure 4.7 from *Atmospheric sciences: an introductory survey* by John M. Wallace and Peter V. Hobbs, copyright © 1977 by Academic Press, reproduced by permission of the publisher; **Figures 7.14 and 7.20**: Figures 4 and 7 from *Descriptive micrometeorology* by R. E. Munn, copyright © 1966 by Academic Press, reproduced by permission of the publisher; **Figures 10.39 and 10.40**: Figure 15.4 from *Atmospheric circulation systems: their structure and physical interpretation* by C. W. Newton and R. Palmen, copyright © 1969 by Academic Press, reproduced by permission of the publisher; **Figure 16.4**: from Meese et al. 1994 in *Science* 266: 1680–1682, published by the American Association for the Advancement of Science; **Figures 11.15, 11.16, 13.24, 17.13 and 17.14**: American Meteorological Society; **Figures 16.10 and 16.11**: from Timothy C. Partridge, 'Cainozoic environmental change in southern Africa, with special emphasis on the last 200 000 years', *Progress in Physical Geography* 1997, 1:3–22, Arnold Publishers. **Figure 13.26**: from S. J. Mason and M. R. Jury, 'Climatic variability and change over southern Africa: a reflection on underlying processes', *Progress in Physical Geography* 1997, 1:23–50, Arnold Publishers; **Figure 16.2**: from *Gondwana Geology*, 1975, Australian National University Press; **Figure 16.19**: from K. Heine (ed.) *Palaeoecology of Africa* 23, 1993, A. A.

Balkema, P. O. Box 1675, Rotterdam, Netherlands; **Figure 9.23**: Lightning Strike on Brixton Tower, Johannesburg, South Africa. Photographer: Don Briscoe APSSA, PPSA; **Figure 10.32**: © British Crown Copyright 1985; **Figure 17.2**: Cambridge University Press, Figure 4.6 from *The Global Climate* by Simmoris and Bengtsson (1984); **Figures 5.7 and 6.13**: Cambridge University Press, Figure 17 from *Clouds, rain and rainmaking* by B. J. Mason (1962); **Figure 9.14**: from Carte, 1981, 'Morphology of persistent storms in the Transvaal on 16/17 October 1978', *Beitrage zur Phyzik der Atmosphare*, 54, 86–100; **Figure 15.15**: from Rayner and Law, 1995, 'A comparison of modelled responses to prescribed CO_2 sources', CSIRO Australia, Division of Atmospheric Research, Technical Paper No. 36, reproduced by permission of CSIRO Australia; **Figures 13.10 and 13.11**: from Defant in *Physical oceanography*, Vol 1, Figures 83 and 332, 1961; **Figures 14.15 and 14.16**: Russel L. DeSouza, Ph.D., Florida State University, 1972; **Figures 5.11 and 7.7**: Figures 6.17 and 3.2 from *Meteorology, the Atmosphere in Motion* by J. R. Eagleman, 1992, Trimedia; **Figure 16.8**: reprinted from *Quaternary Science Reviews*, 16, Partridge et al., in 'Orbital fording of climate over South Africa: a 2000-year rainfall record from the Pretoria Saltpan', pages 1125–1133, copyright 1997 with permission from Elsevier Science; **Figure 13.6**: reprinted from *Deep Sea Research*, 18, Bang, in 'The southern Benguela Current region on February 1966, Part II: Bythythermography and air-sea interactions', pages 209–224, copyright 1971, with permission from Elsevier Science; **Figure 16.4**: used by permission from the Geological Society of America; **Figure 7.10**: Figure 8.1 from Hess, *Introduction to Theoretical Meteorology* (1959); **Figures 2.5, 7.30, 16.18, 16.19, 17.7, 17.9, 17.10, 17.11 and 17.12**: IPCC Secretariat; **Figures 7.15 and 8.11**: Hewson and Longley, Figures 24 and 117 from *Meteorology Theoretical and Applied* (1944); **Figure 13.29**: Kluwer Academic Publishers, from *Boundary Layer Meteorology*, 83, 1997, 311–330, 'Atmospheric boundary-layer fluxes and structure across a land-sea transition zone in south-eastern Africa', Jury et al., with kind permission from Kluwer Academic Publishers; **Figure 16.9**: from *Climatic Change*, 10, 1987, 1–21, 'Late Quaternary environmental changes inferred from small mammalian fauna, southern Africa', Thackaray, with kind permission from Kluwer Academic Publishers; **Figure 9.23**: © Gary Ladd 1972; **Figure 7.12**: Longman Group, Figure 2.9 from *Contemporary Climatology* by Sellers and Robinson (1986); **Figure 13.16**: Marine Technology Society from *Marine Technology Society Journal*, 16, 3–10; **Table 2.1 and Figure 2.1**: Miller and Thompson, Table 1.1 and Figure 1.1 from *Elements of Meteorology* (1970); **Figure 7.22, Figure 7.24, Figure 7.25, Figure 7.25, Table 14.1, Table 7.3, Figure 14.22, Figure 14.4, Figure 14.8, Figure 14.9 and Figure 7.21**: Methuen & Co., respectively Figure 2.4, Figure 2.6, Figures 5.1 and 5.2, Table 8.4, Table A2.3, Figures 8.6 and 9.3, Figure 2.9, Figures 8.1 and 8.2, Figure 8.1(b), and Figure 1.12 from *Boundary Layer Climates* by Oke (1978); **Figure 16.32**: Nature. Reprinted with permission from *Nature* 365: 143–147, Bond et al., Copyright 1993 Macmillan Magazines Limited; **Figure 5.10**: J. F. R. McIlveen, Figure 6.14 in *Basic Meteorology*; **Figures 9.1 and 10.31**: Ludlam, Frank., *Clouds and Storms*, University Park: The Pennsylvania State University Press, 1980, Figures 6.2 and 9.14, Copyright 1980 by The Pennsylvania State University. Reproduced by permission of the publisher; **Figure 7.16 and Table 2.2**: Routledge, Figure 2.7 and Table 1.2 from *Atmosphere, weather and climate* by Barry and Chorley (1998); **Figure 13.12**: George Retseck; **Figure 9.16**: Royal Meteorological Society; **Figures 12.50 and 12.51**: Society of South African Geographers; **Figures 17.5 and 17.16**: *South African Journal of Science*; **Figure 13.23**: from Springer-Verlag, 'SOI and QBO signals in the African region', *Theoretical and Applied Climatology*, 50, 103–115, Jury et al., 1994; **Figure 7.5**: Figure 8.4 © Arthur N. Strahler, published by John Wiley & Sons, Inc.; **Figure 7.10**: Figure 7.3 from *The science and wonders of the atmosphere*, Gedzelman, copyright © 1980. Reprinted by permission of John Wiley and Sons, Inc.; **Figure 13.17**: World Meteorological Organization.

While every effort has been made to trace and acknowledge copyright holders, this has not always been possible. Should any infringement have occurred, apologies are tendered and any omissions will be rectified in the event of a reprint of this book.

Index